Botany: The Science of Plant Life

Botany: The Science of Plant Life

Edited by **Austin Balfour**

R CALLISTO
REFERENCE

New York

Published by Callisto Reference,
106 Park Avenue, Suite 200,
New York, NY 10016, USA
www.callistoreference.com

Botany: The Science of Plant Life
Edited by Austin Balfour

International Standard Book Number: 978-1-63239-105-6 (Hardback)

Printed in the United States of America.

Contents

Preface

Human beings are predominantly dependent on plants for their survival. They are crucial to the future of human society as they are the primary source for food, oxygen, medicine, and other material products. Hence it can be said that the study of plants, or Botany, is important as it is a step towards understanding nature that supports all forms of life and moreover, to help improve the conditions of human existence.

Botany is the study of plant life and is a part of biology. It first originated as herbalism, one of the oldest branches of science, where man's interaction with and study of plants was merely limited to the spheres of food and medicine. It was with the emergence of botanical gardens that the study of plants took an academic turn. Since then the study of plants has expanded manifold with the advancements in the techniques of research using genetic analysis and DNA sequences to help classify plants more accurately.

The book's aim is to engage in the importance of plants in the survival of beings as the primary producers of oxygen, water, energy, carbon. Additionally, it will be addressing various environmental concerns and how botany can help tackle them.

This book is a product of the combined efforts of many researchers and scientists. I would like to thank them all for their contribution. I would also like to thank my publisher for the constant support and trust in me.

Editor

Overexpression of a Novel Component Induces *HAK5* and Enhances Growth in *Arabidopsis*

Eri Adams,[1] Celine Diaz,[1,2] Minami Matsui,[1] and Ryoung Shin[1]

[1] *RIKEN Center for Sustainable Resource Science, 1-7-22 Suehirocho, Tsurumi-ku, Yokohama, Kanagawa 230-0045, Japan*
[2] *Center for Research in Agricultural Genomics (CRAG), Universitat Autònoma de Barcelona, Cerdanyola del Vallès, 08193 Barcelona, Spain*

Correspondence should be addressed to Ryoung Shin; ryoung.shin@riken.jp

Academic Editors: C. Chang, I. Paponov, and C. Xu

Plants have developed mechanisms to adapt to the potassium deficient conditions over the years. In *Arabidopsis thaliana*, expression of a potassium transporter HAK5 is induced in low potassium conditions as an adaptive response to nutrient deficiency. In order to understand the mechanism in which *HAK5* is regulated, the full-length cDNA overexpressor gene hunting system was employed as a screening method. Of 40 genes recovered, At4g18280 was found to be dramatically induced in response to potassium-deficiency and salt stress. Plants overexpressing this gene showed higher *HAK5* expression and enhanced growth. These plants were also less sensitive to potassium-deficiency in terms of primary root growth. Taken together, these data suggest that this novel component, At4g18280, contributes to regulation of *HAK5* and, consequently, tolerance to potassium-deficiency in plants.

1. Introduction

Potassium (K^+) is an essential plant macronutrient and the most abundant inorganic cation in higher plants. K^+ functions as a major osmoticum which regulates turgor pressure and membrane potentials to drive plant growth. K^+ is also important for a variety of vital physiological processes in plants such as enzyme activities, phloem transport, photosynthesis, and stress response. Therefore, K^+-deficiency is observed as growth arrest due to a variety of physiological retardations including reduced photosynthesis and impaired phloem transport of sucrose, reviewed in [1–3]. One commonly observed adaptation mechanism providing plants with better access to limited nutrients is alteration of the root architecture such as root hair formation, primary root growth, and lateral root formation [4].

In order to respond to the wide range of K^+ concentrations in the soil, plants have also developed sophisticated K^+ uptake systems. In *Arabidopsis*, the major K^+ channel functioning in the roots under the K^+-sufficient conditions is assumed to be K^+ TRANSPORTER1 (AKT1) [5]. In response to K^+-deficiency, expression of high-affinity K^+ transporters, members of the K^+ uptake permease (KUP) family, is induced

such as *HIGH-AFFINITY K^+ TRANSPORTER5* (*HAK5*) and *KUP3* [6, 7] such that the ability to absorb K^+ is increased under the limited conditions. Of these transporters, HAK5 is probably the most well-studied high-affinity K^+ transporter in *Arabidopsis*. Upon K^+-deficiency, ethylene production is rapidly induced followed by reactive oxygen species (ROS) accumulation to upregulate expression of *HAK5* [8, 9]. Recently, it has been reported that expression of *HAK5* is also induced in response to cesium, which belongs to the same alkali metal group as K^+ and competes with K^+ in terms of uptake in plants, although it is not known whether *HAK5* induction is due to K^+-deficiency caused by cesium [10]. Several transcription factors are suggested to regulate *HAK5* expression including RAP2.11, DDF2, JLO, TF_IIA, and bHLH121 [11, 12]. It is important to understand detailed mechanisms by which plants maintain their survival under K^+-deficiency, especially from an agricultural point of view. However, molecular mechanisms of plant response to K^+-deficiency are not well known.

To investigate the upstream factors which regulate the expression of *HAK5*, the FOX (full-length cDNA overexpressor) gene hunting system in the background of *Arabidopsis* plants carrying the *HAK5* promoter fused to the luciferase reporter construct (*HAK5pro::LUC*) was created. The *Arabidopsis* FOX collection in *Agrobacterium tumefaciens* contains around 10,000 independent normalised full-length cDNA from *Arabidopsis* under the control of CaMV 35S promoter [13]. The FOX hunting system is a gain-of-function system and is a powerful tool to screen the responsible genes involved in a phenotype of interest. Using this system, we found a series of genes which induce expression of *HAK5* and some of these are induced themselves in response to K^+-deficiency. Of those, a gene At4g18280 was strongly induced by K^+-deficiency as well as salt stress. Plants overexpressing this gene showed better growth and resistance to K^+-deficiency but their K^+ and sodium (Na^+) concentrations were not altered. Taken together, we propose that this protein of unknown function At4g18280, in coordination with HAK5, may regulate efficient use and allocation of nutrients.

2. Materials and Methods

2.1. Plant Materials, Luciferase Imaging, and Growth Conditions. The *Arabidopsis thaliana* accession Col-0 was used as a wild type. *Agrobacterium* library containing *Arabidopsis* FOX [13] was transformed into 128 plants each of three independent *HAK5promoter::LUC* lines, *HAK5pro::LUC*-2, *HAK5pro::LUC*-8 and *HAK5pro::LUC*-12 [9,12]. The T1 plants were selected for their resistance to hygromycin. A total of 877 plants were recovered, transferred onto control plates without antibiotics, and grown for 7 days before imaging.

Plants were sprayed with luciferin (Duchefa Biochemie, Haarlem, The Netherlands) and kept in the dark for a few minutes prior to imaging. NightSHADE LB 985 (Berthold, Bad Wildbad, Germany) was used to image luciferase chemiluminescence. Those plants with higher luciferase activities compared to the background lines were selected for further analysis. The second, third, and fourth screenings were performed in subsequent generations to narrow down the candidates.

Seeds were sown on media containing 1.25 mM KNO_3, 1.5 mM $Ca(NO_3)_2$, 0.5 mM KH_2PO_4, 0.75 mM $MgSO_4$, 50 μM H_3BO_3, 10 μM $MnCl$, 2 μM $ZnSO_4$, 1.5 μM $CuSO_4$, 0.075 μM $NH_4Mo_7O_{24}$, and 74 μM Fe-EDTA, pH 5.8, with $Ca(OH)_2$, 1% (w/v) sucrose, and 0.5% (w/v) phyto agar (Duchefa Biochemie) for screening. For physiological and qRT-PCR assays, media containing 1.75 mM KCl, 2 mM $Ca(NO_3)_2$, 0.5 mM phosphoric acid, 0.75 mM $MgSO_4$, 50 μM H_3BO_3, 10 μM $MnCl$, 2 μM $ZnSO_4$, 1.5 μM $CuSO_4$, 0.075 μM $NH_4Mo_7O_{24}$, and 74 μM Fe-EDTA, pH 5.8, with $Ca(OH)_2$, 1% sucrose, and 1% SeaKem agarose (Lonza, Basel, Switzerland) were used as control. For non-potassium (−K) media, KCl was excluded. For non-phosphorus (−P) media, phosphoric acid was excluded. For non-nitrogen (−N) media, $Ca(NO_3)_2$ was replaced with $CaCl_2$. For NaCl media, 100 mM NaCl was added. After stratification for 3 to 4 days at 4°C, plants were placed vertically in a growth cabinet at 22°C in a 16-hour light/8-hour dark photocycle with a light intensity of 70–90 μmol/m^2/s.

2.2. Confirmation of Inserted cDNA and Sequencing. The selected seedlings were transferred into soil and a few rosette leaves were flash-frozen in liquid N_2. The samples were ground using a mixer mill and genomic DNA was extracted. A single insertion of a cDNA fragment was confirmed by PCR using a primer set: F2 5′-CATTTATTCGGAGAGGTACGT-AT and R2 5′-GGATTCAATCTTAAGAAACTTTATTGC-CAA [14]. The PCR products were purified using a Wizard SV Gel and PCR Clean-Up System (Promega, Wisconsin, USA) and sequenced using a F6 primer: 5′-CCCCCCCCCCCCD (A or G or T).

2.3. qRT-PCR Analysis. For *HAK5* and FOX gene expression in Col-0, seedlings were grown on control media for 3 days, transferred onto −K, −N, −P, or 100 mM NaCl media, and grown for 7 days. The samples were flash-frozen in liquid N_2 and ground using a mixer mill. Total RNA was extracted, treated with DNaseI (Invitrogen, California, USA), and synthesised into cDNA using SuperScript III (Invitrogen). Quantitative real-time reverse transcription-PCR (qRT-PCR) was performed using THUNDERBIRD SYBR qPCR mix (TOYOBO, Osaka, Japan) and a Mx3000P qPCR system (Agilent Technologies, California, USA). The amplification conditions were 95°C for 15 seconds and 60°C for 30 seconds. The cycle was repeated 40 times, preceded by 95°C for 1 minute and followed by a dissociation programme to create melting curves. Three biological replicates with three technical replicates for each treatment were run. The β-tubulin gene (*TUB2*) was used as a reference gene. The primers used are summarised in Supplemental Table 1 as shown in supplementary material available online at http://dx.doi.org/10.1155/2014/490252. Primer sequences of HAK5 and TUB2 have been previously published [15]. One-way ANOVA with Dunnett's multiple comparison posttest ($P < 0.05$) or t-test with Welch's correction ($P < 0.05$) was performed using Prism (GraphPad Software, California, USA) to determine the statistical significance.

2.4. Root Growth Assay. FOX lines and Col-0 were grown on control media for 3 days, transferred onto media with or without K^+ and grown for 11 days. Plates were scanned and primary root lengths were measured using ImageJ (http://rsbweb.nih.gov/ij/). Roots and shoots were separated and weighted on a precision balance. One-way ANOVA with Bonferroni's multiple comparison posttest was performed using Prism to determine the statistical significance.

2.5. Potassium and Sodium Concentration Analysis. FOX lines and Col-0 were grown on control media for 3 days, transferred onto media with or without K^+ and grown for 11 days. Seedlings were washed in MilliQ water, dried on a piece of paper towel, placed in a paper envelope, and dried in an oven at 65°C for 3 to 4 days. Approximately 2 mg of dried samples were extracted in 1 mL of 60% (v/v) HNO_3 at 125°C for 1 hour followed by 1 mL of 30% (v/v) H_2O_2 and diluted with MilliQ water to get total volume of 10 mL. The samples were further diluted 10 or 100 times with 6% (v/v) HNO_3. K^+ and Na^+ concentrations were measured

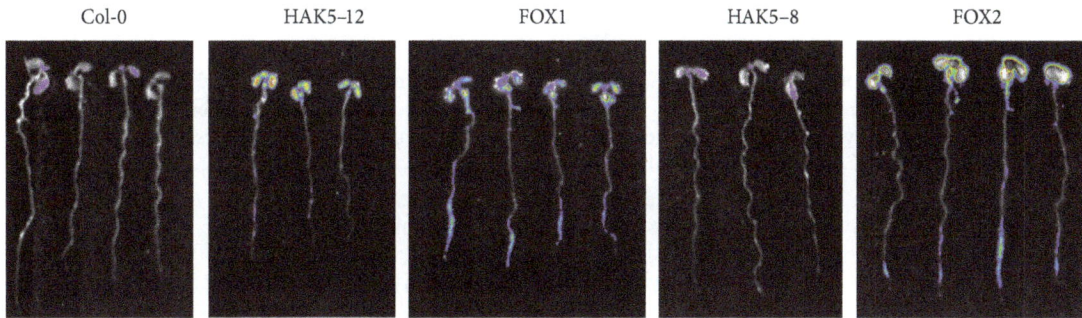

FIGURE 1: Chemiluminescence imaging of Col-0 and two FOX lines with their parental lines (*HAK5pro::LUC*-12 for FOX1, *HAK5pro::LUC*-8 for FOX2) grown for 7 days under the control condition. Pseudocolour represents the intensity of chemiluminescence. FOX1 expresses At4g18280; FOX2 expresses At2g36370.

on a flame atomic absorption spectrometer AAnalyst 200 (PerkinElmer). Concentrations were calculated against each standard curve and one-way ANOVA with Bonferroni's multiple comparison posttest ($P < 0.05$) was performed using Prism to determine the statistical significance.

3. Results

3.1. Screening and Selection of Candidate Genes which Induce HAK5 Expression. The high-affinity K^+ transporter, HAK5, is a crucial component for plant growth and survival especially under the K^+-deficient conditions. Expression of *HAK5* is known to be rapidly induced in response to K^+-deficiency but promptly reverts to basal levels upon resupply of K^+ however the detailed mechanism of this exquisite regulation is not well known. In order to seek out the components involved in regulation of *HAK5* expression, *Arabidopsis* plants carrying the *HAK5pro::LUC* reporter constructs were transformed with *Agrobacterium* library containing *Arabidopsis* FOX [13]. The T1 population was segregated on selection media containing hygromycin. The resistant plants were transferred onto control media which contained sufficient amounts of K^+. In this condition, expression of *HAK5* is not induced; therefore, luciferase activity is minimal for the parental *HAK5pro::LUC* lines. Out of 877 transformants, upon multiple screenings of subsequent generations, 47 candidate lines were selected for enhanced luciferase activity compared to their parental lines (Figure 1). Genomic DNA for each line was extracted, PCR was performed to confirm the insertion of FOX genes and purified PCR fragments were sequenced. A total of 40 genes were recovered (Table 1), of which 9 genes were recovered from two independent plant lines and two plant lines hosted two FOX genes. Endogenous *HAK5* expression was also determined to be induced in the selected FOX lines (Table 1).

Since the candidate genes were expected to be involved in induction of *HAK5* expression which is known to be induced under K^+-deficiency, we tested whether expression of these genes was also differentially regulated under K^+-deficiency. Of 11 candidate genes selected, 5 genes were found to be induced in response to K^+-deficiency in the roots and 3 genes in the shoots also (Table 2). These results are consistent with the predominant role of HAK5 in the roots.

3.2. Involvement of FOX Genes in Other Nutrient Deficiency and Salt Stress. Five FOX genes that were induced in response to K^+-deficiency in the roots, At4g18280, At5g47790, At5g22360, At5g55310, and At2g36370 (Table 2), were also analysed under nitrogen-deficiency, phosphorus-deficiency, and salt stress in order to investigate whether they are general nutrient stress-response factors. Moderate increases ($P < 0.05$) of At5g47790 in nitrogen-deficiency, At5g22360 in phosphorus-deficiency (Supplemental Figure 1) and a dramatic increase ($P < 0.01$) of At4g18280 under salt stress (Figure 2(a)) were observed. Some K^+ transporter genes including *HAK5* are known to be induced in response to salt stress [12, 16, 17], possibly causing similar effects as K^+-deficiency to plants. Our data might indicate that At4g18280 could regulate expression of *HAK5* not only under K^+-deficiency but also under salt stress.

3.3. Phenotype of FOX Lines in K^+-Deficiency. Five FOX lines expressing the genes that are induced in response to K^+-deficiency (Table 2) were investigated for their phenotype. The FOX lines expressing At5g47790, At5g22360, At5g47790, and At2g36370 (FOX2) did not show altered growth in terms of primary root length compared to wild type (Col-0) under K^+-deficiency. However, the FOX line expressing At4g18280 (FOX1) showed enhanced growth in the K^+-deficient condition (Figures 2(b) and 2(c)). FOX1 contains full-length At4g18280 and its expression levels are 47.5 times more in average than one in its parental line, *HAK5pro::LUC*-12 ($\log_2$ fold change 5.57). FOX1 also contains a second FOX gene At5g59160, tandem with At4g18280 but expression of this gene was not enhanced ($\log_2$ fold change −0.01).

Further analysis of FOX1 revealed that its lateral root number, lateral root density, and both root and shoot weight were increased in the control condition (Figure 3), suggesting that induction of At4g18280 expression renders enhanced growth to plants.

3.4. K^+ and Na^+ Accumulation in FOX1. According to our data, a protein annotated as a glycine-rich cell wall protein (At4g18280) seems to be involved in regulation of *HAK5* expression. Since HAK5 is a K^+ transporter which absorbs exogenous K^+ into the root system, predominantly working

TABLE 1: List of the genes that were identified in the *HAK5pro::LUC* reporter lines transformed with FOX library for increased luciferase activity. Number of the lines that were recovered for each gene is given as a hit number, if multiple. Endogenous *HAK5* expression levels in the roots for selected lines are also given as log_2 fold change compared to their parental *HAK5pro::LUC* lines under the control condition.

AGI	Gene description	Hit	HAK5
At1g17620	Late embryogenesis abundant (LEA) hydroxyproline-rich glycoprotein family	2	
At1g23310	ALANINE-2-OXOGLUTARATE AMINOTRANSFERASE 1 (AOAT1), GLUTAMATE:GLYOXYLATE AMINOTRANSFERASE 1 (GGAT1)	2	
At1g29050	Trichome birefringence-like 38 (TBL38)		
At1g34000	One-helix protein 2 (OHP2)		
At1g48630	Receptor for activated C kinase 1B (RACK1B)		7.845
At1g51590	Alpha-mannosidase 1 (MNS1)		
At1g56580	Smaller with variable branches (SVB)		
At1g62380	ACC oxidase 2 (ACO2)	2	
At1g69530	Expansin 1 (EXP1)		
At2g01970	Endomembrane protein 70 protein family	2	
At2g21660	Cold, circadian rhythm, and RNA binding 2 (CCR2), glycine-rich RNA-binding protein 7 (GRP7)	2	
At2g34830	WRKY DNA-binding protein 35 (WRKY35), maternal effect EMBRYO ARREST 24 (MEE24)		
At2g36370	Ubiquitin-protein ligase		0.906
At2g36895	Unknown protein		
At3g07430	Embryo defective 1990 (EMB1990)		
At3g53730	Histone superfamily protein		
At4g03190	Auxin signaling f-box protein1 (AFB1), GRR1-like protein 1 (GRH1)		
At4g03960	Plant, and fungi atypical dual-specificity phosphatase 4 (PFA-DSP4)		
At4g13195	Clavata3/ESR-related 44, CLE44		
At4g18280	Glycine-rich cell wall protein related		1.581
At4g22190	unknown (EAR motif)		
At4g25240	GPI-anchored SKU5-like protein (SKS1)		
At4g28860	Casein kinase-like 4 (ckl4);		
At5g01820	SnRK3.5, CIPK14		
At5g03040	IQ-DOMAIN 2 (IQD2)		6.856
At5g09760	pectinesterase family protein		
At5g14740	Carbonic anhydrase 2 (CA2)		
At5g15410	CYCLIC NUCLEOTIDE GATED CHANNEL2 (CNGC2), defense no death 1 (DND1)		
At5g22360	Vesicle-associated membrane protein 714 (VAMP714)		0.902
At5g44160	Nutcracker (NUC)		
At5g44750	Homologous to Y-family DNA polymerase (REV1)	2	
At5g47790	SMAD/FHA domain-containing protein		
At5g52880	F-Box family protein	2	
At5g54585	Unknown protein		
At5g55310	Topoisomerase 1 (TOP1)	2	6.104
At5g57360	FKF1-like protein 2 (FKL2)		
At5g59160	Type one serine/threonine protein phosphatase 2 (TOPP2)		
At5g59320	Lipid transfer protein 3 (LTP3)	2	
At5g59500	Unknown protein		
At5g61250	Glucuronidase 1 (GUS1)		

TABLE 2: List of the genes that were analysed for their response to potassium deficiency ($-$K). Expression levels of each candidate gene under $-$K in roots and shoots of wild type plants (Col-0) are given as $\log_2$ fold change compared to the ones under the control condition. Statistically significant difference is marked in bold. b.d. indicates below detection.

AGI	Description	Root	Shoot
At5g47790	SMAD/FHA domain-containing protein	**1.14**	b.d.
At5g03040	IQ-domain 2 (iqd2)	−0.13	—
At5g55310	Topoisomerase 1 (TOP1)	**0.64**	b.d.
At5g22360	Vesicle-associated membrane protein 714 (VAMP714)	**0.44**	**−0.86**
At4g18280	glycine-rich cell wall protein related	**4.14**	**5.19**
At2g21660	cold, circadian rhythm, and RNA binding 2 (CCR2), glycine-rich RNA-binding protein 7 (GRP7)	−0.21	**1.17**
At2g36370	Ubiquitin-protein ligase	**1.53**	**1.09**
At4g03960	plant and fungi atypical dual-specificity phosphatase 4 (PFA-DSP4)	b.d.	—
At5g44160	Nutcracker (NUC)	b.d.	0.26
At5g01820	SnRK3.5, CIPK14	0.06	—
At1g48630	receptor for activated C kinase 1B (RACK1B)	**−0.65**	—

under the K$^+$-deficient conditions, the role of At4g18280 in this process was investigated. FOX1 expressing At4g18280 was grown in K$^+$-sufficient and -deficient conditions and K$^+$ concentrations in the seedlings were determined. In both sufficient and deficient conditions, K$^+$ levels were comparable to those in Col-0 (Figure 4(a)), indicating that, although At4g18280 is involved in induction of HAK5 expression, it does not contribute to K$^+$ accumulation.

Accumulation of Na$^+$ was also determined in FOX1 under K$^+$-deficiency and salt stress. Na$^+$ concentrations in FOX1 were also comparable to those in Col-0 under the conditions tested (Figure 4(b)), suggesting that At4g18280 alone is perhaps not sufficient for activation of HAK5 for K$^+$ or Na$^+$ uptake.

4. Discussion

It is important for plants to efficiently absorb a macronutrient K$^+$ especially when K$^+$ is not readily available. Upon K$^+$-deficiency, *Arabidopsis* quickly induces expression of a high-affinity K$^+$ transporter HAK5 and increases the efficiency of K$^+$ uptake. In order to understand the detailed regulatory mechanism of HAK5 expression and components involved in this process, plants carrying the HAK5 promoter fused to a luciferase reporter gene, HAK5pro::LUC, were transformed with the FOX library and analysed. After multiple screenings, 47 plants out of 877 were selected for increased luciferase activities under the non-inducing condition. Sequencing of the FOX genes expressed in each FOX plant revealed 40 genes which might be responsible for inducing HAK5 expression. Randomly selected plant lines were tested for endogenous expression of HAK5 to confirm the integrity of the system. Nine of the genes recovered were represented in two independent FOX lines, suggesting importance of those genes and validity of the screening. These include At1g23310, ALANINE-2-OXOGLUTARATE AMINOTRANSFERASE1 (AOAT1)/GLUTAMATE:GLYOXYLATE AMINOTRANS-FERASE1 (GGAT1), At1g62380, ACC OXIDASE2 (ACO2),

and At5g44750, DNA polymerase REV1. It has been reported that ethylene production in response to K$^+$-deficiency is required for ROS accumulation and consequent HAK5 expression [9]. AOAT1/GGAT1 is suggested in 1-aminocyclopropane-1-carboxylate (ACC) biosynthesis, the first step of ethylene biosynthesis, according to the GO term and ACO2 is an enzyme which converts ACC into ethylene. Another member of the ACO family, ACO1 was previously shown to be induced in response to K$^+$-deficiency [8]. Moreover, REV1 is suggested to be involved in biosynthesis of hydrogen peroxide, a form of ROS, according to the GO term. Our results reinforce previous findings that overexpression of ethylene or ROS biosynthesis genes can mimic K$^+$-deficiency response in terms of HAK5 induction. Other genes are represented in one plant line such as At5g15410, CYCLIC NUCLEOTIDE GATED CHANNEL2 (CNGC2) and At4g03190, AUXIN SIGNALING F-BOX PROTEIN1 (AFB1). CNGC2 was reported to function as a monovalent cation channel and conduct K$^+$ but not Na$^+$ [18], and therefore it is possible that CNGC2 and HAK5 are coregulated in response to K$^+$-deficiency. The relationship between auxin and K$^+$-deficiency was also previously reported [19]. The gene list also includes a few genes related to protein degradation such as F-box proteins and ubiquitin-protein ligase, suggesting involvement of this process in K$^+$-deficiency response.

In order to find novel components that are coregulated with HAK5, transcript levels of 11 selected genes in response to K$^+$-deficiency was tested in roots and shoots. Of those, five genes in roots and three genes in shoots showed statistically significant induction in reference to the K$^+$-sufficient condition. Induction of a gene encoding a protein annotated as glycine-rich cell wall protein (At4g18280) and a gene encoding an ubiquitin-protein ligase (At2g36370) were particularly high in both tissues. Previous microarray results also showed increased transcript levels of At4g18280 in shoots of K$^+$-starved *Arabidopsis* plants [20]. Upon physiological studies of the FOX lines, only FOX1 with At4g18280 was

(a)

(b)

Col-0   FOX1

(c)

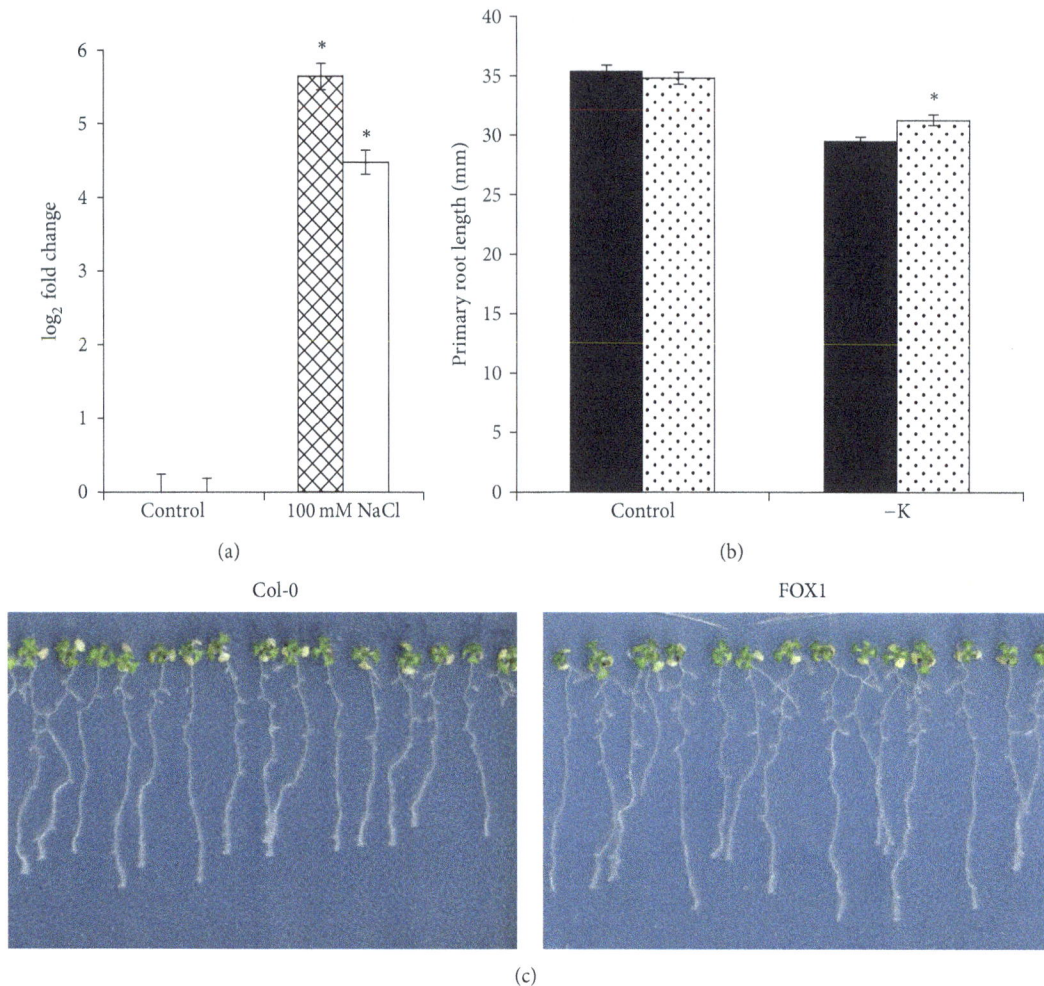

FIGURE 2: Role of At4g18280 in salt stress and K^+-deficiency. (a) Gene expression of *HAK5* (crisscross bars) and *At4g18280* (white bars) under salt stress. Three-day-old Col-0 seedlings germinated on control media were transferred onto control media or media containing 100 mM NaCl and grown for 7 days prior to analysis. Values are $\log_2$ ratios relative to the control condition. Error bars indicate standard error for three biological replicates with three technical replicates. Each sample contained more than 15 seedlings. Statistically significant differences ($P < 0.01$) compared to each control sample are indicated as asterisks. (b) Primary root lengths of Col-0 (black bars) and FOX1 (dotted bars). Three-day-old seedlings germinated on control media were transferred onto control media or media lacking K^+ (−K) and grown for 11 days prior to measurement. Error bars indicate standard error ($n \geq 78$). Statistically significant differences ($P < 0.01$) compared to Col-0 are indicated as an asterisk. (c) Image of Col-0 and FOX1 on −K. Three-day-old seedlings germinated on control media were transferred onto −K and grown for 11 days prior to imaging.

associated with enhanced primary root growth under the K^+-deficient condition. Although primary root lengths of FOX1 were comparable to those of Col-0 in the control condition, lateral root number, lateral root density, and root and shoot weight were increased in FOX1 in the control condition, but not to the same extent in the K^+-deficient condition. These data indicate that induced *HAK5* expression caused by overexpression of At4g18280 or overexpression of At4g18280 alone renders enhanced general growth to the plants. These data also indicate that overexpression of At4g18280 causes less sensitivity to K^+-deficiency in terms of primary root growth but not in terms of secondary root growth or shoot growth. This lower sensitivity of FOX1 to K^+-deficiency was

not due to changes in K^+ accumulation in the plant body, suggesting that At4g18280 is unlikely to be directly involved in K^+ uptake.

Transcriptional regulation of the five genes induced in the roots in response to K^+-deficiency was also analysed under nitrogen-deficiency, phosphorus-deficiency, and salt stress. Both *HAK5* and At4g18280 were shown to be dramatically induced in response to salt stress. This might suggest that At4g18280 was a general transcriptional regulator for *HAK5*; however, its expression was not induced by cesium treatment ($\log_2$ fold change 0.18) which could induce *HAK5* expression [10]. Na^+ concentrations in FOX1 were comparable to those in Col-0 under the control condition, K^+-deficiency, or salt

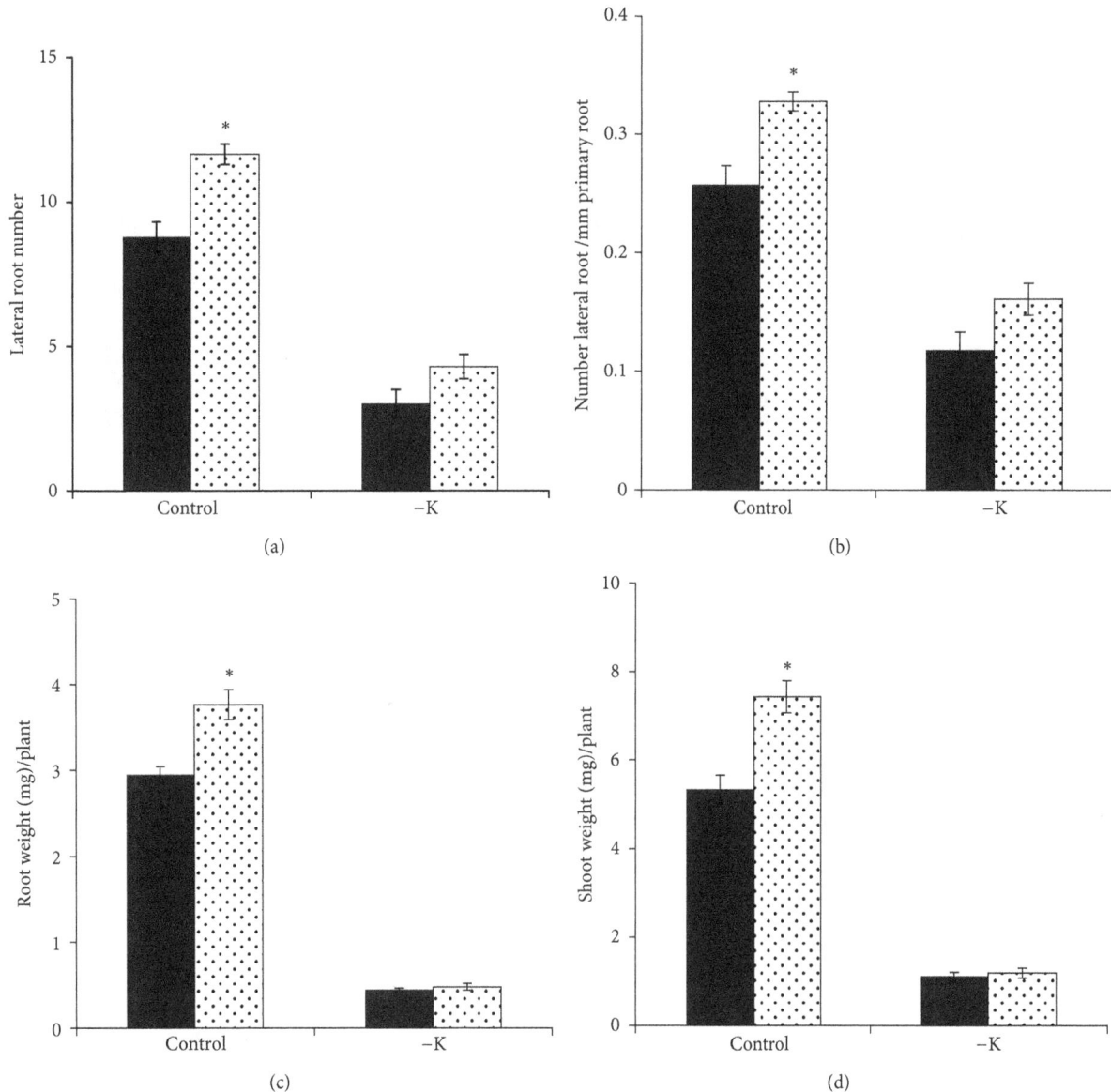

FIGURE 3: Physiological analysis of FOX1. (a) Lateral root number, (b) lateral root density, (c) fresh root weight, and (d) fresh shoot weight of Col-0 (black bars) and FOX1 (dotted bars). Three-day-old seedlings germinated on control media were transferred onto control media or media lacking K^+ (−K) and grown for 11 days prior to measurement. Error bars indicate standard error. $n = 14$ for (a) and (b), $n = 6$ with each replicate containing 7 seedlings for (c) and (d). Statistically significant differences ($P < 0.01$) compared to Col-0 are indicated as asterisks.

stress. Also, no morphological phenotype was observed in FOX1 under salt stress. Taken together, At4g18280 is not likely to have a direct role in Na^+ uptake.

According to the eFP Browser (http://bar.utoronto.ca/efp/cgi-bin/efpWeb.cgi), it is shown that At4g18280 is predominantly expressed in reproductive organs such as siliques and seeds under normal condition, where HAK5 is also expressed. At4g18280 is annotated as a protein related to glycine-rich cell wall protein according to the TAIR website (http://www.arabidopsis.org/); however, it lacks conventional glycine-rich domains [21]. BLAST (http://blast.ncbi.nlm.nih.gov/) search reveals no known protein with high similarity and a protein domain search engine SMART (http://smart.embl-heidelberg.de/) does not identify any

conserved domains or motifs. A gene encoding a glycine-rich protein from alfalfa was reported to be induced by salt and drought stresses and abscisic acid treatment [22]. *Arabidopsis* glycine-rich protein GRP9, whose transcripts are salt-inducible, is suggested to be involved in lignin synthesis in response to salt stress [23]. Although these previous findings correspond conveniently with our data, At4g18280 is unlikely to be a glycine-rich cell wall protein.

5. Conclusion

An approach combining promoter reporter constructs and the FOX hunting system successfully revealed a list of genes that might be capable of regulating *HAK5* expression. Of

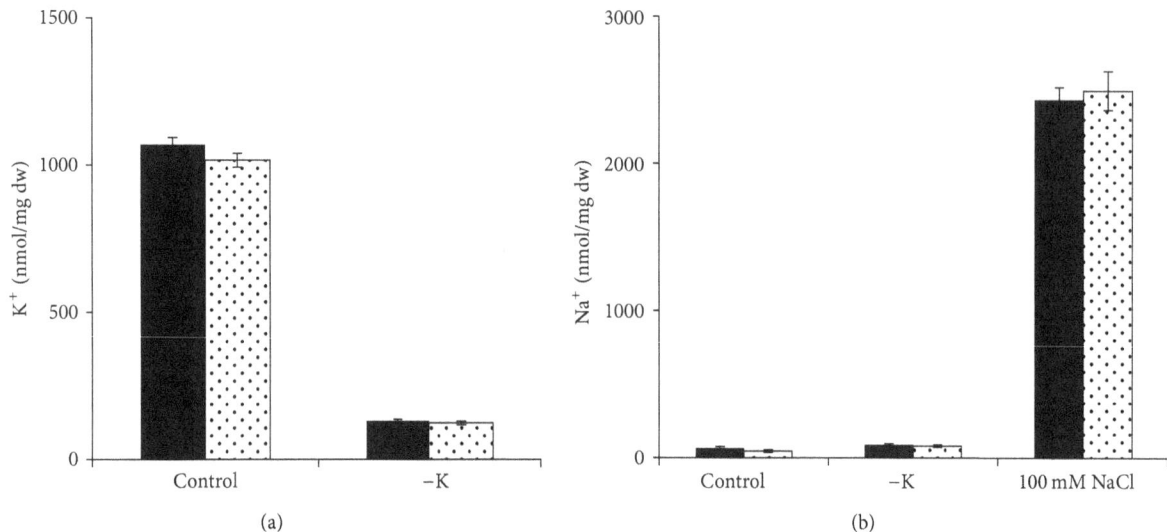

FIGURE 4: K^+ and Na^+ concentrations in FOX1. (a) K^+ and (b) Na^+ concentrations in Col-0 (black bars) and FOX1 (dotted bars). Three-day-old seedlings germinated on control media were transferred onto control media, media lacking K^+ ($-K$), or media containing 100 mM NaCl and grown for 11 days prior to analysis. Error bars indicate standard error for four to six biological replicates. Each replicate contained more than 15 seedlings.

those, we have shown that a novel component At4g18280 is coregulated with *HAK5* in K^+-deficiency and salt stress response, and overexpression of this gene induces *HAK5* expression even under the K^+-sufficient conditions in *Arabidopsis*. This coexpression is likely specific to K^+ and Na^+ responses rather than a general monovalent cation response since expression of At4g18280 is not induced in response to cesium, which induces *HAK5* expression. However, K^+ and Na^+ concentrations do not increase in the plants which overexpress At4g18280, suggesting that other components are required to activate HAK5 for K^+, and possibly Na^+, uptake. Overexpression of At4g18280 enhances general plant growth and provides a mechanism through which plants can adapt to K^+-deficiency, possibly due to efficient use of nutrients. Further functional study of At4g18280 is awaited to reveal the entire picture of *HAK5* regulation through this novel component.

Conflict of Interests

The authors declare that there is no conflict of interests regarding the publication of this paper.

Acknowledgments

This research was financially supported by RIKEN. The authors thank Ms. Takae Miyazaki for taking care of the plants and Dr. Michael Adams for comments and discussion on the paper.

References

[1] A. Amtmann, J. P. Hammond, P. Armengaud, and P. J. White, "Nutrient sensing and signalling in plants: potassium and phosphorus," *Advances in Botanical Research*, vol. 43, pp. 209–257, 2005.

[2] F. Alemán, M. Nieves-Cordones, V. Martínez, and F. Rubio, "Root K^+ acquisition in plants: the *Arabidopsis thaliana* model," *Plant and Cell Physiology*, vol. 52, no. 9, pp. 1603–1612, 2011.

[3] Y.-F. Tsay, C.-H. Ho, H.-Y. Chen, and S.-H. Lin, "Integration of nitrogen and potassium signaling," *Annual Review of Plant Biology*, vol. 62, pp. 207–226, 2011.

[4] J. López-Bucio, A. Cruz-Ramírez, and L. Herrera-Estrella, "The role of nutrient availability in regulating root architecture," *Current Opinion in Plant Biology*, vol. 6, no. 3, pp. 280–287, 2003.

[5] M. Basset, G. Conejero, M. Lepetit, P. Fourcroy, and H. Sentenac, "Organization and expression of the gene coding for the potassium transport system AKT1 of *Arabidopsis thaliana*," *Plant Molecular Biology*, vol. 29, no. 5, pp. 947–958, 1995.

[6] Z. Qi, C. R. Hampton, R. Shin, B. J. Barkla, P. J. White, and D. P. Schachtman, "The high affinity K^+ transporter AtHAK5 plays a physiological role *in planta* at very low K^+ concentrations and provides a caesium uptake pathway in *Arabidopsis*," *Journal of Experimental Botany*, vol. 59, no. 3, pp. 595–607, 2008.

[7] E. J. Kim, J. M. Kwak, N. Uozumi, and J. I. Schroeder, "*AtKUP1*: an *Arabidopsis* gene encoding high-affinity potassium transport activity," *The Plant Cell*, vol. 10, no. 1, pp. 51–62, 1998.

[8] R. Shin and D. P. Schachtman, "Hydrogen peroxide mediates plant root cell response to nutrient deprivation," *Proceedings of the National Academy of Sciences of the United States of America*, vol. 101, no. 23, pp. 8827–8832, 2004.

[9] J.-Y. Jung, R. Shin, and D. P. Schachtman, "Ethylene mediates response and tolerance to potassium deprivation in *Arabidopsis*," *The Plant Cell*, vol. 21, no. 2, pp. 607–621, 2009.

[10] E. Adams, P. Abdollahi, and R. Shin, "Cesium inhibits plant growth through jasmonate signaling in *Arabidopsis thaliana*," *International Journal of Molecular Sciences*, vol. 14, no. 3, pp. 4545–4559, 2013.

[11] M. J. Kim, D. Ruzicka, R. Shin, and D. P. Schachtman, "The *Arabidopsis* AP2/ERF transcription factor RAP2. 11 modulates

plant response to low-potassium conditions," *Molecular Plant*, vol. 5, no. 5, pp. 1042–1057, 2012.

[12] J.-P. Hong, Y. Takeshi, Y. Kondou, D. P. Schachtman, M. Matsui, and R. Shin, "Identification and characterization of transcription factors regulating *Arabidopsis HAK5*," *Plant and Cell Physiology*, vol. 54, no. 9, pp. 1478–1490, 2013.

[13] T. Ichikawa, M. Nakazawa, M. Kawashima et al., "The FOX hunting system: an alternative gain-of-function gene hunting technique," *The Plant Journal*, vol. 48, no. 6, pp. 974–985, 2006.

[14] M. Higuchi, Y. Kondou, T. Ichikawa, and M. Matsui, "Full-length cDNA overexpressor gene hunting system (FOX hunting system)," *Methods in Molecular Biology*, vol. 678, pp. 77–89, 2011.

[15] S. J. Ahn, R. Shin, and D. P. Schachtman, "Expression of *KT/KUP* genes in *Arabidopsis* and the role of root hairs in K^+ uptake," *Plant Physiology*, vol. 134, no. 3, pp. 1135–1145, 2004.

[16] F. J. M. Maathuis, "The role of monovalent cation transporters in plant responses to salinity," *Journal of Experimental Botany*, vol. 57, no. 5, pp. 1137–1147, 2006.

[17] H. Su, D. Golldack, C. Zhao, and H. J. Bohnert, "The expression of HAK-type K^+ transporters is regulated in response to salinity stress in common ice plant," *Plant Physiology*, vol. 129, no. 4, pp. 1482–1493, 2002.

[18] Q. Leng, R. W. Mercier, B.-G. Hua, H. Fromm, and G. A. Berkowitz, "Electrophysiological analysis of cloned cyclic nucleotide-gated ion channels," *Plant Physiology*, vol. 128, no. 2, pp. 400–410, 2002.

[19] R. Shin, A. Y. Burch, K. A. Huppert et al., "The *Arabidopsis* transcription factor MYB77 modulates auxin signal transduction," *The Plant Cell*, vol. 19, no. 8, pp. 2440–2453, 2007.

[20] C. R. Hampton, H. C. Bowen, M. R. Broadley et al., "Cesium toxicity in *Arabidopsis*," *Plant physiology*, vol. 136, no. 3, pp. 3824–3837, 2004.

[21] A. Mangeon, R. M. Junqueira, and G. Sachetto-Martins, "Functional diversity of the plant glycine-rich proteins superfamily," *Plant Signaling & Behavior*, vol. 5, no. 2, pp. 99–104, 2010.

[22] R. Long, Q. Yang, J. Kang et al., "Overexpression of a novel salt stress-induced glycine-rich protein gene from alfalfa causes salt and ABA sensitivity in *Arabidopsis*," *Plant Cell Reports*, vol. 32, no. 8, pp. 1289–1298, 2013.

[23] A.-P. Chen, N.-Q. Zhong, Z.-L. Qu, F. Wang, N. Liu, and G.-X. Xia, "Root and vascular tissue-specific expression of glycine-rich protein AtGRP9 and its interaction with AtCAD5, a cinnamyl alcohol dehydrogenase, in *Arabidopsis thaliana*," *Journal of Plant Research*, vol. 120, no. 2, pp. 337–343, 2007.

2

The Effect of Phosphorus Reduction and Competition on Invasive Lemnids: Life Traits and Nutrient Uptake

Joëlle Gérard and Ludwig Triest

Plant Biology and Nature Management (APNA), Vrije Universiteit Brussel, Pleinlaan 2, 1050 Brussels, Belgium

Correspondence should be addressed to Joëlle Gérard; jgerard@vub.ac.be

Academic Editors: R. B. Peterson and T. L. Weir

Introduction of invasive macrophytes often leads to competition with native species or with already established invasive species. Competition between invasive species in multiple-invaded systems is expected to be particularly high, especially when they share growth form and position in the water column. We performed indoor experiments between invasive free-floating *Lemna minuta* and *Landoltia punctata* in monocultures and mixtures under a phosphorus gradient concurring with hypereutrophic, eutrophic, mesotrophic, and oligotrophic conditions. Our results showed that a phosphorus reduction from hypereutrophic to eutrophic had important negative impacts on the relative growth rate (RGR) of both species. A further reduction to mesotrophic condition did not alter either species RGR. However, species strategies and nutrient uptake differed. Both intra- and interspecific interference occurred; however, the intensity differed between phosphorus concentrations. Difference in RGR (RGRD) showed *L. minuta* to gain at high phosphorus levels, while a reduction favoured *L. punctata*. In oligotrophic condition, either species hardly produced new daughter fronds. Our results are useful to (1) understand the effects of phosphorus and setting target values in the process of eutrophication reduction and (2) diminish the impacts of invasive lemnids since a water column phosphorus reduction would prevent large impacts.

1. Introduction

In the last 100 years, a variety of invasive aquatic plants have been introduced to Europe due to increasing travel and trade [1]. These introductions often lead to competitive interactions with native species and with already established alien species [2]. Numerous studies on competition have demonstrated competitive superiority of alien plants over native plants due to their higher growth rates [3, 4]. However, competition between alien species has rarely been explored. A high level of competition might be expected, especially between related species with shared growth form, occurring in the same position in the water column [5, 6].

Eutrophication of aquatic ecosystems is a second major threat to freshwater biodiversity. It plays an important facilitating role in the invasion process [7, 8], since it can increase the invasibility of water bodies [9] and change the competitive balance between plant species leading to changes in species composition [10]. Eutrophication of most freshwater ecosystems is enhanced by phosphorus inputs [11], since P is often the limiting element for freshwater macrophytes [12]. P can occur as soluble reactive P (SRP), particulate organic P, and soluble organic P. P is delivered to aquatic ecosystems as a mixture of these forms, but aquatic plants only take up P in inorganic form, primarily as orthophosphate [13].

Several studies have shown that a high resource availability boosts the performance of invasive species [7, 8, 14], since successful invaders are capable of using limiting resources more efficiently than less invasive species, leading to higher relative growth rates (RGRs) [15]. However, most of these studies were performed only in high nutrient levels [16] because species with high competitive performance are known to occur in nutrient-rich areas and not in low-nutrient areas [17]. Reducing phosphorus levels is a powerful tool to reduce eutrophication [12] and might help in controlling the productivity and competitiveness of invasive macrophytes, possibly leading to changes in plant community composition. It is crucial that the latter is confirmed by competition experiments in low and high nutrients to examine whether invasiveness of plants is reduced after a water column nutrient reduction. Since competitiveness for nutrients should incorporate a plant's capability to use that nutrient for growth and its ability to store nutrients when its concentration is low [18], we studied morphology and life-traits, as well as nutrient storage in plant tissues.

Many alien plant species owe their invasive success to traits related to morphology, physiology, and reproduction [19]. Various studies have attempted to identify common traits associated with invasiveness, to better understand and predict the level of species invasiveness [20, 21]. Even though it is apparent that one set of traits is not applicable to every invasive plant, the search for traits is still crucial [19]. However, studies should focus on related species with the same growth form, since similar species may use different traits to promote invasiveness [19, 22].

Lemnids, commonly known as duckweed, are small fast growing aquatic plants and are known to be a symptom of high-nutrient concentration in small water bodies [23]. Lemnids develop dense mats when enough nutrients and light are available. These mats reduce submersed plant abundance by eliminating sunlight penetration and interfere with gaseous exchange, reducing fish populations [24, 25]. Duckweeds are P-hyperaccumulators and can use internally stored phosphorus for growth when it is no longer available [26, 27]. Hyperaccumulation can present advantages for the invasion of lemnids and can lead to higher competition with other invasive lemnid species. In this study, we used two invasive lemnid species: *Lemna minuta* Kunth (least duckweed) and *Landoltia punctata* Les & Crawford (dotted duckweed). *Lemna minuta* is native to North and South America and is currently an invasive alien species in Europe including Belgium [28]. *Landoltia punctata* originates from Australia and Southeast Asia and has invaded several European countries [29]. The species is not present in Belgium but has been reported in The Netherlands [30]. Since lemnids rely on the water column for nutrients, studies on water column nutrient change might help in understanding the effects of phosphorus and setting target values in the process of eutrophication reduction [31]. These results will show whether *L. minuta* and *L. punctata* are able to invade habitats ranging in trophic condition, or only high nutrient loaded water columns. Results will also show whether *L. punctata* should be considered for the Belgian watch or alert list for invasive species.

The present study aimed at investigating competitive outcomes between the two invasive species *Lemna minuta* and *Landoltia punctata* growing in varying P concentrations representing hypereutrophic, eutrophic, mesotrophic, and oligotrophic conditions. The main purposes of this study are to investigate (1) the intraspecific and interspecific competition between the invasive lemnids, (2) the life-history traits that enhance the invasive potential of the most invasive species, and (3) the effect of a phosphorus reduction on plant growth, competition, and plant strategies.

2. Materials and Methods

Lemna minuta was collected from a nature area "Kalkense Meersen" in Belgium, and *L. punctata* was obtained from the National Botanical Garden. A single plant was cultured for each species in full strength Hoagland's solution to obtain stock cultures. Plants were cultivated in glass aquaria in a temperature-controlled growth room at $25 \pm 2°C$ by day, $19 \pm 2°C$ by night under controlled photoperiod (16/8 hr day/night), and PAR approximately between 63 and $72 \, \mu mol \, m^{-2} \, s^{-1}$.

We performed four indoor experiments, corresponding to the trophic states as recognized in limnology [32]: oligotrophic ($0–10 \, \mu g \, P \, L^{-1}$), mesotrophic ($10–30 \, \mu g \, P \, L^{-1}$), eutrophic ($30–100 \, \mu g \, P \, L^{-1}$), and hypereutrophic ($>100 \, \mu g \, P \, L^{-1}$), with the latter being used as a control. The experiments were performed from August 2010 to March 2011 in a single temperature-controlled room at the Vrije Universiteit Brussel. For the hypereutrophic experiment, a standard Hoagland and Arnon [33] solution was used ($P = 30.97 \, mg \, L^{-1}$). For the other experiments, we used Hoagland and Arnon solution lacking phosphorus [33] and added KH_2PO_4 such that the phosphorus concentration concurred with the maximum value of each trophic state. To assess the importance of intra- and interspecific interactions, each experiment followed a complete additive design [34]. The initial frond number for $L. \, minuta : L. \, punctata$ was $0:14$, $0:28$, $14:28$, $14:14$, $28:28$, $28:14$, $28:0$, $14:0$, and each combination replicated 5 times. A density of 14 consisted of 4 similar sized mother fronds, each with 2 daughter fronds, and 1 mother frond with 1 daughter frond (Figure 1). A density of 28 contained twice these groups. Each combination of plants was placed in a 600 mL glass beaker filled with 250 mL solution, resulting in 40 beakers. Beakers were randomly placed in a growth room under identical conditions as during cultivation. Nutrient solutions were renewed every other day during 20 days, after which all plants were separated according to species. At the end of the eutrophic, mesotrophic, and oligotrophic experiments, digital images were taken to estimate the total frond area of each species. Plants were placed in a black container to minimize background effects, filled with an equal amount of water, placed at a constant distance from the camera. This method of estimation was chosen since it is noninvasive or damaging [35] and allows different parameters to be determined using appropriate user-friendly software [36]. Plant pixels were isolated from the total digital image in Adobe Photoshop 7.0, and the total frond number was counted in Image J.

In addition, we determined species traits by randomly selecting 15 clusters of fronds per species in each beaker and determining the number of fronds per cluster and the longest root per cluster. For the oligotrophic experiments all frond clusters were analysed since the final number of clusters was lower than 15. Plants were then oven-dried at $70°C$ for 48 hours and weighed.

Particulate organic nitrogen (PON) and carbon (POC) concentrations in plant tissue were analyzed simultaneously using a Flash EA 1112 (Thermo) elemental analyzer [37], at the Vrije Universiteit Brussel. Known amounts of samples ($\pm 10 \, mg$) are packed in tin cups and placed sequentially in the sampler of the analyzer. Samples are first injected under

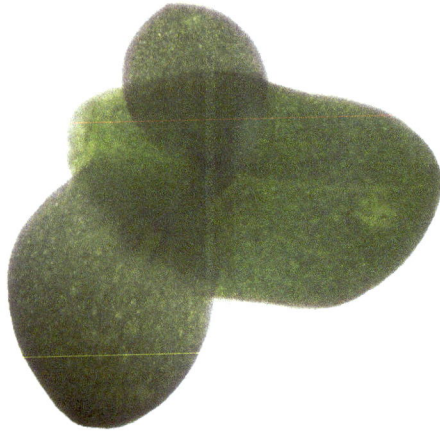

FIGURE 1: Underside of *L. punctata* group of fronds consisting of a mother frond and two clonally produced daughter fronds. The length of the mother frond is approximately 0.4–0.7 cm.

a continuous He flow in a combustion oven (1020°C) with a pulse of oxygen (flash combustion) where organic C and N are converted to CO_2, N_2, and nitrogen oxide gasses. Gasses then sequentially pass through a reduction oven filled with Cu (640°C) where all nitrogen oxides are reduced to N_2, through a water trap, and through a GC column where CO_2 and N_2 gasses are separated before reaching a thermal conductivity detector. An acetanilide standard was used to make 6-point standard curves. Samples are analyzed randomly with blanks (empty tin cups) and standards.

2.1. Statistical Analysis. The differences total frond area, number of fronds per cluster, total number of fronds, root length, and tissue nutrient concentration between species and between nutrient levels were analyzed with nonparametric Mann-Whitney U test and Kruskal-Wallis ANOVA. To determine the effect of species, water phosphorus and species × phosphorus on biomass, N%, and C%, a two-way ANOVA with aligned rank transformed (ART) data was performed.

The relative growth rate (RGR) of each species was calculated according to the formula: RGR $= \ln(Y/y)/t$, where y is the species initial stand biomass and Y the species stand biomass at the end of the experimental period t. The effects of competition and phosphorus reduction on RGR were analyzed using multiple linear regression leading to the equation

$$\mathrm{RGR}_i = a_{i0} + a_{ii}y_i + a_{ij}y_j + \varepsilon. \tag{1}$$

The coefficient a_{ii} determines the intraspecific effects of species i on its own RGR, whereas a_{ij} shows the interspecific effects of species j on the RGR of species i. The constant a_{i0} indicates a constant RGR for species i.

We calculated and modelled relative growth rate difference (RGRD) as described by Connolly and Wayne [38]: $\mathrm{RGR}_2 - \mathrm{RGR}_1$ (2 = *L. punctata*, 1 = *L. minuta*). If all coefficients are zero, then no change in biomass composition occurs. If RGRD > 0, *L. punctata* gains more than *L. minuta* and vice versa if RGRD < 0. The species with the higher RGR is

considered more efficient over the course of the experiment, having a higher output per unit input [39].

3. Results

3.1. Relative Growth Rate. The RGRs of both species in monocultures and mixtures (Table 1) were reduced significantly when lowering nutrients from hypereutrophic to eutrophic condition. However, no significant differences were observed for either species' RGR between eutrophic and mesotrophic condition. In oligotrophic condition both species hardly grew and both species' RGRs were consequently much lower than in other trophic conditions (*L. punctata* mixtures: H (3, $N = 80$) = 67,06 $P < 0.0001$, monocultures: H (3, $N = 40$) = 32,98 $P < 0.0001$; *L. minuta* mixtures: H (3, $N = 80$) = 67,72 $P < 0.0001$, monocultures: H (3, $N = 40$) = 33,00 $P < 0.0001$).

Lemna minuta always performed better in monocultures than mixed cultures except in oligotrophic condition (Mann-Whitney U test; H $P < 0.0001$, E $P < 0.0001$, M $P < 0.001$), while *L. punctata* showed no difference between monocultures and mixtures except in hypereutrophic condition (Mann-Whitney U test: $P < 0.0001$).

A comparison of the RGR between species indicated that *L. minuta* grew faster than *L. punctata* in monocultures except in oligotrophic (Mann-Whitney U test; H $P < 0.001$, E $P < 0.01$, M $P < 0.01$). In mixtures, however, both species performed similarly except in hypereutrophic condition, where *L. minuta* grew faster (Mann-Whitney U test: $P < 0.0001$).

3.2. Strategy

3.2.1. Plant Traits

Difference between Species. We found that in monocultures, both species covered a similar total area in eutrophic and mesotrophic conditions, while in oligotrophic condition, *L. punctata* covered a larger total area (Mann-Whitney U test; $P < 0.001$). In mixtures, *L. punctata* produced a higher total area than *L. minuta* in all experiments (Mann-Whitney U test; E: $P < 0.0001$, M: $P < 0.05$; O: $P < 0.0001$). Both species produced equal total amounts of fronds except in mixtures in oligotrophic condition, where *L. punctata* produced more fronds and in eutrophic monocultures, where *L. minuta* produced the most fronds. (Mann-Whitney U test; E monocultures: $P < 0.01$; O mixtures: $P < 0.05$). *L. punctata* possessed clusters consisting of more fronds than *L. minuta* in all experiments, in monocultures and mixtures, except in mesotrophic condition in mixture (Mann-Whitney U test; E: mixtures $P < 0.0001$, monocultures $P < 0.05$; M: monocultures $P < 0.01$; O: mixtures and monocultures $P < 0.0001$). At the end of each experiment, *L. punctata* had longer roots than *L. minuta* in mixtures and monocultures (Mann-Whitney U test, M: mixtures and monocultures $P < 0.01$, O: mixtures and monocultures $P < 0.0001$).

Effect of Nutrient Reduction. A nutrient reduction from eutrophic to mesotrophic condition did not change the

TABLE 1: RGR (mean ± SE) (g g^{-1} d^{-1}) of *L. punctata* and *L. minuta* in monocultures and mixtures in hypereutrophic, eutrophic, mesotrophic, and oligotrophic conditions.

	Hypereutrophic	Eutrophic	Mesotrophic	Oligotrophic
L. punctata				
Monocultures	0.283 ± 0.003	0.133 ± 0.008	0.131 ± 0.005	0.006 ± 0.002
Mixtures	0.255 ± 0.002	0.134 ± 0.005	0.129 ± 0.003	0.005 ± 0.001
L. minuta				
Monocultures	0.329 ± 0.003	0.164 ± 0.003	0.157 ± 0.005	0.004 ± 0.001
Mixtures	0.290 ± 0.003	0.141 ± 0.002	0.126 ± 0.006	0.006 ± 0.001

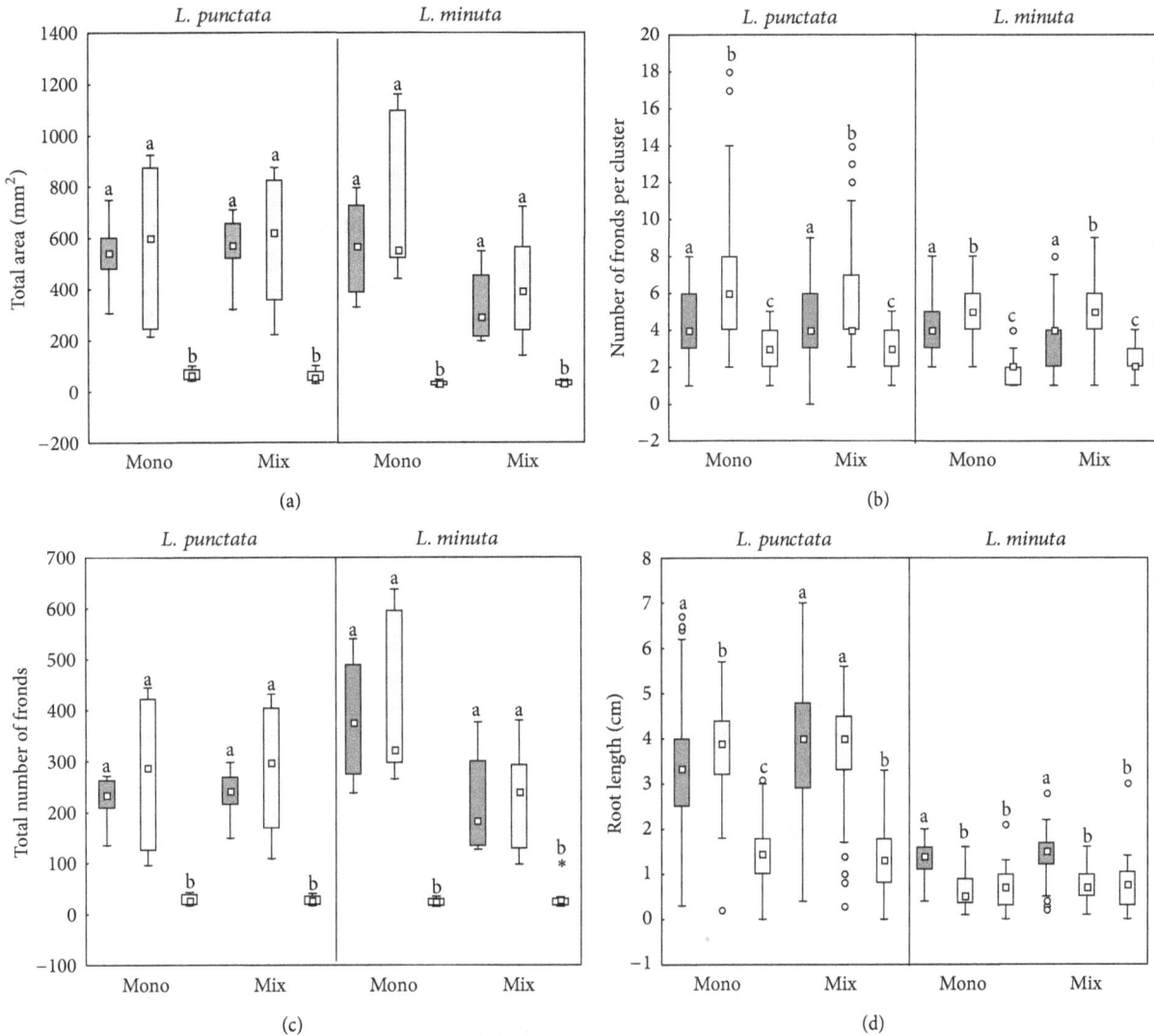

FIGURE 2: *L. punctata* and *L. minuta* traits after cultivation in eutrophic (dark grey), mesotrophic (light grey), and oligotrophic (white) conditions. Data were pooled for monocultures and mixtures at each phosphorus concentration. Different letters indicate significant differences between nutrient levels. □ represents the median, the box represents the interquartile range, and the vertical lines indicate the minimum and the maximum non-outlier values. Outliers are represented by ○, extremes by ∗.

species total frond area or total frond number, in mixtures and in monocultures (Figures 2(a) and 2(c)). However, both species produced clusters of fronds consisting of a higher numbers in mesotrophic condition, in monocultures and mixtures (Figure 2(b)). In oligotrophic condition, both

species produced few new fronds, thereby generating a lower frond area. A comparison of the longest root length (Figure 2(d)) revealed that in mixtures, *L. punctata* produced roots of similar length in eutrophic and mesotrophic conditions, while in oligotrophic condition, the length is significantly reduced. In monocultures, however, roots were significantly higher in mesotrophic condition than in eutrophic and oligotrophic conditions. *L. minuta* produced longer roots in eutrophic conditions and shorter roots of similar length in mesotrophic and oligotrophic conditions, in monocultures and in mixtures.

3.2.2. Tissue Nutrient Content

Difference between Species. When compared between species, the N% was significantly higher for *L. punctata* than for *L. minuta* in all experiments except in oligotrophic condition (H: 7.23 ± 0.05 versus 6.62 ± 0.06 $P < 0.0001$, E: 2.80 ± 0.08 versus 2.46 ± 0.04 $P < 0.01$, M: 2.78 ± 0.04 versus 2.37 ± 0.04 $P < 0.0001$, O: 1.89 ± 0.04 versus 1.85 ± 0.04). The overall tissue C% was significantly higher for *L. punctata* than for *L. minuta* in all experiments (H: 41.08 ± 0.09 versus 40.26±0.65 $P < 0.0001$, E: 40.91±0.50 versus 39.88±0.46 $P < 0.05$, M: 42.66 ± 0.23 versus 40.98 ± 0.27 $P < 0.0001$, O: 38.77 ± 0.52 versus 37.88 ± 0.51 $P < 0.05$).

Effect of Nutrient Reduction. The overall tissue N% followed the pattern that was observed for biomass (Figures 3(a) and 3(b)). In both species the tissue N% was significantly lower when the phosphorus concentration in the medium was reduced from hypereutrophic to eutrophic condition. However, no difference was observed between eutrophic and mesotrophic conditions. The oligotrophic condition, however, impeded both species N uptake and resulted in a tissue N% even lower than 2%.

The tissue C% showed a different pattern (Figure 3(c)). *L. punctata* tissue C% was similar in hypereutrophic and eutrophic conditions but increased in mesotrophic condition. In oligotrophic condition, however, the P in the medium was so low that C uptake was very low. For *L. minuta*, C% remained similar in hypereutrophic, eutrophic, and mesotrophic conditions but was lower in oligotrophic level.

Two-way ANOVA (Table 2) showed that water phosphorus concentration had a significant influence on biomass, N%, and C%, while species only had an effect on C% and biomass. Species x phosphorus concentration had a significant influence on all.

3.3. Influence of Nutrient Reduction and Intra- and Interspecific Effects on Species' RGR.
Both species showed a declining constant growth rate when phosphorus was reduced (Table 3). In hypereutrophic and eutrophic condition, the RGRs of *L. minuta* and *L. punctata* were affected by negative inter- and intraspecific interactions. This indicates that an increase in either species initial biomass would decrease both species RGR. In mesotrophic condition, the RGR of *L. minuta* was also influenced by negative intra- and interspecific effects. The opposite, however, was observed for *L. punctata* RGR, which was influenced by positive intra- and interspecific effects, indicating that an increase of either

TABLE 2: Effects of the water phosphorus level (P_{H_2O}), species, and their interaction ($P_{H_2O}*$ species) on the species N and C tissue content and biomass. Figures are *F*-ratios and levels of significance, based on two-way ANOVA with aligned rank transformed (ART) data.

	N	C	Biomass
P_{H_2O}	119.36[****]	72.19[****]	63.08[****]
Species	0.90[n.s.]	30.84[****]	419.58[****]
P_{H_2O} × species	9.32[****]	3.04[*]	12.05[****]

n.s.: nonsignificant, [*]$P < 0.05$, [****]$P < 0.0001$.

species initial biomass has positive effects on the *L. punctata* RGR.

3.4. Change in Biomass Composition.
The results of the RGRD model showed that a change in biomass composition occurred when *L. punctata* and *L. minuta* were grown together, in all conditions except in oligotrophic condition. The negative constant confirmed that a hypereutrophic and eutrophic condition favoured *L. minuta*, while a nutrient reduction to mesotrophic condition shifted the composition in favour of *L. punctata* (constant H: −0.0357, E: −0.0064, M: 0.0030).

4. Discussion

4.1. Effect of Nutrient Reduction on Growth.
In this study, both species' growth was negatively influenced by the reduction in phosphorus concentration. The average RGRs found in this study are similar to ranges in the literature [40–42]. Several authors observed an effect on plant RGR after a change in nutrients [43–45]. In this study, however, a reduction in phosphorus from eutrophic to mesotrophic condition had no effect on species' biomass production. One reason could be the small difference in phosphorus concentration between the two treatments. This was a possible explanation in the study of Hastwell et al. [46], who observed no change in biomass when reducing nutrients (0.06 versus 0.12 mg $P L^{-1}$, 0.76 versus 1.03 kjeldahl $N L^{-1}$). However, in other studies biomass production did change when reducing water nutrients to similar concentrations [47, 48]. This indicates that a reduction in phosphorus from hypereutrophic to eutrophic condition could have a substantial impact on both invasive species growth and spread, though an additional reduction to mesotrophic condition would have no further effect.

4.2. Competition.
We hypothesized that the outcome of competition would change with varying phosphorus concentration. *Lemna minuta* clearly benefited from high phosphorus concentrations, while a reduction to mesotrophic condition favoured *L. punctata*. A change in species competitive outcome with similar growth form after nutrient reduction was found in several studies [7, 45].

Our results show that competition occurred in every experiment except in oligotrophic condition. In hypereutrophic condition, the RGR of *L. punctata* was influenced by

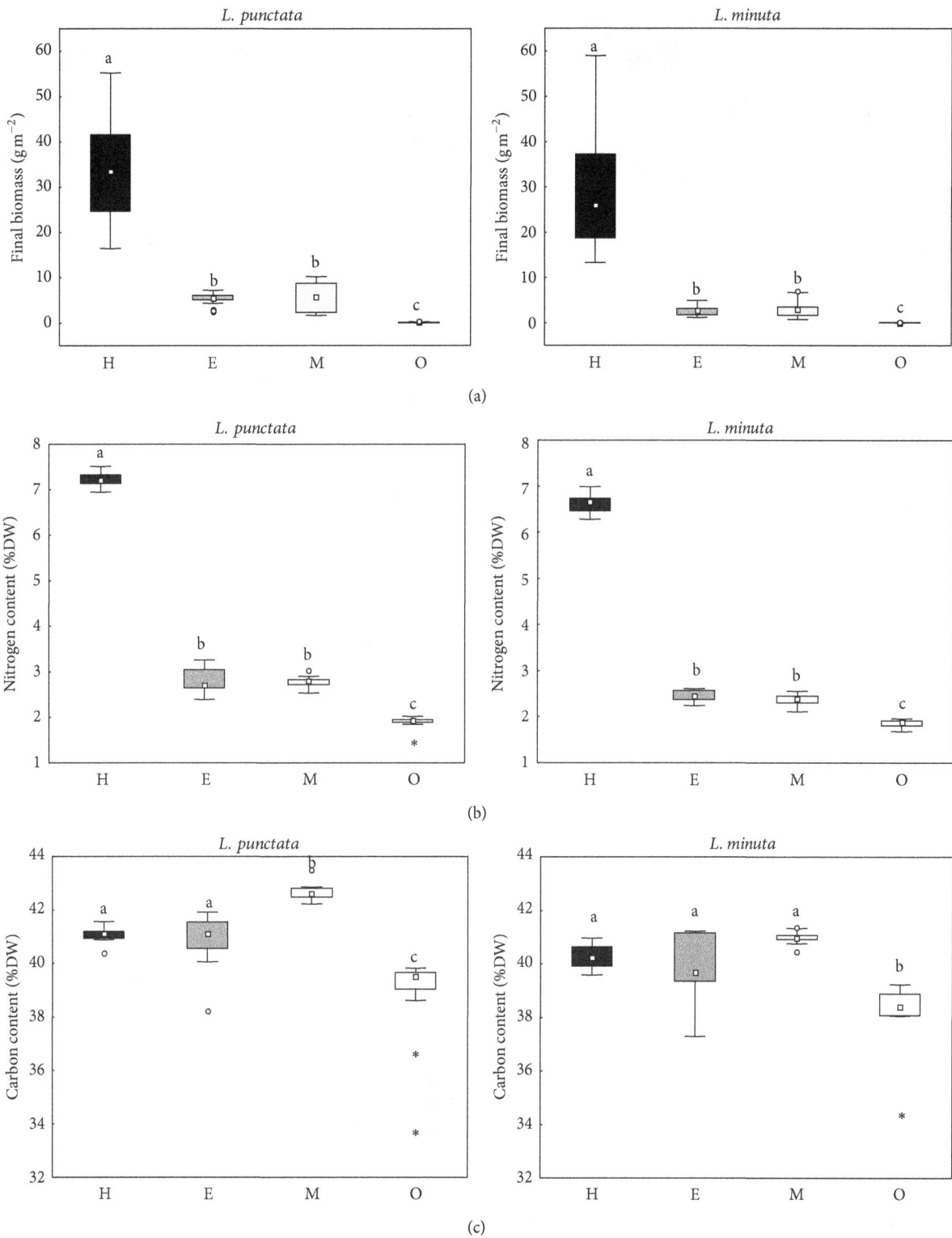

FIGURE 3: Final biomass and plant N and C content in *L. punctata* and *L. minuta* in hypereutrophic (black), eutrophic (dark grey), mesotrophic (light grey), and oligotrophic (white) conditions. □ represents the median, the box represents the interquartile range, and the vertical lines indicate the minimum and the maximum non-outlier values. Outliers are represented by ○, extremes by ∗. Different letters indicate significant differences between nutrient levels for each species.

TABLE 3: Linear equations for relative growth rate (RGR) of *L. punctata* (p) and *L. minuta* (m) in hypereutrophic (H), eutrophic (E), mesotrophic (M), and oligotrophic (O) conditions. (y_1) represents *L. minuta* and (y_2) represents *L. punctata*. Values in bold indicate significant coefficients $^{****}P < 0.0001$, $^{***}P < 0.001$, $^{**}P < 0.01$, $^{*}P < 0.05$, n.s.: nonsignificant.

Nutrient condition	Linear model	R^2	F
H	$RGR_p = \mathbf{0.2594}^{****} - 27.0565(y_1) - \mathbf{19.8473}^{****}(y_2)$	0.69	$F_{2,17} = 18.94^{****}$
	$RGR_m = \mathbf{0.2903}^{****} - \mathbf{64.9527}^{***}(y_1) - \mathbf{20.1793}^{****}(y_2)$	0.72	$F_{2,17} = 21.9^{****}$
E	$RGR_p = \mathbf{0.1345}^{****} - \mathbf{19.6726}^{**}(y_1) - \mathbf{18.7755}^{****}(y_2)$	0.77	$F_{2,17} = 29.25^{****}$
	$RGR_m = \mathbf{0.1408}^{****} - 11.9127(y_1) - 2.5873(y_2)$	0.19	$F_{2,17} = 1.99^{n.s.}$
M	$RGR_p = \mathbf{0.1294}^{****} + \mathbf{15.0709}^{****}(y_1) + \mathbf{9.0791}^{****}(y_2)$	0.84	$F_{2,17} = 43.37^{****}$
	$RGR_m = \mathbf{0.1264}^{****} - \mathbf{24.1459}^{*}(y_1) - \mathbf{15.6399}^{***}(y_2)$	0.58	$F_{2,17} = 11.58^{***}$
O	$RGR_p = \mathbf{0.0041}^{*} + 3.5400(y_1) - 1.1187(y_2)$	0.15	$F_{2,17} = 1.55^{n.s.}$
	$RGR_m = \mathbf{0.0059}^{***} + 2.22600(y_1) - 1.26294(y_2)$	0.15	$F_{2,17} = 1.55^{n.s.}$

intraspecific competition. The lack of interspecific competition seems evident since competition is expected to occur when nutrients are in short supply [49]. Since fronds are clustered in groups, competition with conspecific neighbours remains possible. A large reduction in phosphorus to eutrophic condition led to intra- and interspecific competition influencing the growth rate of *L. punctata*. In mesotrophic conditions, positive effects of both species have been observed on the RGR of *L. punctata*, which could indicate facilitation. Since only few lemnid competition experiments exist, no facilitative interactions are known in function of nutrients. However, facilitation has been observed by Driever et al. [42] in *L. minor*, producing mats in which temperature increases, thereby helping other species grow. For *L. minuta*, if any effects occurred, these were always negative.

4.3. Strategies. Even though *L. punctata* and *L. minuta* possess many similarities, their behaviour differed when facing competition. *L. punctata* did not change its strategy when facing competition. *L. minuta*, however, produced less total fronds, thereby covering a smaller total area. Even though a phosphorus reduction from eutrophic to mesotrophic condition induced no change in either RGR, frond morphology, or number, both species produced clusters consisting of more fronds. Although biomass allocation to roots is often observed at low nutrient supply [50], the root length of both species decreased when phosphorus was reduced. This can be explained by the fact that growth of meristematic tissue and especially root growth is associated with P and that increasing P promotes root growth [51]. In addition, lemnids are able to take up nutrients by the leaves [26]. Clusters consisting of more fronds could result in a larger contact surface with the water to maximize nutrient uptake.

4.4. Effect of Nutrient Reduction on Tissue Nutrient Content. The result that high phosphorus concentrations favour *L. minuta*, while lowering the concentration favours *L. punctata*, is supported by the nutrient tissue content, showing that in all conditions except for oligotrophic, *L. punctata* was able to store more N and C than *L. minuta*. Our N% and C% results concur with percentages found in other studies for *L.*

minuta and *L. punctata*, as well as other related lemnid species [50, 52, 53].

We found that P stress resulted in a lowered uptake of N and a stable or slight increase in C. Interactions between nutrients in tissue have been observed by other authors [54–56]; however, few elucidated on this. According to Xu et al. [50] an increase in carbon content in low nutrients could be due to the duckweed's starch production. However, Chapin [57] found similar results in N-stressed plants with lowered tissue P and increased C. Our results can similarly be explained by the fact that a nutrient limitation leads to a growth reduction and therefore a lower carbon need. Since photosynthesis continues, an accumulation of unused carbohydrates occurs. To match the plant's lower C requirement, photosynthesis declines, thereby reducing the amount of N needed which lowers the N uptake [57].

5. Conclusions

Lemnids are mostly controlled by mechanical removal; however, these tiny plants are never removed entirely and can rapidly cover a pond again due to their high turnover rates. Environmental control such as nutrient reduction is therefore needed to affect the species growth rates. Our results show that *L. minuta* and *L. punctata* are able to invade a wide variety of habitats ranging in trophic condition; however, a water column phosphorus reduction from hypereutrophic to eutrophic condition already induces a significant decrease in growth rates. Water quality targets should therefore be implemented and further studied [31]. The study of nutrient targets however should be species-based [58] to assure an effective control of individual freshwater species since different species can react differently to changed nutrient conditions. We can conclude that a $100\,\mu g\,P\,L^{-1}$ is a realistic goal for pond managers to achieve and to reduce *Lemna* cover.

We also found *L. punctata* to outcompete the very invasive *L. minuta* at lower nutrient conditions, emphasizing that the former species might become even more abundant if not carefully monitored. However, since its ecology hardly differs from *L. minuta* and native duckweed, this species will probably not become invasive [30]. Due to the widespread nonnative distribution of *L. minuta* and the species' impact, we advise that *L. minuta* should be added to Invasive Alien Species lists of European countries, as is the case for Belgium, and considered for management.

Conflict of Interests

The authors declare that there is no conflict of interests regarding the publication of this paper.

Acknowledgments

The authors would like to thank D. Maes, N. Brion, T Sierens, and K. Van Eyndonck. This Ph.D. research is funded by a scholarship provided by the government agency for Innovation by Science and Technology and by the Vrije Universiteit Brussel (BAS42).

References

[1] C. S. James, J. W. Eaton, and K. Hardwick, "Competition between three submerged macrophytes, *Elodea canadensis* Michx, *Elodea nuttallii* (Planch.) St John and *Lagarosiphon major* (Ridl.) Moss," *Hydrobiologia*, vol. 415, pp. 35–40, 1999.

[2] M.-H. Barrat-Segretain and A. Elger, "Experiments on growth interactions between two invasive macrophyte species," *Journal of Vegetation Science*, vol. 15, no. 1, pp. 109–114, 2004.

[3] J. H. Burns, "A comparison of invasive and non-invasive dayflowers (Commelinaceae) across experimental nutrient and water gradients," *Diversity and Distributions*, vol. 10, no. 5-6, pp. 387–397, 2004.

[4] E. Grotkopp and M. Rejmánek, "High seedling relative growth rate and specific leaf area are traits of invasive species: phylogenetically independent contrasts of woody angiosperms," *American Journal of Botany*, vol. 94, no. 4, pp. 526–532, 2007.

[5] B. Gopal and U. Goel, "Competition and allelopathy in aquatic plant communities," *The Botanical Review*, vol. 59, no. 3, pp. 155–210, 1993.

[6] D. M. Cahill, J. E. Rookes, B. A. Wilson, L. Gibson, and K. L. McDougall, "Turner review no. 17. *Phytophthora cinnamomi* and Australia's biodiversity: impacts, predictions and progress towards control," *Australian Journal of Botany*, vol. 56, no. 4, pp. 279–310, 2008.

[7] T. K. Van, G. S. Wheeler, and T. D. Center, "Competition between *Hydrilla verticillata* and *Vallisneria americana* as influenced by soil fertility," *Aquatic Botany*, vol. 62, no. 4, pp. 225–233, 1999.

[8] C. C. Daehler, "Performance comparisons of co-occurring native and alien invasive plants: implications for conservation and restoration," *Annual Review of Ecology, Evolution, and Systematics*, vol. 34, pp. 183–211, 2003.

[9] M. A. Davis, J. P. Grime, and K. Thompson, "Fluctuating resources in plant communities: a general theory of invasibility," *Journal of Ecology*, vol. 88, no. 3, pp. 528–534, 2000.

[10] V. H. Smith, G. D. Tilman, and J. C. Nekola, "Eutrophication: impacts of excess nutrient inputs on freshwater, marine, and terrestrial ecosystems," *Environmental Pollution*, vol. 100, no. 1–3, pp. 179–196, 1998.

[11] D. W. Schindler, "Evolution of phosphorus limitation in lakes," *Science*, vol. 195, no. 4275, pp. 260–262, 1977.

[12] A. N. Sharpley, T. Daniel, T. Sims, J. Lemunyon, R. Stevens, and R. Parry, *Agricultural Phosphorus and Eutrophication*, U.S. Department of Agriculture, Agricultural Research Service, 2nd edition, 2003.

[13] D. L. Correll, "The role of phosphorus in the eutrophication of receiving waters: a review," *Journal of Environmental Quality*, vol. 27, no. 2, pp. 261–266, 1998.

[14] E. C. Adair, I. C. Burke, and W. K. Lauenroth, "Contrasting effects of resource availability and plant mortality on plant community invasion by *Bromus tectorum* L.," *Plant and Soil*, vol. 304, no. 1-2, pp. 103–115, 2008.

[15] J. L. Funk and P. M. Vitousek, "Resource-use efficiency and plant invasion in low-resource systems," *Nature*, vol. 446, no. 7139, pp. 1079–1081, 2007.

[16] G. Fogarty and J. M. Facelli, "Growth and competition of *Cytisus scoparius*, an invasive shrub, and Australian native shrubs," *Plant Ecology*, vol. 144, no. 1, pp. 27–35, 1999.

[17] C. L. Gaudet and P. A. Keddy, "Competitive performance and species distribution in shoreline plant communities: a comparative approach," *Ecology*, vol. 76, no. 1, pp. 280–291, 1995.

[18] C. Garbey, K. J. Murphy, G. Thiébaut, and S. Muller, "Variation in P-content in aquatic plant tissues offers an efficient tool for determining plant growth strategies along a resource gradient," *Freshwater Biology*, vol. 49, no. 3, pp. 346–356, 2004.

[19] P. Pysek and D. M. Richardson, "Traits associated with invasiveness in alien plants: where do we stand?" in *Biological Invasions*, W. Nentwig, Ed., vol. 193, pp. 97–125, Springer, Berlin, Germany, 2007.

[20] M.-H. Barrat-Segretain, A. Elger, P. Sagnes, and S. Puijalon, "Comparison of three life-history traits of invasive *Elodea canadensis* Michx. and Elodea nuttallii (Planch.) H. St. John," *Aquatic Botany*, vol. 74, no. 4, pp. 299–313, 2002.

[21] Y. Feng, J. Wang, and W. Sang, "Biomass allocation, morphology and photosynthesis of invasive and noninvasive exotic species grown at four irradiance levels," *Acta Oecologica*, vol. 31, no. 1, pp. 40–47, 2007.

[22] Y.-L. Feng and G.-L. Fu, "Nitrogen allocation, partitioning and use efficiency in three invasive plant species in comparison with their native congeners," *Biological Invasions*, vol. 10, no. 6, pp. 891–902, 2008.

[23] R. Portielje and R. M. Roijackers, "Primary succession of aquatic macrophytes in experimental ditches in relation to nutrient input," *Aquatic Botany*, vol. 50, no. 2, pp. 127–140, 1995.

[24] R. A. Janes, J. W. Eaton, and K. Hardwick, "The effects of floating mats of *Azolla filiculoides* lam. and *Lemna minuta* Kunth on the growth of submerged macrophytes," *Hydrobiologia*, vol. 340, no. 1-3, pp. 23–26, 1996.

[25] M. E. Hernandez and W. J. Mitsch, "Deepwater macrophytes and water quality in two experimental constructed wetlands at Olentangy River Wetland Research Park," in *Deepwater Vegetation and Water Quality*, pp. 45–50, 2004.

[26] E. Landolt and R. Kandeler, *The Family of Lemnaceae—A Monographic Study, Volume 2*, Veröffentlichungen des Geobotanischen Institutes der Eidg. Techn. Hochschule, Stiftung Rübel , Zurich, Switzerland, 1987.

[27] J. M. Novak and A. S. K. Chan, "Development of P-hyperaccumulator plant strategies to remediate soils with excess P concentrations," *Critical Reviews in Plant Sciences*, vol. 21, no. 5, pp. 493–509, 2002.

[28] J. Lambinon, J. E. de Langhe, L. Delvosalle, and J. Duvigneaud, *Flora van België, het Groothertogdom Luxemburg,*

Noord-Frankrijk en de aangrenzende gebieden (Pteridofyten en Spermatofyten), National Botanical Garden, Meise, Belgium, 1998.

[29] E. Landolt, *The Family of Lemnaceae—A Monographic Study. Volume 1*, Veröffentlichungen des Geobotanischen Institutes der Eidg. Techn. Hochschule, Stiftung Rübel, Zurich, Switzerland, 1986.

[30] J. L. C. H. van Valkenburg and R. Pot, "*Landoltia punctata* (G.Mey.) D.H.Les & D.J.Crawford (Smal kroos), nieuw voor Nederland," *Gorteria*, vol. 33, pp. 41–49, 2008.

[31] P. A. Chambers, C. Vis, R. B. Brua, M. Guy, J. M. Culp, and G. A. Benoy, "Eutrophication of agricultural streams: defining nutrient concentrations to protect ecological condition," *Water Science and Technology*, vol. 58, no. 11, pp. 2203–2210, 2008.

[32] C. Bronmark and L. A. Hanssen, *The Biology of Lakes and Ponds*, Oxford University Press, Oxford, UK, 2nd edition, 2005.

[33] R. Hoagland and I. Arnon, "The water-culture method for growing plants without soil," *Circular*, vol. 347, pp. 1–32, 1950.

[34] C. J. T. Spitters, "An alternative approach to the analyses of mixed cropping experiments. 1. Estimation of competition effects," *Netherlands Journal of Agricultural Science*, vol. 31, pp. 1–11, 1983.

[35] H. E. Nilsson, "Remote sensing and image analysis in plant pathology," *Annual Review of Phytopathology*, vol. 33, pp. 489–527, 1995.

[36] L. C. Purcell, "Soybean canopy coverage and light interception measurements using digital imagery," *Crop Science*, vol. 40, no. 3, pp. 834–837, 2000.

[37] J. Nieuwenhuize, Y. E. M. Maas, and J. J. Middelburg, "Rapid analysis of organic carbon and nitrogen in particulate materials," *Marine Chemistry*, vol. 45, no. 3, pp. 217–224, 1994.

[38] J. Connolly and P. Wayne, "Assessing determinants of community biomass composition in two-species plant competition studies," *Oecologia*, vol. 142, no. 3, pp. 450–457, 2005.

[39] B. C. Hwang and W. K. Lauenroth, "Effect of nitrogen, water and neighbor density on the growth of *Hesperis matronalis* and two native perennials," *Biological Invasions*, vol. 10, no. 5, pp. 771–779, 2008.

[40] L. T. Valderrama, C. M. del Campo, C. M. Rodriguez, L. E. De-Bashan, and Y. Bashan, "Treatment of recalcitrant wastewater from ethanol and citric acid production using the microalga *Chlorella vulgaris* and the macrophyte *Lemna minuscula*," *Water Research*, vol. 36, no. 17, pp. 4185–4192, 2002.

[41] N. Cedergreen and T. V. Madsen, "Light regulation of root and leaf NO3- uptake and reduction in the floating macrophyte Lemna minor," *New Phytologist*, vol. 161, no. 2, pp. 449–457, 2004.

[42] S. M. Driever, E. H. van Nes, and R. M. M. Roijackers, "Growth limitation of *Lemna minor* due to high plant density," *Aquatic Botany*, vol. 81, no. 3, pp. 245–251, 2005.

[43] I. Woo and J. B. Zedler, "Can nutrients alone shift a sedge meadow towards dominance by the invasive *Typha × glauca*?" *Wetlands*, vol. 22, no. 3, pp. 509–521, 2002.

[44] S. Szabó, R. Roijackers, M. Scheffer, and G. Borics, "The strength of limiting factors for duckweed during algal competition," *Archiv für Hydrobiologie*, vol. 164, no. 1, pp. 127–140, 2005.

[45] J. Njambuya, I. Stiers, and L. Triest, "Competition between *Lemna minuta* and *Lemna minor* at different nutrient concentrations," *Aquatic Botany*, vol. 94, no. 4, pp. 158–164, 2011.

[46] G. T. Hastwell, A. J. Daniel, and G. Vivian-Smith, "Predicting invasiveness in exotic species: do subtropical native and invasive exotic aquatic plants differ in their growth responses to macronutrients?" *Diversity and Distributions*, vol. 14, no. 2, pp. 243–251, 2008.

[47] G. Rubio, J. Zhu, and J. P. Lynch, "A critical test of the two prevailing theories of plant response to nutrient availability," *American Journal of Botany*, vol. 90, no. 1, pp. 143–152, 2003.

[48] R. M. Wersal and J. D. Madsen, "Comparative effects of water level variations on growth characteristics of *Myriophyllum aquaticum*," *Weed Research*, vol. 51, no. 4, pp. 386–393, 2011.

[49] R. Aerts, "Interspecific competition in natural plant communities: mechanisms, trade-offs and plant-soil feedbacks," *Journal of Experimental Botany*, vol. 50, no. 330, pp. 29–37, 1999.

[50] J. Xu, W. Cui, J. J. Cheng, and A.-M. Stomp, "Production of high-starch duckweed and its conversion to bioethanol," *Biosystems Engineering*, vol. 110, no. 2, pp. 67–72, 2011.

[51] S. I. Tisdale, W. L. Nelson, and J. D. Beaton, *Soil Fertility and Fertilizers*, Macmillan, New York, NY, USA, 1985.

[52] B. A. Fulton, R. A. Brain, S. Usenko, J. A. Back, R. S. King, and B. W. Brooks, "Influence of nitrogen and phosphorus concentrations and ratios on Lemna gibba growth responses to triclosan in laboratory and stream mesocosm experiments," *Environmental Toxicology and Chemistry*, vol. 28, no. 12, pp. 2610–2621, 2009.

[53] L. Kufel, M. Strzałek, U. Wysokińska, E. Biardzka, S. Oknińska, and K. Ryś, "Growth rate of duckweeds (*Lemnaceae*) in relation to the internal and ambient nutrient concentrations û testing the droop and monod models," *Polish Journal of Ecology*, vol. 60, pp. 241–249, 2012.

[54] T.-H. Kim, W.-J. Jung, B.-R. Lee, T. Yoneyama, H.-Y. Kim, and K.-Y. Kim, "P effects on N uptake and remobilization during regrowth of Italian ryegrass (Lolium multiflorum)," *Environmental and Experimental Botany*, vol. 50, no. 3, pp. 233–242, 2003.

[55] S. Güsewell, "Nutrient resorption of wetland graminoids is related to the type of nutrient limitation," *Functional Ecology*, vol. 19, no. 2, pp. 344–354, 2005.

[56] M. A. Hussaini, V. B. Ogunlela, A. A. Ramalan, and A. M. Falaki, "Mineral composition of dry season maize (*Zea mays* L.) in response to varying levels of nitrogen, phosphorus and irrigation at Kadawa, Nigeria," *World Journal of Agricultural Sciences*, vol. 4, pp. 775–780, 2008.

[57] F. S. Chapin, "Integrated responses of plants to stress," *BioScience*, vol. 41, pp. 29–36, 1991.

[58] I. Clarke, Z. Stokes, and R. Wallace, "Habitat Restoration Planning Guide for Natural Resource Managers," 2010, http://www.environment.sa.gov.au/Home.

3

Ecological Implications of Acorn Size at the Individual Tree Level in *Quercus suber* L.

Soledad Ramos,[1] **Francisco M. Vázquez,**[2] **and Trinidad Ruiz**[3]

[1] *Departamento de Ingeniería del Medio Agronómico y Forestal, Escuela de Ingenierías Agrarias, Universidad de Extremadura, Avendia Adolfo Suárez s/n, 06007 Badajoz, Spain*
[2] *Departamento de Producción Forestal, Centro de Investigación La Orden-Valdesequera, Consejería de Infraestructura y Desarrollo Tecnológico, Junta de Extremadura, Apartado de Correos 22, 06080 Badajoz, Spain*
[3] *Departamento de Biología Vegetal, Ecología y Ciencias de la Tierra, Área de Botánica, Facultad de Ciencias, Universidad de Extremadura, Avendia de Elvas s/n, 06006 Badajoz, Spain*

Correspondence should be addressed to Soledad Ramos; solesuber@yahoo.es

Academic Editors: S. Ogita, S.-W. Park, and T. Vogt

Few studies have determined the influence of acorn size on germination and predation percentage at tree level. To evaluate the seed size influence at individual tree level, trees producing two different sizes of acorn were chosen. Our results show that smaller acorns were significantly more infested (49.6–75.3%) than larger ones (11.0–27.33%). About germination, big acorns achieved the best germination percentage compared to the smaller ones (18% in infested and 76% in sound acorns for the small acorn group versus 69.3% in infested and 93.3% in sound acorns belonging to the big acorn group). We also found that there was a difference in behaviour between big and small seeds at tree level. The same size belonging to different functional groups presented a difference at the behavioural level per tree. Infested small acorns from trees 8 and 10 had only 33 and 13% germination, while big acorns from trees 2, 3, and 6 (there was no difference between both sizes) presented 67, 97, and 83%, respectively. These results indicate that the production of acorns with two different sizes could be a strategy for species regeneration, producing each size for a different purpose.

1. Introduction

Quercus L. is a genus that produces a great acorn-size variation within the species and even at the individual tree level. Some Mediterranean *Quercus* species, such as holm oak (*Q. ilex* L.) and cork oak (*Q. suber* L.), frequently bloom twice a year. Male flowering, female flowering, or both may happen in spring and again in autumn [1–3]. Most of these species need only one season to complete the cycle (from flower until seed maturity) whereas others require two years. Both annual and biennial acorns are found in various species of the subgenus *Cerris*. Cork oak, however, is the only known oak species with annual and biennial acorns on the same tree [4–6]. Many female flowers show an annual cycle, with anthesis in spring or autumn and mature fruits during the same year. Several trees also produce biennial acorns with anthesis in spring or autumn of the first year and fruit set in the second year. All of this could be the origin of acorn size variability because spring flowering has more time to grow than autumn ones in annual cycle. Furthermore, the biennial flowering has more time than annual ones. All of these reasons make cork oak a very interesting species that produces acorn from September to January with different sizes and different origins.

Researchers have been interested in that variability, producing many works about the relationship between acorn size and germination or seedling production in cork oak [7–9] and in other species [10–14]. However, all of those studies of this trait have been carried out at the stand level not at the individual tree level. Only Tecklin and McCreary [15], working with blue oak at this level, found that the different behaviours from diverse acorn sizes were due not only to the size but also for each mother tree. This variability could be a reproductive strategy, because it is well known that one size of fruit is not favourable from an evolutionary point of view. Predators usually are adapted to attack a narrow range of fruit size [16, 17].

In accordance with previously published work, which aimed to determine the reproductive behaviour of cork oak [18], we proposed to understand the reason that this species produces different acorn sizes at the individual tree level. Specifically, the goals of this work were (a) to study biometrically the acorns to justify the existence of both big and small acorns; (b) to analyze the relationship between acorn size and germination; and (c) to analyze the relationship between acorn size and infestation percentage.

2. Materials and Methods

2.1. Plant Material. The study was conducted in a mountainous area in South-Western Spain—Sierra de Jerez de los Caballeros located at 500–693 m above sea level (UTM 29SQC94)—that is covered by holm oak and cork oak forming dehesa systems.

Acorns were collected from the ground beneath trees during November (at this time most of the acorns are on the ground [19]). To evaluate the influence of acorn size at the individual tree level, we selected seventeen cork oak that fulfilled the following conditions: they had two clearly different acorn sizes (big and small), without intermediate sizes, and their crowns did not overlapp any tree to avoid picking up acorns belonging to another individual. To carry out the different studies, 200 acorns per individual were randomly collected, 100 per each class of acorn size. Samples were identified by a number and letter according to their mother tree and class size and placed in plastic bags to be transported and stored at 4°C in the refrigerator.

2.2. Acorn Biometry Determination and Infestation Study. In order to ratify the variability of acorn size at the individual tree from the 200 stored acorns in the refrigerator, 50 acorns were randomly chosen, 25 from each class of acorn size (850 acorns in total). In these samples, after removing the cupules, length and width were measured to calculate the acorn volume according to the formulae designed for this kind of acorns by Ramos [18]. The acorns without holes could present larvae inside for that those acorns, after being measured, were carefully dissected and examined for the presence of predators (insects) or their remains.

2.3. Acorn Germination. The germination test was carried out two days after to be collected, in order to avoid loss of viability. Germination was studied in 3 replicas of 10 acorns each one from each collected sample, 30 were infested (acorns with insect holes) and the other 30 were apparently sound (without insect holes) with a total of 60 acorns studied per individual. In order to verify the absence of larvae, cotyledons were removed after seedling emergence. Whole sets were stratified on wet sand and were placed in a room simulating the environmental conditions of the natural population when the acorn germination is carried out (oscillation day/night aprox. 3°C to 18°C). Germination was recorded every 2-3 days for three months, and the acorns were watered when necessary. They were considered germinated when the radicle protruded about 2 mm.

TABLE 1: Nested ANOVA of the variation of the acorn volume according to acorn class (small versus big) nested within individual tree in *Quercus suber.*

	DF	MS	F	P
Tree	16	48.930	60.437	0.000
Acorn class (tree)	17	137.694	170.077	0.000
Error	811	0.810		

DF: degree of freedom; MS: mean square based on type III sums of squares.

2.4. Statistical Analysis. The statistical analyses were chosen according to Sokal and Rohlf [20] and were carried out with Statistical Software SPSS 15.0 for Windows [21]. After testing normality of the acorn volume variable using the Kolmogorov-Smirnov test with Lilliefors correction and Homoskedasticity using Levene's test, comparison of means was analyzed by a nested ANOVA with individual as principal factor and acorn class nested within the individual. Post hoc comparisons were analyzed using Tukey's HSD. The relationship between small and big acorn classes within individuals was analyzed using Pearson linear correlation. The comparison of acorn infestation and germination percentage was analyzed using the Pearson chi-square test by means of a two-factor (Row x Column) contingency table.

3. Results

3.1. Acorn Biometry. Acorn biometry was significantly different among individuals and class acorn nested within individuals (Table 1). The volume average varied between 1.65 and 3.91 cc in the small group and from 3.49 to 8.56 cc in the big group (Table 2). The individuals 8 and 10 could be stadistically separated of the other trees. However, the small acorn class of the 8 and 10 individuals was overlapping with the big acorn class belonging to other individuals, mainly to 1, 2, 3, 6, and 16 individuals. In spite of that, those small acorns had a different behaviour compared with those small acorns as we will comment hereinafter (Table 3).

By other hand, a significant positive correlation was found between the two acorn sizes at the individual tree level ($r = 0.845$, $P < 0.001$), that is, the higher acorn size of the big class, the higher acorn size of the small class. For this reason, it could be possible to consider trees 8 and 10 different from the rest (Figure 1).

3.2. Acorn Infestation and Germination. At population level, significant differences among acorn size were observed in the percentage of infested acorns ($\chi^2 = 154.87$, $P < 0.001$, Table 2). The mean percentage in the small class varied between 16 and 100%, while this variable in the big class ranged from 0 to 68%. At the individual tree level, there were significant differences in infestation between small and big acorns, being always lower infestation in big acorns, even some trees (8 and 9) presented 0% in this class (Table 2).

Infestation caused a significant reduction in germination percentage in both acorn classes, but this phenomenon was most intense and common in small acorns (Table 2, i.e, tree 2 small sound 43.33% versus small infested 0% and big

TABLE 2: Mean values of volume (cc), infestation, and germination percentage (sound versus infested) of both small and big acorn classes at individual tree level. In parentheses is shown the value of chi-square test and significance when comparing small versus big acorns.

Tree	Class	Volume[a]	Infestation (χ^2)	Germination Sound (χ^2)	Germination Infested (χ^2)	$\chi^{2\mathrm{b}}$
1	Small	2.06 ± 0.43***	16 (0.76ns)	100 (0.00ns)	63.33	13.47**
	Big	3.49 ± 0.59	8	100	—	—
2	Small	1.78 ± 0.54***	96 (6.64**)	43.33 (8.53**)	0 (30.00***)	16.60***
	Big	4.36 ± 0.92	68	80	66.67	1.36ns
3	Small	1.65 ± 0.37***	68 (13.88***)	96.67 (0.00ns)	10 (45.27***)	45.27***
	Big	4.66 ± 0.95	16	96.67	96.67	0.00ns
4	Small	1.99 ± 0.61***	100 (25.76***)	56.67 (10.76**)	13.33 (32.27***)	12.38***
	Big	5.55 ± 1.24	32	93.33	86.96	0.74ns
5	Small	2.32 ± 0.68***	80 (8.33**)	66.67 (12.00**)	16.67 (39.10***)	15.43***
	Big	6.71 ± 1.55	40	100	96.67	1.02ns
6	Small	2.16 ± 0.49***	44 (8.42**)	90 (0.22ns)	36.67 (13.61***)	18.37***
	Big	4.48 ± 0.95	8	93.33	83.33	1.46ns
7	Small	1.87 ± 0.52***	44 (2.88ns)	80 (2.31ns)	16.67 (42.86***)	24.09***
	Big	5.89 ± 1.57	20	93.33	100	2.07ns
8	Small	3.75 ± 0.73***	44 (14.10***)	93.33 (4.09*)	33.33	23.25***
	Big	8.56 ± 0.86	0	100	—	
9	Small	2.86 ± 0.75***	88 (39.29***)	83.33 (12.86***)	40	10.00**
	Big	6.59 ± 0.60	0	100	—	
10	Small	3.91 ± 0.67***	72 (24.53***)	93.33 (1.55ns)	13.33	38.57***
	Big	8.56 ± 0.91	4	98.33	—	
11	Small	2.74 ± 0.71***	84 (5.88*)	71.43 (5.46**)	16.67 (10.34***)	17.38***
	Big	5.64 ± 0.95	52	93.33	56.67	10.76**
12	Small	2.61 ± 0.65***	72 (6.52*)	73.33 (6.41**)	20 (7.18**)	17.14***
	Big	5.35 ± 0.98	36	96.67	53.33	15.02***
13	Small	2.55 ± 0.72***	52 (4.16*)	93.33 (0.35ns)	0 (10.59***)	52.50***
	Big	5.53 ± 1.68	24	96.67	29.63	28.71***
14	Small	1.87 ± 0.88***	76 (3.13ns)	36.67 (27.81***)	0 (7.935**)	13.47***
	Big	4.90 ± 1.03	52	100	23.33	37.30***
15	Small	2.16 ± 0.57***	32 (0.94ns)	80 (2.59ns)	3.33	36.27***
	Big	5.04 ± 0.95	20	63.33	—	
16	Small	2.01 ± 0.47***	76 (2.73***)	63.33 (5.43*)	3.33	24.30***
	Big	4.18 ± 1.35	8	85	—	
17	Small	1.78 ± 0.77***	72 (24.53***)	66.67 (9.02**)	13.33	17.78***
	Big	5.27 ± 1.09	4	96.67	—	0.00ns

[a]Significant differences between big and small acorns within the tree after nested ANOVA with class acorn nested in individual.
[b]Chi-square value and significance to compare germination between sound and infested acorns for each class per tree.
—indicate those trees that didn't present acorns apparently infested to be tested to germination.
ns: not significant; *$P < 0.05$; **$P < 0.01$; ***$P < 0.001$.

sound 80% versus big infested 66.67%). When considering exclusively infested or sound acorns, the germination percentage was always higher or sometimes similar in big acorns, although this difference was stronger inside the infested acorns (Table 2, i.e, tree 2 small sound 43.33% versus big sound 80% and small infested 0% versus big infested 66.67%).

The volume of the small class to trees 8 and 10 was overlapped with the big class of trees 1, 2, 3, 6, and 16. In spite of this the infestation percentage is significantly higher in the small acorn of those individuals than in the big acorns of the other trees (Table 3). In addition the reduction of germination percentage is higher in small acorns of individuals 8 and 10. In the case of sound acorns these differences disappear, only tree 1 showing 100% of germination differs significantly.

4. Discussion

4.1. Individual Tree Importance. One of the most relevant results obtained in this study is the role played by the individual tree. A significant difference was found between the two nearest sizes belonging to different volume class (i.e., 8 and 10S versus 1, 2, 3, 6, and 16B) in germination and infestation levels. These results indicate that size is important even at the individual tree level. Had it not been differentiated by tree

TABLE 3: Acorn infestation and germination percentage of small acorn class (**S**) belonging to individuals **8** and **10** (bold type) and big acorn class (B) belonging to individuals 1, 2, 3, 6, and 16 which size are statistically the same (mean ± S.E.).

Individual	Class acorn	Infestation	Germination	
			Sound	Infested
8	**S**	**44**	**93.33**	**33.33**
1	B	$12 (8.42^{**})$	$100 (4.09^{*})$	—
2	B	$68 (2.92^{ns})$	$80 (2.31^{ns})$	$66.67 (6.67^{**})$
3	B	$16 (4.67^{*})$	$96.67 (0.35^{ns})$	$96.67 (26.45^{***})$
6	B	$8 (8.42^{**})$	$93.33 (0.00^{ns})$	$83.33 (15.43^{***})$
16	B	$8 (8.42^{**})$	$85 (1.30^{ns})$	—
10	**S**	**72**	**93.33**	**13.33**
1	B	$12 (21.33^{***})$	$100 (4.09^{*})$	—
2	B	$68 (0.10^{ns})$	$80 (2.31^{ns})$	$66.67 (17.78^{***})$
3	B	$16 (15.91^{***})$	$96.67 (0.35^{ns})$	$96.67 (42.09^{***})$
6	B	$8 (21.33^{***})$	$93.33 (0.00^{ns})$	$83.33 (29.43^{***})$
16	B	$8 (21.33^{***})$	$85 (1.30^{ns})$	—

ns: no significant; $^{*}P < 0.05$; $^{**}P < 0.01$; $^{***}P < 0.001$.

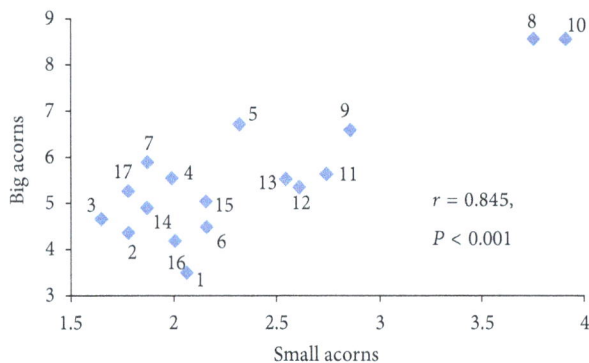

FIGURE 1: Pearson correlation between small and big acorns per individual.

collection, these seeds would have been included in the same volume group. The scarcity of references about the importance of mother plant in other species [15, 22] is peculiar.

In this study, the measurements were performed at the individual tree level because it gives us information on tree-to-tree variability within a population. Some trees (8 and 10) would be classified as bigger, because their acorns were placed at the top of the distribution by size shown in Figure 1.

Bigger acorns tended to present higher germination and lower acorn infestation. This fact is very clear at the individual tree level for those trees which presented the same size in different volume classes (8 and 10S versus 1, 2, 3, 6, and 16B) (Table 3). If we collected acorns regardless of tree but only the size. All of them would have included in the same class. However, by looking at their behaviour as found, clear differences can be seen in Table 3. This could indicate that each tree, genetic or physiological characteristics, produces a certain size, but also their sizes oriented towards a specific purpose.

4.2. Effect of Size. This study contributes to an understanding of the role of predispersal seed predation by insects and seed size in plant recruitment. Thus, larger seeds also satiated

the weevil larvae, and so a larger acorn size increased the likelihood of embryo survival because it has more mass to be consumed [12, 14, 23, 24]. These experiments reinforce the results obtained by others [12, 14, 24] about the importance of the satiation mechanism by means of larger seeds.

It is well known that predispersive acorn predation by insects is one of the constraints for the sexual regeneration of oak [13], so our results show that the endogenous consumption of seeds gives seed size an important role in the chance of seed's surviving predation.

A negative relationship between seed size and insect-infested acorns ratio was found in our study. The infested percentage was 23.29% for larger seeds and 65.41% for small ones, similar to those shown by Espelta et al. [13] for *Q. ilex* and *Q. humilis*.

Our study showed a reduction in germination percentage from sound acorns to those damaged by weevils, as observed by other authors [25]. However, infested larger seeds have shown a better germination ratio than infested smaller acorns according to the results found in other experiments [14, 22, 24]. To understand this result we should be aware of two facts: first, that a larger acorn size means it has larger cotyledons and second that insects oviposit on the basis of acorn, near the cupule [26]. Thus, the larvae began to consume the cotyledons from the base to the embryonic apex. So, the higher size acorn should have greater probability that its embryo will not be damaged, unlike what happens with small seed.

4.3. Ecological Implications. Our results show that cork oak has not only large intraspecific seed-size variability but also per tree. This variability is regarded as an important aspect in the evolution of angiosperms [17]. The seed size of a species represents the amount of maternal investment in the individual offspring [16]. This variability, together with its advantages, could be considered at the species level or at the ecosystem level. At the individual tree level, it presents a strategy to improve the seedling success, because larger seeds have more resources [17]. Thus the seed size variability at the individual level tree could be a reproductive strategy. Our study showed how not only important acorn size but also size within the tree (8 and 10S versus 1, 2, 3, 6, and 16B).

From another point of view, it could be possible that each tree produces different sizes with different roles, for example, small sizes to satisfy its predators; this would agree with those authors who find that birds choose infested acorns [27]. This situation also maintains the genetic variability in the population. On the other hand, larger seed sizes could be more appreciated by rodents as showed by Muñoz and Bonal [28].

In summary, we can conclude that it is important to distinguish acorn size at the individual tree level, not just at the stand level, because they show different behaviours. Furthermore, we postulate that cork oak represents a good reproductive strategy based on two ways that are not contradictory. Each tree could produce two acorn sizes in order to obtain different objectives: smaller ones could be produced for pest satiation and larger ones to produce seedlings. The same strategy may be followed by the whole stand, increasing the genetic variability at stand level.

Acknowledgments

The authors thank the staff members of the forestry department of the SIDT Junta de Extremadura for assistance with technical work and Mr. Manolo Méndez for permission to work with the cork oak placed in his dehesa. The first author also thanks Dr. J. López, Dr. A. Ortega-Olivencia, and Dr. T. Rodríguez-Riaño from Botany Área of Extremadura University for considerably improving this paper. This study was financed by INIA Projects Ph.D. Thesis "Biología reproductiva de una masa de alcornoque (*Quercus suber* L.) en el Sur de Badajoz (España)."

References

[1] R. Corti, "Ricerche sul ciclo riproduttivo di specie del gen. *Quercus* della flora italiana I. Osservazioni sul ciclo riproduttivo in *Quercus coccifera* L.," *Annali Accademia Italiana di Scienze Forestali*, vol. 2, pp. 235–264, 1954.

[2] P. Bianco and B. Schirone, "On *Quercus coccifera* L. s.l.: variation in reproductive phenology," *Taxon*, vol. 34, no. 3, pp. 436–439, 1985.

[3] R. Currás and E. Laguna, "Datos sobre la fenología de algunas especies forestales valencianas," *Montes*, vol. 10, pp. 50–52, 1986.

[4] R. Corti, "Ricerche sul ciclo riproduttivo di specie del gen. *Quercus* della flora italiana II. contributo alla biologia ed alla sistematica di *Quercus suber* L. in particulare delle forme a sivilupo biannale della ghianda," *Annali Accademia Italiana di Scienze Forestali*, vol. 4, pp. 55–136, 1955.

[5] J. A. Elena-Roselló, J. M. de Rio, J. L. García, and I. G. Santamaría, "Ecological aspects of the floral phenology of the cork-oak (*Q. suber* L.): why do annual and biennial biotypes appear?" *Annals of Forest Science*, vol. 50, pp. 114s–121s, 1993.

[6] P. M. Díaz, *Variabilidad de la fenología y del ciclo reproductor de Quercus suber L. en la península Ibérica [Ph.D. thesis]*, 2000, http://oa.upm.es/814/.

[7] J. M. Montoya, "Efectos de la profundidad de siembra y del tamaño de la bellota en el repoblado de *Quercus suber* L.," *Anales INIA: Serie Forestal*, vol. 6, pp. 9–16, 1982.

[8] J. L. García Valdecantos, "Influence of seed weight on the early growth of *Quercus suber* L. seedlings," *The International Plant Propagators Society*, vol. 35, pp. 432–436, 1985.

[9] M. Pardos, I. Cañellas, and A. Bachiller, "Influencia del tamaño de bellota y del régimen de riego en la calidad de planta de alcornoque cultivada en vivero. II Congreso Forestal Español," *Pamplona*, vol. 3, pp. 23491-37496, 1997.

[10] C. Bonfil, "The effects of seed size, cotyledon reserves, and herbivory on seedling survival and growth in *Quercus rugosa* and *Q. Laurina* (Fagaceae)," *American Journal of Botany*, vol. 85, no. 1, pp. 79–87, 1998.

[11] G. Ke and M. J. A. Werger, "Different responses to shade of evergreen and deciduous oak seedlings and the effect of acorn size," *Acta Oecologica*, vol. 20, no. 6, pp. 579–586, 1999.

[12] R. Bonal, A. Muñoz, and M. Díaz, "Satiation of predispersal seed predators: the importance of considering both plant and seed levels," *Evolutionary Ecology*, vol. 21, no. 3, pp. 367–380, 2007.

[13] J. M. Espelta, P. Cortés, R. Molowny-Horas, and J. Retana, "Acorn crop size and pre-dispersal predation determine interspecific differences in the recruitment of co-occurring oaks," *Oecologia*, vol. 161, no. 3, pp. 559–568, 2009.

[14] X. F. Yi and Y. Q. Yang, "Large acorns benefit seedling recruitment by satiating weevil larvae in *Quercus aliena*," *Plant Ecology*, vol. 209, no. 2, pp. 291–300, 2010.

[15] J. Tecklin and D. D. McCreary, "Acorn size as a factor in early seedlings growth of blue oaks," in *Proc. Symp. Oak Woodlands and Hardwood Rangeland Management*, R. B. Standiford, Ed., vol. 126 of *USFS Gen. Tech. Rep. PSW*, pp. 48–53, 1991.

[16] M. R. Leishman, I. J. Wright, A. T. Moles, and M. Westoby, "The evolutionary ecology of seed size," in *Seeds: The Ecology of Regeneration in Plant Communities*, M. Fenner, Ed., pp. 31–57, CABI Publishing, Wallingford, UK, 2nd edition, 2000.

[17] S. M. Mandal, D. Chakraborty, and K. Gupta, "Seed size variation: influence on germination and subsequent seedling performance in *Hyptis suaveolens*, Lamiaceae," *Research Journal of Seed Science*, vol. 1, pp. 26–33, 2008.

[18] S. Ramos, *Biología reproductiva de una masa de alcornoque, Quercus suber L., en el sur de Badajoz [Ph.D. thesis]*, Universidad de Extremadura, Badajoz, Spain, 2003, http://dialnet.unirioja.es/servlet/tesis?codigo=525.

[19] G. Montero and I. Cañellas, *Manual de Reforestaciñon y Cultivo de Alcornoque (Quercus suber L.)*, Instituto Nacional de Investigaciones Agrarias y Alimetnarias, Madrid, Spain, 1999.

[20] R. R. Sokal and F. J. Rohlf, *Biometry. The Principles and Practice of Statistics in Biological Research*, W. H. Freeman, New York, NY, USA, 3rd edition, 2000.

[21] M. Ferrán, *SPSS Para Windows. Análisis Estadístico*, McGraw-Hill, Madrid, Spain, 2001.

[22] M. J. Leiva and R. Fernández-Alés, "Holm-oak (*Quercus ilex* subsp. *Ballota*) acorns infestation by insects in Mediterranean dehesas and shrublands: Its effect on acorn germination and seedling emergence," *Forest Ecology and Management*, vol. 212, no. 1–3, pp. 221–229, 2005.

[23] A. T. Moles and M. Westoby, "Seedling survival and seed size: a synthesis of the literature," *Journal of Ecology*, vol. 92, no. 3, pp. 372–383, 2004.

[24] Z. Xiao, M. K. Harris, and Z. Zhang, "Acorn defenses to herbivory from insects: implications for the joint evolution of resistance, tolerance and escape," *Forest Ecology and Management*, vol. 238, no. 1–3, pp. 302–308, 2007.

[25] J. A. Lombardo and B. C. McCarthy, "Seed germination and seedling vigor of weevildamaged acorns of red oak," *Canadian Journal of Forest Research*, vol. 39, no. 8, pp. 1600–1605, 2009.

[26] F. J. Soria and M. E. Ocete, "Principales Tortrícidos perforadores del fruto del alcornoque en la Sierra Norte de Sevilla," *Boletín de Sanidad Vegetal Plagas*, vol. 22, no. 1, pp. 63–69, 1996.

[27] W. C. Johnson, L. Thomas, and C. S. Adkisson, "Dietary circumvention of acorn tannins by blue jays—implications for oak demography," *Oecologia*, vol. 94, no. 2, pp. 159–164, 1993.

[28] A. Muñoz and R. Bonal, "Are you strong enough to carry that seed? Seed size/body size ratios influence seed choices by rodents," *Animal Behaviour*, vol. 76, no. 3, pp. 709–715, 2008.

4

Factors Affecting Seedling Emergence and Dry Matter Characteristics in *Musa balbisiana* Colla

A. B. Nwauzoma[1,2] **and K. Moses**[1]

[1] *Department of Applied & Environmental Biology, Rivers State University of Science & Technology, PMB 5080, Port Harcourt 500001, Nigeria*
[2] *Embrapa Agroenergia, PqEB-Final W3 Norte, Asa Norte, 7077091 Brasilia, DF, Brazil*

Correspondence should be addressed to A. B. Nwauzoma; drnwabarth@yahoo.com

Academic Editors: F. A. Culianez-Macia and N. Rajakaruna

The effects of storage duration (0, 2, 4, 6, 8, and 10 days), sterilization with sodium hypochlorite (0, 5, 10, 15, and 20%), and weaning media on seedling characteristics and dry matter content in *Musa balbisiana* seedlings were studied. The experiment was factorial in a completely randomized design with five replicates. The result indicates that increase in NaOCl concentration and number of days in storage significantly ($P = 0.5$) increased the period of seedling emergence. Also, soaking in NaOCl for 20 min had significant effect on average seedling emergence at 15 and 20% concentrations, compared to 10 min soaking at the same concentrations. The combined effects of storage duration and sterilization resulted in a decrease in the duration of seedling emergence. Seeds previously sterilized with either water or NaOCl had no significant effect on seedling growth, leaf and corm dry weight, but affected almost all the dry matter traits. A mixture of poultry manure, top soil, and river sand as weaning media gave better seedling growth and increased dry matter characteristics. We conclude that *M. balbisiana* seeds require after-ripening treatment to enhance germination, sterilizing seeds with 5% NaOCl for 10 min and air-drying under ambient condition for 2–6 days were found most appropriate, and a mixture of poultry manure, top soil, and river sand is recommended as weaning medium for growth and dry matter composition in *M. balbisiana* seeds.

1. Introduction

Bananas and plantains (*Musa* spp.) are giant perennial herbs that grow in the humid agroecological zones of the tropics where they are important staples and contribute significantly to the incomes of the rural dwellers that grow them in compound or home gardens [1]. Bananas and plantains are derived from intra- and interspecific crosses between the diploid wild species *Musa acuminata* Colla (genome A) and *M. balbisiana* Colla (genome B) [2]. Most cultivated *Musa* are triploids ($2n = 3x = 33$) with the following genomic constitutions: dessert bananas predominantly AAA, plantains AAB, and cooking bananas ABB [2]. The wild progenitors (*M. acuminata* and *M. balbisiana*) of the edible bananas produce seeds, while most of the edible clones are seedless with a few exception such as "Pisang awak" subgroup ABB [3].

A major constraint to *Musa* breeding is the scarcity of healthy planting materials, due to low seed-set and viability [4]. Intractable fertilization barriers such as moderate to high levels of female sterility and triploidy make genetic improvements of parthenocarpic *Musa* clones slow and technically difficult [5]. Hybrid plant production in the most common triploid clones is further complicated by low seed-set and germination caused by endosperm failure [4]. Seeds are excellent dispersal units which have emerged in the course of plant evolution and are important genetic delivery systems essential for sustainable agriculture. *Musa* seeds are orthodox seeds; that is, the seeds can be dried to very low level moisture content (below 7%), stored at subzero temperatures, and are employed only for propagation in breeding programs. The possibility of conserving bananas in the form of seed is envisaged for a long-term preservation of *Musa* sp. Under favorable environmental conditions, at least 3–6 weeks is required for the initiation of germination in soil, and germination may occur either in a flush or intermittently over a period of 3–15 weeks. Germination percentages differ

Table 1: Relative proportion (%) of weaning substrates.

Media	TP	PM	SD	RS
1	25	25	25	25
2	33.3	33.3	0	33.3
3	33.3	50	33.3	0
4	50	50	0	0
5	0	50	50	0
6	50	33.3	16.7	0

Spent sawdust (SD), river sand (RS), topsoil (TP), and Poultry manure (PM).

Table 2: Effects of sterilization (NaOCl) and storage duration on seedling emergence in *M. balbisiana*.

Conc. (%) NaOCl	Days of storage/duration of seedling emergence (days)					
	0	2	4	6	8	10
0 (control)	16.7	17.6	18.8	20.7	21.6	22.2
5	16.0	17.1	18.8	20.2	21.3	21.7
10	18.3	19.8	21.6	22.3	25.1	27.3
15	19.6	23.4	36.2	27.0	27.7	26.5
20	20.2	22.9	24.2	27.2	28.8	28.7
LSD$_{(0.05)}$	1.72	1.59	9.06	1.69	1.95	1.72

Table 3: Effects of soaking time and NaOCl concentrations on days to seedling emergence in *M. balbisiana*.

Soaking time (min)	NaOCl concentration/seedling emergence time (days)				
	0	5	10	15	20
10	17.6	19.3	20.4	22.6	23.6
20	17.6	20.0	22.4	24.2	26.2
30	19.3	21.1	23.8	23.8	24.9
LSD$_{(0.05)}$	1.33	1.23	1.42	1.31	1.51

International Institute of Tropical Agriculture (IITA), Onne station (lat. 4° 43 N, long 7° 01 E, and 10 m altitude above sea level). *Musa balbisiana* seeds were extracted from matured fruits obtained from the field genebank of IITA Onne. *M. balbisiana* Colla, a diploid ancestor of commercial parthenocarpic varieties, was used because of the availability of large quantities of seeds. The weaning substrates sawdust, river sand, topsoil, and poultry manure were obtained from Port Harcourt and Onne towns. JIK (3.5% a.i) a Nigerian commercial product of sodium hypochlorite (NaOCl) used for sterilization was obtained from the market. A total of 1000 freshly harvested and viable *Musa balbisiana* seeds were selected using floatation method. Moisture Content (MC) of the seeds was also determined [14]. The seeds were air-dried, and only properly filled ones were selected for use.

markedly between harvest lots [6–8], depending on the maturity of the fruit at time of seed harvest, postharvest age of the seed, and method of storage [6, 7].

Little is known about the factors that affect seed germination in *Musa*, except that germination is extremely variable and relatively difficult to obtain under natural conditions [5]. The use of *Musa* seedlings as research tools and the increased emphasis on banana breeding programs require improved germination rates. Understanding the germination process will help in the conservation strategies and breeding programs [9]. Seed germination is considered as a critical phase in the reproductive cycle and is of great importance for species fitness, and variation in germination percentage has been interpreted as an adaptation to ecological conditions [10, 11]. Several reports have indicated the limited and variable seed germination exhibited by *Musa* [6, 12]. The work in [5] used an optimized *in vitro* culture of zygotic embryos to identify factors affecting germination and seedling growth in *Musa acuminata* subspecies *malaccensis*. The work in [13] employed *in vitro* embryo culture technique to produce hybrid germplasm in *Musa balbisiana*. Therefore, the objective of this study is to evaluate the effects of seed treatment chemicals, storage conditions, and weaning media on seedling emergence, growth, and dry matter content in *Musa balbisiana*.

2. Materials and Method

2.1. Experimental Site and Collection of Samples. This work was carried out at the Rivers State University of Science and Technology, Port Harcourt, Nigeria (Department of Applied and Environmental Biology), and formerly at the

2.2. Experimental Layout and Treatments. The experiment was factorial in a completely randomised design. The factors considered were number of days in storage after extraction (0, 2, 4, 6, 8, and 10 days) and sterilization with sodium hypochlorite (NaOCl, 3.5% active ingredient), with the following concentrations 5, 10, 15, and 20% and 0% as control (seeds soaked in tap water). After sterilization, seeds were washed under running tap water and air-dried under ambient room condition before planting. However, seeds with no-storage treatment (0) were planted immediately, while the remaining seeds were and stored in unsealed thin cellophane from where subsequent seeds were taken and planted accordingly.

There were five replicates and 10 seeds per replicate were sown in black nursery bags measuring 21 × 11 cm filled with sawdust. Medium sterilization was as described by [15] and watered with deionized water. Data on the effects of NaOCl concentration on days to seedling emergence after extraction was recorded. A seed was considered to have germinated when the radical had pierced through the testa up to 2 mm in height.

The effect of weaning media on growth parameters of *M. balbisiana* seeds was studied with six different weaning media, compounded in different ratios, namely, spent sawdust (SD), river sand (RS), topsoil (TP), and decomposed poultry manure (PM) (Table 1). Thereafter, 5-week-old seedlings (10 seedlings/weaning substrate) were planted in black nursery bags of 21 × 11 cm. Substrate sterilization and watering were as previously described, and data were taken at 4 and 8 weeks after transplanting on number of photosynthetically active leaves, height (cm) of seedling from

TABLE 4: Interaction between days after extraction (DAE) and varying concentrations of NaOCl on seedling emergence in *M. balbisiana*.

	Days to percentage seedling emergence								
	10%	20%	30%	40%	50%	60%	70%	80%	90%
DAE									
2	18.4	21.2	23.5	25.6	27.8	30.8	31.7	26.7	24.2
4	22.3	24.0	25.3	27.5	27.4	29.1	31.4	34.2	24.2
6	18.6	20.4	21.6	23.2	25.0	27.1	26.4	27.4	25.0
8	19.0	20.9	22.4	22.9	24.4	24.9	24.9	29.9	23.6
10	22.8	25.1	26.0	26.4	26.2	24.2	25.8	17.9	24.2
NaOCl Conc. (%)									
0 (control)	17.4	19.4	21.3	21.9	23.0	23.8	24.6	24.9	25.1
5	18.3	20.6	23.1	23.9	26.9	25.0	27.7	23.4	24.2
10	20.8	23.0	23.8	25.6	25.0	28.0	27.3	34.6	24.2
15	21.0	22.5	23.1	25.4	26.5	27.1	29.5	23.9	14.2
20	23.7	26.0	27.4	28.7	29.4	32.1	31.1	29.4	33.5
LSD$_{(0.05)}$	1.25	1.38	1.32	1.28	1.41	1.52	1.45	1.48	NS

DAE: Days after extraction.

the base of the pseudostem ("soil level") to the level of the last foliage, excluding the cigar leaves; plant girth/circumference (cm) taken from the base of the pseudostem above ground level and leaf area [16]. Leaf area = $(L \times W)0.8$, where L = length of leaf and W = width of leaf.

Destructive sampling was done at the end of 8 weeks on five randomly selected seedlings from each weaning medium, and data on the number of roots, length of longest roots (cm), and fresh weight (g) of corms, roots, leaves, and pseudostem, as well as their respective dry weights at 70°C, were recorded. Data were analyzed with GENSTAT Discovery Edition 1 [17] in a factorial completely randomized design. The least significant difference (LSD) at the 5% probability level was used to separate the means and detect the effects of storage, sterilization, weaning media, and interaction between storage and concentrations of sterilization medium on *M. balbisiana* seeds.

3. Results

Sterilization with sodium hypochlorite (NaOCl) at higher concentration significantly (P = 0.05) influenced the number of days to seedling emergence in *M. balbisiana*. The shortest duration in seedling emergence was observed in 5% concentration NaOCl which was not significantly different from the control (Table 2). There was no significant difference between seeds sterilized with 15 and 20% NaOCl, except in day 4 where seeds with 15% NaOCl sterilization took longer days (36.2 days) to emerge compared to 24.2 days in 20% NaOCl (Table 2). However, significant differences were observed between 20 and 10% concentration in all of the treatments. Thus, increase in both NaOCl concentration used in seed sterilization and number of days in storage resulted in an increase in the number of days to seedling emergence.

Soaking in NaOCl for 20 min had significant effect on average seedling emergence in *M. balbisiana* at 15 and 20%

concentrations, compared to 10 min soaking at the same concentrations (Table 3). Also, soaking for 30 min at 20% NaOCl resulted in longer days to seedling emergence in *M. balbisiana* at various concentrations (Table 3).

There was a significant interaction effect between days after extraction and varying concentrations of NaOCl on seedling emergence, except at 90% emergence (Table 4). Seeds soaked in water (0%), which was the control, emerged earlier than other treatments, except in 90% germination, while seeds sterilized with 20% NaOCl took longer days to emerge. At two, 6, and 8 DAE had the shortest period to emergence at 10%, 20%, and 30%, respectively, while 8 DAE had the shortest seedling emergence period in 40–70%. At 10% emergence, 2 and 10 DAE had the shortest (18.4) and longest (22.8) days to emergence, while at 30% the maximum (26.0) and minimum (21.6) days to emergence were at 10 and 6 DAE. At 70% emergence, 8 and 2 DAE had the shortest (24.9) and longest (31.7) days to emergence. It was further observed that 8 and 6 DAE had the longest (25.0) days, while an equal number of days were noted for 2, 4, and 10 DAE to seedling emergence, which was the shortest. Thus, increase in days after extraction causes decrease in the number of days to emergence (Table 4).

Seeds previously sterilized with either water or NaOCl had no significant effect on number of leaves, seedling height, leaf and corm dry weight, and corm dry matter (Table 5). However, significant effects were observed on total dry matter, stem dry weight, and almost all the dry matter traits (Table 5). Although seeds soaked in water resulted in taller seedlings, seeds sterilized with sodium hypochlorite had higher percentage root and leaf dry matter.

Table 6 shows the effect of weaning media on seedling height, number of leaves, and dry matter yield and distribution in *M. balbisiana* seedlings. Seedlings grown in a mixture of poultry manure, top soil, and river sand (PM/TP/RS; weaning media 2) as well as those on poultry

TABLE 5: Effects of previous seed treatments on growth, dry matter yield, and distribution in *M. balbisiana* seedlings.

Previous seed treatment	Ht (cm)	NL	Dry weight (g)					Dry matter (%)			
			TDW	Root	Stem	Leaf	Corm	Root	Stem	Leaf	Corm
Water	48.6	7.7	22.1	2.04	9.43	9.22	1.44	10.3	37.8	45.6	6.39
NaOCl	47.8	7.5	16.1	2.12	4.31	8.62	1.03	14.6	26.7	52.4	6.37
LSD$_{(0.05)}$	NS	NS	4.40	2.93	2.93	NS	NS	2.91	4.7	4.6	NS

Ht: height of seedling, NL: number of leaves, and TDW: total dry weight.

TABLE 6: Effect of weaning media on growth, dry matter yield, and distribution in *M. balbisiana* seedlings.

Weaning media	Ht (cm)	NL	Dry weight (g)					Dry matter (%)			
			TDW	Root	Stem	Leaf	Corm	Root	Stem	Leaf	Corm
1	51.4	7.0	25.53	2.85	11.66	9.76	1.25	12.4	39.73	42.31	5.53
2	56.2	7.6	25.20	2.42	9.10	11.42	2.26	10.1	32.15	49.58	8.14
3	45.0	7.1	12.08	1.58	4.42	5.41	0.67	15.0	33.10	46.12	5.82
4	52.2	8.0	20.64	2.57	5.63	11.05	1.39	12.4	28.00	53.10	6.51
5	34.6	8.2	8.06	1.28	2.28	3.99	0.52	17.1	27.47	48.94	6.54
6	49.8	7.5	23.8	1.78	8.13	11.86	1.31	7.62	32.91	53.77	5.69
LSD$_{(0.05)}$	6.42	0.8	7.62	0.76	5.06	2.97	1.01	5.04	8.14	7.88	2.47

Weaning media: 1: PM/TP/SD/RS; 2: PM/TP/RS; 3: PM/TP/SD; 4: PM/TP; 5: PM/SD; 6: PM/TP/SD; PM: poultry manure; TP: top soil; SD: sawdust and RS: river sand; Ht: height of seedling; NL: number of leaves; TDW: total dry weight.

manure/topsoil/sawdust/river sand (PM/TP/SD/RS; weaning media 1) and a mixture of poultry manure/top soil (PM/TP; weaning media 6) were significantly taller, compared to other treatments (Table 6). At least seven leaves were produced in each weaning medium, although significantly higher numbers were observed in weaning media 4 and 5. There were differences in dry weight traits; for instance, total dry weight was lowest in weaning media 5 (8.06 g) and weaning media 4 (20.64 g). The other weaning media had total dry weight (TDW) ranging from 23.8 g to 25.5 g. Root dry weight was higher in weaning media 1, 2, and 4 than in 3, 5, and 6. Stem dry weight was significantly different in weaning media 1, 2, and 6 with 11.66, 9.10, and 8.13 g, respectively. The least values for this trait were observed in weaning media 5 (2.28 g), 3 (4.42 g), and 4 (4.53 g). Higher leaf dry weight was obtained in weaning media 1, 2, 4, and 6, while the same for corm was highest in weaning medium 2 (2.26 g), followed by media 4, 6, and 1, that is, 1.39 g, 1.31 g, and 1.25 in that order. Percentage dry matter in root was higher in weaning media 5 and 3 and the least in 6 (7.62%). In stem, percentage of dry matter was highest in weaning medium 1 (39.73) and lowest in 5 (27.47). Significantly higher, percentage of leaf dry matter was produced in weaning medium 6 with 53.77, compared to weaning medium 1 which had 42.31. Similar observations were made in % corm dry matter.

4. Discussion

There are reports that storage duration [9, 18–20], soaking in water [6, 21, 22], and sterilization with sodium hypochlorite [20] affect seed germination characteristics in plants. The present work confirms that increase in NaOCl concentration and number of days in storage resulted in an increase in the number of days to seedling emergence in *Musa balbisiana*.

Similar results were obtained by [9] on germination and seedling characteristics of *Periploca angustifolia* Labill after long duration in storage. This is because as seeds get older the membrane system gradually becomes permeable allowing many electrolytes to flow out of cells leading to loss of vigor in such seeds. Similar effects were also observed when seeds were soaked in higher concentrations of NaOCl for a long time. Though sterilization with NaOCl is reported to reduce competition with pathogens which could impede the rate of growth and development of seedlings, [20], also it prolongs seedling emergence. It could be that the seeds may have to be resterilized after 10 days to reduce the time for seedling emergence. Soaking in water had better seedling emergence than NaOCl. Our findings further show that delayed planting for 10 days after seed extraction reduced seedling emergence which could be attributed to loss of seed viability as a result of loss in moisture. It was evident from the study that soaking duration influenced earliness to emergence and the total number of seedlings that emerged, while sterilization significantly affected the quality of seedlings produced. Seedlings arising from previously sterilized seeds had less total dry weight accumulation than water soaked seeds. This could be as a result of long duration of soaking seeds in NaOCl concentration. According to [23], the percentage germination in *Musa balbisiana* is highly variable depending on factors such as fruit maturity at seed harvest, the postharvest age of the seed, and the method of storage. The work in [21] reported that embryo germination was achieved in *Musa balbisiana* seeds after soaking in water for five days, prior to embryo excision. Similarly, Simmonds [6] and [23] also reported that the soaking of *Musa* seeds before sowing was either deleterious or ineffective to germination.

A mixture of poultry manure, top soil, and river sand as a weaning medium resulted in better seedling growth and dry matter characteristics, which is supposed to be a reflection of the chemical properties of the medium. It should be noted that 5 out of the 6 weaning media formulations had the same proportion of poultry manure as nutrient source. Lower bulk density enhances better root substrate relation [22], provided such media have the ability to hold the plant firmly in place and at the same time have adequate drainage and aeration [24]. A combination of poultry manure, top soil, sawdust, and river sand (PM/TP/SD/RS) with poultry manure, top soil, and river sand (PM/TP/RS) had higher bulk density and were more aggregated than sawdust and poultry manure (SD/PM), thus having relatively greater pore spaces for proper aeration. Media compaction and in situ media decomposition could cause an undesirable decrease in drainage and aeration leading to impediment of root growth. The decomposition of SD causes a depletion of available nitrogen within the potting mixture [25]. Thus, poor plant performance of SD/PM in the traits studied might be due to nutrient deficiency, especially nitrogen immobilization.

In conclusion, *M. balbisiana* seed needs after-ripening treatment to enhance seedling germination. Also, sterilizing *M. balbisiana* seeds with 5% dilution of sodium hypochlorite (NaOCl) at 10 min duration and air-drying under ambient tropical room condition for 2–6 days were found most appropriate. Finally, a mixture of poultry manure, top soil, and river sand is most appropriate as weaning media for growth and dry matter composition in *M. balbisiana*.

References

[1] F. I. Nweke, J. E. Njoku, and G. F. Wilson, "Productivity and limitations of plantain (*Musa* spp. Cv. AAB) production in compound gardens in southeastern Nigeria," *Fruits*, vol. 43, pp. 161–166, 1998.

[2] N. W. Simmonds, "Bananas," in *Evolution of Crop Plants*, J. Smart and N. W. Simmonds, Eds., pp. 370–374, John Wiley & Sons, New York, NY, USA, 1995.

[3] N. W. Simmonds, *Bananas*, Longmans, Green and Co., London, UK, 1966.

[4] K. J. Shepherd, L. L. Dantas, and B. Alves, "Banana breeding in Brazil," in *Banana and Plantain Breeding Strategies*, G. J. Persley and E. A. De Langhe, Eds., pp. 78–83, ACIAR, Canberra, Australia, 1987.

[5] M. J. Asif, C. Mak, and R. Y. Othman, "In vitro zygotic embryo culture of wild *Musa acuminata ssp.* malaccensis and factors affecting germination and seedling growth," *Plant Cell, Tissue and Organ Culture*, vol. 67, no. 3, pp. 267–270, 2001.

[6] N. W. Simmonds, "The germination of banana seeds," *Journal of Tropical Agriculture*, vol. 29, pp. 35–49, 1952.

[7] N. W. Simmonds, "Experiments on the germination of banana seeds," *Tropical Agriculture*, vol. 36, pp. 259–273, 1959.

[8] R. H. Stover, "A Rapid and simple pathogenicity test for detecting virulent clones of *Fusarium oxysporum f. cubense* using seedlings of *Musa balbisiana*," *Nature*, vol. 184, no. 4698, pp. 1591–1592, 1959.

[9] R. Abdellaoui, A. Souid, D. Zayoud, and M. Neffati, "Effects of natural long storage duration on seed germination characteristics of *Periploca angustifolia* Labill," *African Journal of Biotechnology*, vol. 12, no. 15, pp. 1760–1768, 2013.

[10] L. Navarro and J. Guitián, "Seed germination and seedling survival of two threatened endemic species of the northwest Iberian peninsula," *Biological Conservation*, vol. 109, no. 3, pp. 313–320, 2003.

[11] L. Rajjou and I. Debeaujon, "Seed longevity: survival and maintenance of high germination ability of dry seeds," *Comptes Rendus*, vol. 331, no. 10, pp. 796–805, 2008.

[12] N. Pancholi, A. Wetten, and P. D. Caligari, "Germination of *Musa velutina* seeds: comparison of in vivo and in vitro system," *In Vitro Cellular and Developmental Biology*, vol. 31, no. 3, pp. 127–130, 1995.

[13] K. Z. Ahmed, S. Remy, L. Sági, and R. Swennen, Germination of *Musa balbisiana* seeds and embryos.XVII Reunioa Internacional da Associacao para Cooperacao nas Pesquisas sobre Banana no Caribe e na America Tropical 15 a 20 de Outubro de 2006-Joinville-Santa Catarina, Brasil, 2006.

[14] ISTA, *The International Seed Testing Association*, Basserdorf, Switzerland, 2003.

[15] C. U. Agbo and C. M. Omaliko, "Initiation and growth of shoots of *Gongronema latifolia* Benth stem cuttings in different rooting media," *African Journal of Biotechnology*, vol. 5, no. 5, pp. 425–428, 2006.

[16] J. C. Obiefuna and T. O. C. Ndubizu, "Estimating leaf area of plantain," *Scientia Horticulturae*, vol. 11, no. 1, pp. 31–36, 1979.

[17] GENSTAT, *GENSTAT 5. 0 Release 4. 23DE*, Lawes Agricultural Trust, Rothamsted Experimental Station, 1st edition, 2003.

[18] M. B. McDonald, "Seed deterioration: physiology, repair and assessment," *Seed Science and Technology*, vol. 27, no. 1, pp. 177–237, 1999.

[19] C. C. Hsu, C. L. Chen, J. J. Chen, and J. M. Sung, "Accelerated aging-enhanced lipid peroxidation in bitter gourd seeds and effects of priming and hot water soaking treatments," *Scientia Horticulturae*, vol. 98, no. 3, pp. 201–212, 2003.

[20] K. P. Baiyeri and B. N. Mbah, "Surface sterilization and duration of seed storage influenced emergence and seedling quality of African breadfruit (*Treculia africana* Decne)," *African Journal of Biotechnology*, vol. 5, no. 15, pp. 1393–1396, 2006.

[21] J. C. Afele and E. de Langhe, "Increasing *in vitro* germination of *Musa balbisiana* seed," *Plant Cell, Tissue and Organ Culture*, vol. 27, no. 1, pp. 33–36, 1991.

[22] V. I. Ayodele, "Substrates for production of ornamental in Nigeria," in *Proceeding 15th HORTSON Conference*, NIHORT Ago Iwoye, April 1997.

[23] O. Stotzky, C. A. Cox, and R. D. Goose, "Seed germination studies in *Musa*. I. Scarification and aseptic germination of Musa balbisiana," *American Journal of Botany*, vol. 49, pp. 515–520, 1962.

[24] A. C. Bunt, *Modern Potting Composts*, Pennsylvania State University Press, 1976.

[25] L. W. Woodstock, "Seed Imbibition: a critical period for successful germination," *Journal of Seed Technology*, vol. 12, pp. 1–15, 1988.

5

Relationship between Carbon Stock and Plant Biodiversity in Collaborative Forests in Terai, Nepal

Ram Asheshwar Mandal,[1] **Ishwar Chandra Dutta,**[2]
Pramod Kumar Jha,[3] **and Siddhibir Karmacharya**[1]

[1] *Tirchandra College, Kathmandu, Nepal*
[2] *Tribhuvan University Commission, Kirtipur, Kathmandu, Nepal*
[3] *Central Department of Botany, Tribhuvan University, Kirtipur, Nepal*

Correspondence should be addressed to Ram Asheshwar Mandal; ram.mandal@gmail.com

Academic Editors: C. Bolle, F. A. Culianez-Macia, J. H. Titus, and J.-K. Weng

Reducing emission from deforestation and forest degradation (REDD+) programme has prime concern to carbon stock enhancement rather than biodiversity conservation. Participatory managed forest has been preparing to get benefit under this programme, and collaborative forest is one of them in Nepal. Hence, this research is intended to assess the relationship between carbon stock and biodiversity. Three collaborative forests (CFMs) were selected as study sites in Mahottari district, Nepal. Altogether 96 sample plots were established applying stratified random sampling. The plot size for tree was 20 m × 25 m. Similarly, other concentric plots were established. Diameter at breast height (DBH) and height were measured, species were counted, and soil samples were collected from 0–0.1, 0.1–0.3, and 0.3–0.6 m depths. The biomass was calculated using equation of Chave et al. and converted into carbon, soil carbon was analyzed in laboratory, and plant biodiversity was calculated. Then, relation between carbon stock and biodiversity was developed. Estimated carbon stocks were 197.10, 222.58, and 274.66 ton ha^{-1} in Banke-Maraha, Tuteshwarnath, and Gadhanta-Bardibas CFMs, respectively. The values of Shannon-Wiener Biodiversity Index ranged 2.21–2.33. Any significant relationship between carbon stock and biodiversity, and was not found hence REDD+ programme should emphasize on biodiversity conservation.

1. Introduction

Halting deforestation single can contribute to reduce about 18% atmospheric CO_2 emission [1]. Thus the forest management has objectively been focused on altering the deforestation and forest degradation targeting to get the benefit from reducing emission from deforestation and forest degradation (REDD+) programme in response to climate change [2]. Community managed forests also have been preparing to be candidate under this [3]. Simultaneously, other important part under this programme is biodiversity conservation and promotion.

Deforestation contributes about 5.9 GtCO_2 annually in the world [4]. The current rate of deforestation, clearing tropical forests could release an additional 87 to 130 GtC of CO_2 to the atmosphere by 2100 [5]. In the base year 1994/1995,

net emissions of CO_2 from all sectors in Nepal were estimated to be 9747 Gg and from the land-use change and forestry sectors were about 8117 Gg [6].

Collaborative forest is the management of the forests by three different collaborators—district forest office, district development institutions, and users including distant users. The participation of distant users in forest management and benefit sharing mechanism is the key feature of collaborative forests. Though the main purpose of collaborative forest management is to meet the demand of forest products of users, this also offers to store and sequestrate forest carbon simultaneously. Statistically, about 305.11 million ha forest was managed by community based institutions specifically indigenous people in 36 countries, and in Asia-Pacific it was about 146.00 million ha [7]. In Nepal, there are about 17 collaborative forests managing 43445 ha of forests areas. As

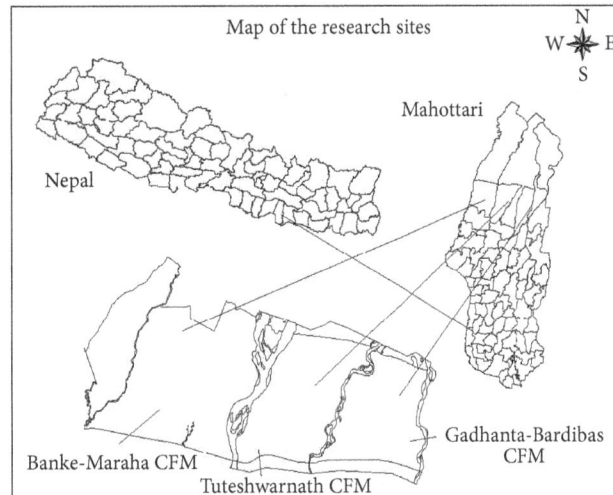

FIGURE 1: Map of the research site.

the carbon stock and biodiversity are intrinsic components in collaborative forests, their assessment is significant. The records of carbon stock and biodiversity are insufficient in collaborative forests, which justify the need of this study.

Approximately 8000 tree species, or 9% of the total number of tree species worldwide, are currently under threat of extinction because of forest decline [8] and impacts of climate change. Deforestation continues at an alarming rate which is consequently affecting the biodiversity in the tropics [9]. So, the climate change, deforestation, forest degradation, and biodiversity are interlinked to each other. Of the world's total land surface area, Nepal covers only 0.1% but harbors 136 ecosystems, about 2% of the flowering plants, 3% of the pteridophytes, and 6% of bryophytes of the world's flora, but 8 species are suspected to be extinct, 1 species is endangered, 7 species are vulnerable, and 31 species fall under the IUCN rare species category [10].

Important elements of REDD+ programme include monitoring, reporting, and verification (MRV) as well as reference emission level (REL) which need sufficient records of carbon stock. Moreover, noncarbon benefit of REDD+ programme has also focused on the biodiversity promotion. Besides, there is another significant growing concern about the relationship between carbon stock and biodiversity: whether working for carbon enhancement through REDD+ programme should include the biodiversity promotion too. Therefore, the research objectives are to assess carbon stock, biodiversity, and their relationship in collaborative forests of Nepal.

2. Material and Method

We selected three collaborative forests, namely, Banke-Maraha, Tuteshwarnath, and Gadhanta-Bardibas CFMs of Mahottari district, Terai (plain areas), which have areas of 2006, 1334, and 1450 ha, respectively. The reason why these forests were selected for this study is because all three are natural forests (Figure 1), and no such studies have been

carried out here. Selected collaborative forests are situated at $26°\ 36'$ to $28°\ 10'$ N and $85°\ 41'$ to $85°\ 57'$ E. The average annual temperature ranges $20–250°C$, and average annual rainfall recorded 1100–3500 mm. The main species of these forests is Sal (*Shorea robusta*), and other species are Saj (*Terminalia tomentosa*), Botdhairo (*Lagerstroemia parviflora*), Harro (*Terminalia chebula*), and Barro (*Terminalia bellirica*).

2.1. Sampling Design and Data Collection. Stratified random sampling was applied to gather the biophysical data. So, three main strata specifically regeneration, pole, and tree based on stage of the forest were delineated on the map of the study areas.

The pilot sampling was carried out to calculate the number of sample plots [11]. For this purpose at least 15 sample plots were taken from each stratum of collaborative forests. In this context, the diameter at breast height and height were measured to determine the minimum number of sample plots based on coefficient of variance [12]. Hence, altogether, 96 samples were collected, out of this, 32 samples from Banke-Maraha CFM, 33 samples from Tuteshwarnath CFM, and 31 from Gadhanta-Bardibas CFM.

Firstly, sample plots were distributed on each stratum on the map, and the coordinates of sample plots were uploaded in GPS. Secondly, concentric sample plots were established in the field by navigating the GPS coordinates. For tree stratum 20 m × 25 m sample plot was laid out, and nested plots for poles (10 m × 10 m), sapling (5 m × 5 m), seedling (5 m × 2 m) and litter, herbs, and grasses (1 m × 1 m) were laid out simultaneously [13]. Similarly, soil sample plot was laid out at the centre of the plot. The height and diameter at breast height of plants having dbh >1 cm were measured. Then, sapling (5 cm < dbh > 1 cm), seedlings, herbs, and shrubs were counted, and fresh weights of their samples were recorded. Moreover, soil samples were collected from three different depths 0–0.10, 0.10–0.30, and 0.30–0.60 m in order to determine the soil carbon. In addition, the list of tree species was prepared to assess the biodiversity.

TABLE 1: Carbon stock in collaborative forests.

Collaborative forests	Above ground C stock $t\,ha^{-1}$		Below ground C stock $t\,ha^{-1}$		Total $t\,ha^{-1}$	Total C t
	LHG	Regeneration + pole + tree	Root	Soil		
Banke-Maraha CFM	4.21	116.72	15.12	61.06	197.10	395398
Tuteshwarnath CFM	3.603	139.8	17.92	61.26	222.58	296927
Gadhanta-Bardibas CFM	6.325	178.88	23.15	66.31	274.66	398268

2.2. Data Analysis

2.2.1. Calculation of Carbon.
It is essential to calculate the forest biomass before determining the carbon except for soil carbon. Therefore, the above ground tree dry biomass (AGTB in kg) was calculated by using AGTB = $0.0509x\rho D^2 H$ [14] for plants dbh >5 cm, where ρ is wood density (g/cc), D is the diameter at breast height (cm), and H is the height of the tree (m).

Biomass of dbh <5 cm was estimated using Tamrakar's [15] equation. This equation only provides the fresh weight, so collected samples were dried in the lab at 105°C until samples showed the constant weight: Ln(AGSB) = $a + b\ln(DBH)$; whereas AGSB is the above ground sapling biomass (kg), Ln is natural log, a and b are constants, and DBH (cm) is diameter at breast height. Similarly, samples of seedling, leaf litter, herbs, and grass (LHG) also dried. Moreover, the root biomass was calculated by using root shoot ratio 0.125. The biomass was converted into carbon by multiplying with 0.47 [11].

Carbon content in the soil was analyzed by Walkley Black Method [16].

Bulk density (BD g/cc) = (oven dry weight of soil)/ (volume of soil in the corer).

SOC = organic carbon content % * soil bulk density (Kg/cc) * thickens of horizon.

Total carbon = total biomass carbon + soil carbon [17].

2.2.2. Biodiversity Calculation.
Biodiversity indices were calculated using following formulae.

Simpson's index $D = \sum n_i(n_i-1)/N(N-1)$, where N is the total number of all organisms and n_i the numbers of individuals of each individual species.

Simpson's diversity index = $1/p_i^2$, where p_i is the total individuals in a species community.

Species richness S is the number of species in the community or sample.

Simpson's evenness $E = D/S$, where D is the Simpson's diversity index and S is the species richness.

Shannon-Wiener Biodiversity Index $H = -\sum_{i=1}^{s}(p_i)(\ln p_i)$, where p_i is the total individuals in a species community [18].

2.2.3. Relationship between Carbon (Biomass) and Biodiversity.
Regression analysis was carried out to find the correlation between carbon stock and biodiversity. For this, only carbon of biomass was used. So, the relationship between carbon and species richness as well as carbon and evenness was developed so that REDD+ policy implication may be worthwhile.

2.2.4. Statistical Analysis.
The collected data set of collaborative forests was tested for normality in order to apply one-way ANOVA, Tukey's test by using software SPSS 17. Similarly, the biodiversity differences were also tested using t-test [19].

3. Results and Discussion

3.1. Comparison of C Stock among Collaborative Forests.
Total carbon stock in collaborative forests varied from site to site. It was found that the highest quantity of carbon stock was 274.66 $t\,ha^{-1}$ in Gadhanta-Bardibas CFM while it was lowest about 197.10 $t\,ha^{-1}$ in Banke-Maraha CFM (Table 1). The reason behind it may be due to various effects of drivers of deforestation and forest degradation. Generally, it was found that loggers of Khayarmara village living near to Banke-Maraha CFM illegally trade the timber and firewood. These types of activities are not so common in other CFMs. In addition, uncontrolled grazing and sudden fire also have been affecting the carbon stock in CFMs. Though, no study was carried out regarding the carbon stock in collaborative forests in Nepal, the pilot study done in Kayarkhola Watershed in community forest showed that 276.5 $t\,C\,ha^{-1}$ in the inventory was done in 2011 [20]. It was found different in studies done in Terai Arc Landscape. In the inventory carried out in 2010, there was 206.15 $t\,C\,ha^{-1}$ in government managed forests, 240 $t\,C\,ha^{-1}$ in community forests, and 274.58 $t\,C\,ha^{-1}$ in protected forests. [21]. These results are very close to the findings of this research.

3.2. Comparison of C Stock Variation in Collaborative Forests.
The ANOVA F test showed that there was variation in C stock in collaborative forests at 5% significant level.

There was clear variation in mean carbon stocks in each collaborative forest. The Tukey's HSD showed that the quantity of C stock of Banke-Maraha, Tuteshwarnath, and Gadhanta-Bardibas CFMs varied with each other at 5% level of significant (Table 2).

3.3. Relationship between Species Richness and Carbon Stock in CFMs.
The result showed that there was positive but

TABLE 2: Multiple comparisons of C stocks using Tukey HSD test.

Variation (I)	Variation (J)	Mean difference ($I - J$)	Std. error	P value
Banke-Maraha CFM	Tuteshwarnath CFM	−25.29*	1.48	0.00
	Gadhanta-Bardibas CFM	−72.33*	1.50	0.00
Tuteshwarnath CFM	Banke-Maraha CFM	25.29*	1.48	0.00
	Gadhanta-Bardibas CFM	−47.04*	1.49	0.00
Gadhanta-Bardibas CFM	Banke-Maraha CFM	72.33*	1.50	0.00
	Tuteshwarnath CFM	47.04*	1.49	0.00

*The mean difference is significant at the .05 level.

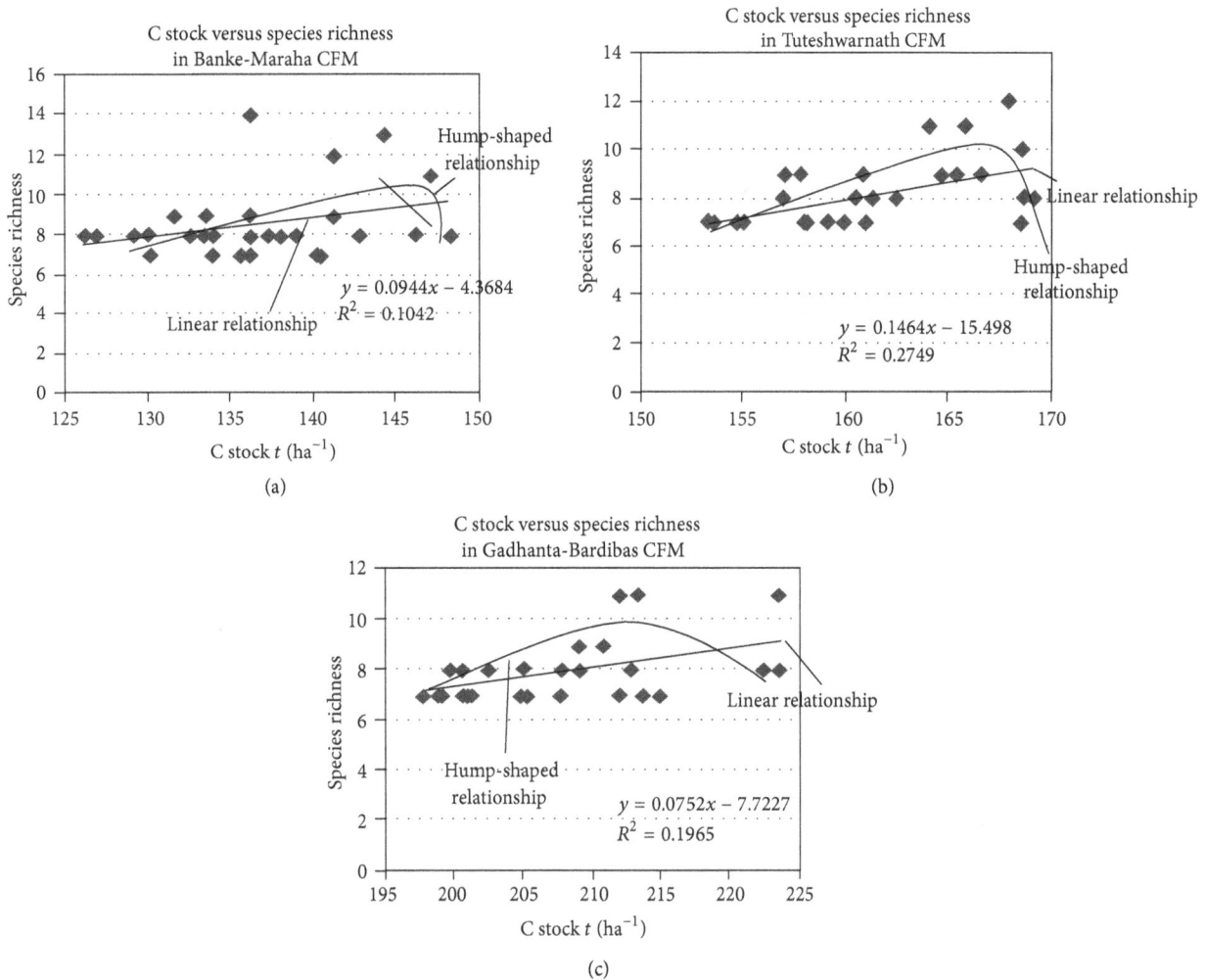

(a)   (b)

(c)

FIGURE 2: (a) Relation between species richness and C stock in Banke-Maraha CFM. (b) Relation between species richness and C stock in Tuteshwarnath CFM. (c) Relation between species richness and C stock in Gadhanta-Bardibas CFM.

very weak relationship between carbon stock and species richness of collaborative forests. However, it showed nearly hump-shaped relationship (Figures 2(a)–2(c)). The values of R^2 of linear regression range from 0.10 to 0.27. Generally, the variation of carbon stock does not depend upon the species diversification. The research done by Karna [22] also supported this idea. He stated that there is positive but weak relationship between carbon stock and biodiversity, though the hump-shaped relationship existed between them [23, 24].

3.3.1. Relationship between Species Richness and Carbon Stock in Banke-Maraha CFM. For more details see Figure 2(a).

3.3.2. Relationship between Species Richness and Carbon Stock in Tuteshwarnath CFM. For more details see Figure 2(b).

3.3.3. Relationship between Species Richness and Carbon Stock in Gadhanta-Bardibas CFM. For more details see Figure 2(c).

TABLE 3: Biodiversity indices in collaborative forests.

Biodiversity indices	Banke-Maraha CFM	Tuteshwarnath CFM	Gadhanta-Bardibas CFM
Shannon-Wiener Biodiversity Index	2.33	2.28	2.21
Simpson's index	0.39	0.41	0.44
Average species richness	8.45	8.12	7.94
Simpson's evenness (mean value)	0.85	0.83	0.79

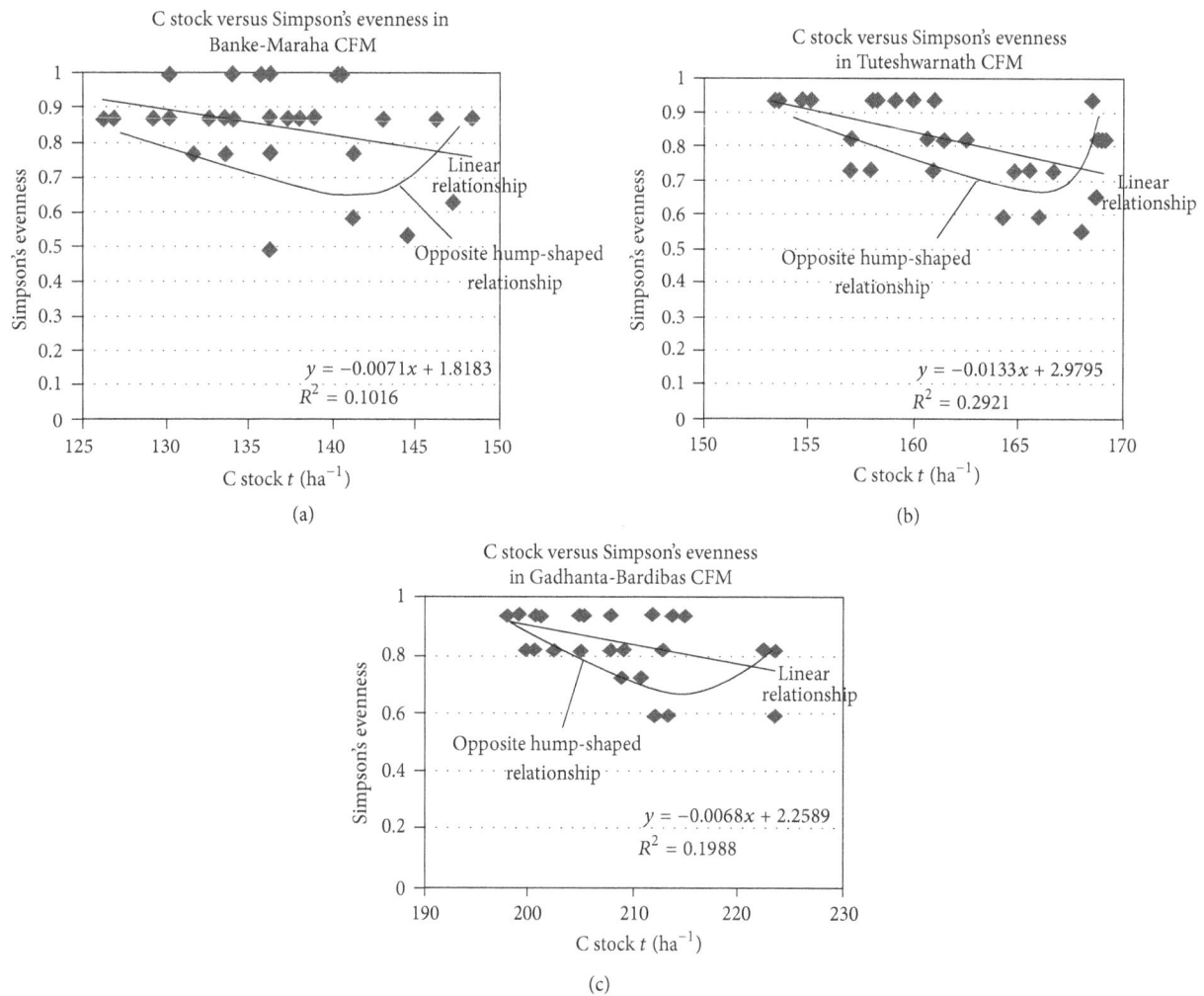

FIGURE 3: (a) Relation between Simpson's evenness and C stock in Banke-Maraha CFM. (b) Relation between Simpson's evenness and C stock in Tuteshwarnath CFM. (c) Relation between Simpson's evenness and C stock in Gadhanta-Bardibas CFM.

3.4. *Relation between Simpson's Evenness and C Stock.* The result showed weak and negative relationship between carbon stock of collaborative forests and Simpson's evenness as the values of R^2 range from 0.10 to 0.29. However, the opposite hump-shaped relationship was found between them (Figures 3(a)–3(c)). This finding is also supported by the study done by Heather et al. [25].

3.4.1. *Relationship between Simpson's Evenness and C Stock in Banke-Maraha CFM.* For more details see Figure 3(a).

3.4.2. *Relationship between Simpson's Evenness and C Stock in Tuteshwarnath CFM.* For more details see Figure 3(b).

3.4.3. *Relationship between Simpson's Evenness and C Stock in Gadhanta-Bardibas CFM.* For more details see Figure 3(c).

3.5. *Variation in Biodiversity in Collaborative Forests.* The values of biodiversity indices varied according to collaborative forests (Table 3). The Shannon-Wiener Biodiversity Index was the highest 2.33 in Banke-Maraha CFM, and it was the lowest 2.21 in Gadhanta-Bardibas CFM. This indicates that the highest biodiversity was in Banke-Maraha CFM.

More species diversity was found in the fringe areas of forest types. There are two rivers, namely, Banke river in the west and Maraha in the east. Because of river rain tropical and *Shorea robusta* mixed forest in Banke-Maraha CFM, it had the highest biodiversity. No management operations except for selection felling have been carried out in collaborative forests. The research done by Sapkota et al. [26] in hill *Shorea robusta* forest showed that values of Shannon-Wiener index and Simpson index were 2.42 and 0.64, respectively, which are close to the value of Shannon Wiener index of Banke-Maraha CFM.

3.6. Comparison of Biodiversity in Collaborative Forests. Statistically, *t*-test showed there was no clear differences in values of Shannon-Wiener indices between Banke-Maraha and Tuteshwarnath CFM, Gadhanta-Bardibas and Tuteshwarnath CFM, and Banke-Maraha and Gadhanta-Bardibas CFM as the *t*-cal < *t*-tab at 5% significant level. So, it can be reasonably confident that there were no clear variation in Shannon Wiener indices value among (i) Banke-Maraha and Tuteshwarnath CFMs, (ii) Tuteshwarnath and Gadhanta-Bardibas CFMs, and (iii) Gadhanta-Bardibas and Banke-Maraha CFMs.

4. Conclusion and Recommendation

The highest quantity of carbon stock was found in Gadhanta-Bardibas CFM while it was lowest in Banke-Maraha CFM. There was positive and very weak relationship between carbon stock and species richness of collaborative forests; it showed nearly hump-shaped relationship. However, the opposite hump-shaped relationship was found between values of Simpson's evenness and carbon stock. So, it indicated that the forest carbon enhancement cannot assure the biodiversity conservation and promotion. Therefore, the REDD+ programme should have parallel focus on biodiversity conservation and promotion.

Conflict of Interests

The authors report no conflict of interests. The authors alone are responsible for the content and writing of the paper.

Acknowledgments

The authors would like to acknowledge the chairperson of Banke-Maraha collaborative, Mr. Raghunath Prashad Yadav, for his sincere help in field data collection. They are also thankful to the staffs of district forest office like Ram Ashis Yadavand and Ram Jiban Yadav for their help during the field work.

References

[1] IPCC, *Fourth Assessment Report of the Intergovernmental Panel on Climate Change*, Intergovernmental Panel on Climate Change, Geneva, Switzerland, 2009.

[2] M. Skutsch and P. E. van Laake, "Redd as multi-level governance in-the-making," *Energy and Environment*, vol. 19, no. 6, pp. 831–844, 2008.

[3] C. Luttrell, L. Loft, M. Gebara, and K. D. Fernada, "Who should benefit and why? Discourses on REDD⁺ benefit sharing," in *Analyzing REDD⁺ Challenges and Choices*, pp. 129–152, CIFOR, Bogor, Indonesia, 2012.

[4] IPCC, *Climate Change 2007: Synthesis Report*, edited by R. K. Pachauri and A. Resinger, Contribution of Working Groups I, II and III to the Fourth Assessment Report of the Intergovernmental Panel on Climate Change, Intergovernmental Panel on Climate Change, Geneva, Switzerland, 2007.

[5] W.G. Ross and P. A. Sheikh, "Deforestation and Climate Change," Congressional Research Service, 2010.

[6] MOPE, "Initial national communication to the conference of the parties of the united nations framework convention on climate change," Tech. Rep., Ministry of Population and Environment, Singhadurbar, Kathmandu, 2004.

[7] RRI and ITTO, "Tropical forest tenure assessment: trends, challenges and opportunities," Rights and Resources Institute and International Tropical Timber Organization, 2009.

[8] S. P. Singh, P. P. Sah, V. Tyagi, and B. S. Jina, "Species diversity contributes to productivity—evidence from natural grassland communities of the Himalaya," *Current Science*, vol. 89, no. 3, pp. 548–552, 2005.

[9] FAO, "Global forest resources assessment," Main Report, FAO Forestry Paper, 2010.

[10] MFSC, *National Biodiversity Action Plan*, Ministry of Forests and Soil Conservation, HMG of Nepal, Global Environment Forum, UNDP, 2000.

[11] K. G. Macdicken, *A Guide to Monitoring Carbon Storage in Forestry and Agroforestry Projects*, Winrock International Institute for Agricultural Development, Arlingt, Va, USA, 1997.

[12] S. D. Moore and G. P. Maccabe, *Introduction to the Practice of Statistics*, M.H. Freeman and Company, New York, NY, USA, 2003.

[13] DOF, *Community Forestry Inventory Guideline*, Community Forests Division, Department of Forests, Kathmandu, Nepal, 2003.

[14] J. Chave, C. Andalo, S. Brown et al., "Tree allometry and improved estimation of carbon stocks and balance in tropical forests," *Oecologia*, vol. 145, no. 1, pp. 87–99, 2005.

[15] P. R. Tamrakar, *Biomass and Volume Tables with Species Description for Community Forest Management*, Ministry of Forests and Soil Conservation, Kathmandu, Nepal, 2000.

[16] A. E. Walkley and J. A. Black, "An examination of the method for determining soil organic method, and proposed modification of the Chromic acid Titration method," *Soil Science*, vol. 37, pp. 29–38, 1958.

[17] E. R. Mini and Y. S. Rao, "An evaluation of soil carbon sequestration in teak and ecalypt plantations," *NeBIO*, vol. 2, no. 3, pp. 9–11, 2011.

[18] J. Barlow, T. A. Gardner, and I. S. Araujo, "Quantifying the biodiversity value of tropical primary, secondary, and plantation forests," *Proceedings of the National Academy of Sciences of the United States of America*, vol. 104, no. 47, pp. 18555–18560, 2007.

[19] J. I. Rocky and C. Miligo, "Regeneration pattern and size-class distribution of indigenous woody species in exotic plantation in Pugu Forest Reserve, Tanzania," *International Journal of Biodiversity and Conservation*, vol. 4, no. 1, pp. 1–14, 2012.

[20] E. Rana, *Processes and Experiences in REDD Pilot Project in Nepal*, International Centre for Integrated Mountain Development, Kathmandu, Nepal, 2011.

[21] Y. Manandhar, *REDD Readiness and Monitoring, Reporting and Verification: Lesson Learned from Field*, World Wildlife Fund, Kathmandu, Nepal, 2010.

[22] Y. K. Karna, *Mapping above ground carbon using world view satellite image and lidar data in relationship with tree diversity*

and forests [M.S. thesis], University of Twente, Enschede, The Netherlands, 2012.

[23] W. Wang, L. Xiangdong, D. K. Daniel, and C. Peng, "Positive relationship between aboveground carbon stocks and structural diversity in spruce-dominated forest stands in brunswick, Canada," *Forest Science*, vol. 57, no. 6, pp. 506–515, 2011.

[24] Q. Guo, "The diversity-biomass-productivity relationships in grassland management and restoration," *Basic and Applied Ecology*, vol. 8, no. 3, pp. 199–208, 2007.

[25] D. V. Heather, R. W. Michael, B. C. Stephen, E. L. Ariel, and N. S. Frederick, "Relationship between abovegroundbiomass and multiple measures of biodiversity in subtropical Forest of Puerto Rico," *Biotropica*, vol. 42, no. 3, pp. 290–299, 2010.

[26] I. P. Sapkota, M. Tigabu, and P. C. Oden, *Tree Diversity and Regeneration of Community-Managed Bhabar Lowland and Hill Sal Forests in Central Region of Nepal*, Swedish University of Agricultural Sciences Southern Swedish Forest Research Centre, Alnarp, Sweden, 2009.

6

Ethnobotanical Studies of Port Harcourt Metropolis, Nigeria

A. B. Nwauzoma[1,2] and Magdalene S. Dappa[1]

[1] *Department of Applied & Environmental Biology, Rivers State University of Science & Technology, PMB 5080,*
 Port Harcourt 500001, Nigeria
[2] *Embrapa Agroenergia-PQEB-Final W3 Norte, Asa Norte, 7077091 Brasilia, DF, Brazil*

Correspondence should be addressed to A. B. Nwauzoma; drnwabarth@yahoo.com

Academic Editors: F. A. Culianez-Macia, G. T. Maatooq, and T. L. Weir

The objective of this study was to show the different ways medicinal herbs are used by the indigenous people in Port Harcourt metropolis (07° 3′ E, 04° 51′ N) in the Niger Delta region, Nigeria. One hundred and fifty structured questionnaires were administered, including oral interviews to herbal practitioners and users located at different parts of the city. Also, three popularly known herbal companies—Emiola naturalist care, Yem-Kem international herbal center, and Abiola medical herbal center—were included as they are healthcare providers, especially the medium income group. The results showed that a total of 83 plant species were recorded and classified according to their family, botanical, common, and local names. Also, the plant part used, mode of preparation, and type of ailment cured were included. The most frequently used plant parts were leaves followed by barks, roots, and fruits and with malaria fever as the most treated ailment. Deforestation, agricultural expansion, and fire were noted as the most important factors threatening the availability of these plants. The authors are of the opinion that paying special attention to the medicinal plants found in the area through conservation may help to amplify their role in the healthcare system, poverty alleviation, and environmental protection.

1. Introduction

Herbal or traditional medicine has been a major aspect of the sociocultural heritage in Africa for hundreds of years even before the advent of conventional medicine. It was once believed to be primitive and wrongly challenged by foreign religions dating back during the colonial rule in Africa and subsequently by the conventional or orthodox medical practitioners [1]. Plant-derived medicines have been part of traditional health care in most parts of the world for thousands of years and there is increasing interest in them as sources in the treatment of diseases [2–4]. The majority of people in developing countries depend on herbalists for their medical care. This is so in Port Harcourt metropolis, which is the hub of oil and gas activities in the Niger Delta region of Nigeria. The city comprises of people from different social and economic strata, ranging from the oil company executive to the water vendor. All these categories of people seek medical care and other social services in the city. Therefore, most of the people especially the poor resort

to herbal medicine because of its affordability, accessibility, and acceptability. The treatment and control of diseases by the use of available medicinal plants in a locality will continue to play significant roles in medical health care implementation in the developing countries of the world [5]. The objective of this survey was to document the herbal plants used by the people in Port Harcourt metropolis, the part used, how they are used, and the type of ailment they cure. Also, the need for the integration of herbal medicine into the formal health care system, complementary relationship amongst health care practitioners and delivery of health services, especially to the low income group in both rural and urban areas is discussed.

2. Materials and Method

2.1. Study Area and Data Collection. Port Harcourt, the capital of the oil rich Rivers State (Figure 1), is located in Southeastern Nigeria (07° 3′ E, 04° 51′ N, and 10 m altitude above sea level) in the humid forest zone of the Niger

FIGURE 1: Map of Rivers State (Nigeria) showing Port Harcourt metropolis.

Delta region, Nigeria. It is densely populated and home to multinational oil and gas companies and as such witnesses the influx of people in search of better living. This population increase often stretches public facilities, including hospitals (which are few) leading to alternative sources like herbal medicine. This makes the ethnobotanical studies of the area very imperative. Therefore, a structured questionnaire was administered to different herbal medicine dealers and users of herbal medicine located at different points in the city—mile 1 market, mile 3 market, flyover area, and three popularly known herbal companies: Emiola naturalist care, Yem-Kem international herbal center, and Abiola medical herbal center, all situated in Port Harcourt metropolis. The herbal companies selected in this study have been in practice for many years in the metropolis and serve as healthcare providers to many people in the city and partner with government during trade fair exhibitions. They did not sponsor this work research or influence the report.

The data in this study were derived from the questionnaires that were administered and oral interviews granted by the individual herbal dealers and the companies. The respondents were both men and women from ages 40 years and above, representing the age group with good knowledge of herbal remedies. The indigenous plants collected during the oral interviews were identified with the aid of floras of the area using [6, 7] and authenticated by Dr. B. O Green (Taxonomist), Department of Applied and Environmental Biology, Rivers State University of Science and Technology, Port Harcourt, Nigeria, where the specimen vouchers were deposited. Ethnomedical confirmations were carried out using [8] and a total of 150 questionnaires were administered and a total of 83 plants species were identified in the study.

3. Results

Table 1 shows the individual plant species, their botanical, common, and local (Yoruba, Igbo, and Hausa) names as well as their families, the use of each plant, and which part(s) of the plant that is being used. Our study shows that the plants have different ethnomedical applications by the people as antiseptic, laxatives, purgative, anticonvulsant, expectorants, anthelmintic, and sedatives in the treatment of malaria, rheumatism, diarrhea, infertility, jaundice, dysentery, gonorrhea, fever, pains, respiratory problem and poultice, and so forth.

3.1. Ailment, Mode of Preparation Dosage, and Administration of Some of the Common Herbs

3.1.1. Skin Diseases, Malaria Fever, Anaemia, Diabetes, and Bronchitis. Fresh leaves and bark of *Mangifera indica* are boiled together with the leaves of Papaya and neem. A glassful of the mixture is taken thrice daily to treat fever, anaemia, and diabetes. It is also used for bathing early in the morning to treat malarial fever.

3.1.2. Stomach Ache, Skin Infection, Diabetes, Loss of Memory, and Prostate Cancer. The tender part of the stem of *Vernonia amygdalina* is used as chewing stick and the bitter water is swallowed daily as remedy for stomach ache. Alternatively, fresh leaves are pounded in a mortar and the juice is pressed out, and a pinch of salt is added to 3 tablespoons of the undiluted juice and taken as a drink 3 times daily to bring immediate relief to stomach ache. For skin infection such as ringworm, itching, rashes, and eczema, the pure undiluted extract of bitter leaf is applied to the affected part daily. For diabetes, 10 handfuls of the fresh leaves are squeezed into 10 liters of water; a glassful is taken 4 times daily for 1 month to reduce sugar level drastically and it also repairs the pancreas. In the case of memory loss, take 1 glass twice daily for at least 2 months, while application of the solution soothes inflamed joints arthritis and eradicates pains.

3.1.3. Malaria Fever, Diabetes, Stomach Ulcer, and Convulsion. The leaves of Papaya are squeezed into one liter of water and a glassful is taken 3 times daily for 7 days to serve as a good treatment for malaria fever and jaundice. Similarly, the green leaves are squeezed into 1 liter of water; one glassful is taken three times daily to treat diabetes and constipation. For intestinal ulcer, unripe Papaya fruit is cut into pieces and the peel and seeds are removed and soaked in five liters of water for 4 days. It is sieved and 1/2 glass is taken 3 times daily for two weeks. The white milky sap of unripe Papaya contains a high percentage of papain which is used for chronic wounds or ulcers. The dry fallen Papaya leaves are washed and ground into powder. Two tablespoonfuls of the ground powder are added into 1/2 glass of palm kernel oil, stirred properly, and rubbed over the body to arrest high fever and convulsion.

3.1.4. Anaemia, Intestinal Ulcer, and Heart Problem. The dried peels of *Musa paradisiaca* are ground into powder; one

TABLE 1: Ethnobotany of some common plant species in Port Harcourt city, Nigeria.

S/no.	Common name	Botanical name	Family name	Local names Yoruba	Local names Ibo	Local names Hausa	Use(s)	Part(s) used
1	Ginger	*Zingiber officinale* (Rose)	Zingiberaceae	Jinga	Chita	Ata-ile	Detoxify liver, bronchitis	corm
2	Edible-stemmed vine	*Cissus quadrangularis* (L.)	Vitaceae	Ogbakiiki	—	Daddor	Caries, dysmenorrhoea, urinary disorders	Tuber, stem
3	Hibiscus	*Hibiscus acetosella*	Malvaceae		Akese	Akese	Dysentery	Leaves
4	Chinese pur/Burweed	*Triumfetta rhomboidea* (Jacaq.)	Tilaceae	Odo	Yanka-dafi	Molanganrsn/ako bolobolo	Gonorrhea	Leaf, Flower Fruit
5	Bush okra/Jew fiber telteria/Jews Mallow	*Corchorus olitorius*	Titiaceae	Ewedu	Ariraa/ulogburu	Lalo	Vegetable, blood purifier	Leaf
6	African star apple	*Chrysophyllum albidum* (G. Don)	Sapotaceae	Odara/udala	Agbalumo	Agwaliba	Delicacy, antinausea	Fruit
7	Fruited gourd	*Coccinia barteri*	Cucurbitaceae			Ewe-oju	Venereal diseases, skin infections, earache	Whole plant
8	Tomatoes	*Solanum lycopersicum*	Solanaceae	Tomati	Tomato	Tomati	Vegetable, Vitamin C	Fruit
9	Pepper, Chili	*Capsicum annuum* L.	Solanaceae	Ata wewe	Ose	Tatashi	Delicacy, seasoning	Fruit
10	Curry leaf	*Thymus vulgaris*	Lamiaceae	Efinrin wewe	Nch-anwu		Antibiotic, carminative	Leaves
11	Orange, Sweet	*Citrus sinensis* (Linn.)	Rutaceae	Osan	Or-oma	Lmu	Dysentery, fever, headache, antimicrobial agent, anthelmintics, toothache, antiscorbutic	Twigs, stem bark, fruits, peel
12	Brimstone tree	*Morinda lucida* (Benth.)	Rubiaceae	Oruwo	Eze-ogu	—	Fever	Leaves
13	African copaiba, balsam tree, niger-opal, maaje	*Daniellia oliveri*	Leguminosae	Iya,	Kadaura,	Ozabwa, Maje	Dysentery, diarrhoea, toothache, urinary infection, astringent, tooth ache	Gum, bark
14	Water leaf	*Talinum triangulare* (Willd.)	Portulacaceae	Gbure	Nte-oka/inene	Alenyruw-a	Vegetable	Leaf
15	Lemon grass	*Cymbopogon citratus*	Poaceae	Koriko-oba	Nch-anwu	—	Malaria	Leaf
16	Hog plum	*Spondias mombin* L.	Anacardiaceae	Iyeye	Ngu-lung-wu	Isada	Infertility	Fruit
17	Cashew nut	*Anacardium occidentale* L.	Anacardiaceae	Kaju	Sas-hu	Kanju	Cough	Bark, Fruit
18	Mango	*Mangifera indica* L	Anacardiaceae	Mongora		Mango	Malaria	Leaf, bark
19	Cocoyam, Wild taro	*Colocasia esculentum* (L.) Schott	Araceae	Koko, kokof-un, kokoibile,	Ede	Gwamba	Anaemia, wounds, rheumatism, poison antidote	Tuber, leaves

TABLE 1: Continued.

S/no.	Common name	Botanical name	Family name	Local names Yoruba	Local names Ibo	Local names Hausa	Use(s)	Part(s) used
20	Giant milk weed/sodom apple	*Calotropis procera*	Asclepeceae	Bom-ubomu		Tumifafiya	Measles	Leaf
21	Bitter leaf	*Vernonia amygdalina* (Del.)	Asteraceae	Ewuro	Onu-gbu, olug-bu	Shiwaka	Pile, lower sugar, vegetable	Leaf
22	Goat weed, floss flower	*Ageratum conyzoides* L.	Compositae	Imi-esu	ula ujula,	Ahenhen	Wounds, ulcers, craw-craw, digestive disturbance, diarrhoea, emetic, skin diseases, antipyretic, gonorrhoea, sleeping sickness, eye wash	Whole plant, leaves, root
23	Siam weed	*Chromolaena odorata* (Linn.)	Compositae	Akintola	Awo-lowo,	Obiarakara	Antimicrobial, dysentery, headache, malaria fever, toothache, haemostatic, skin diseases	Leaves, stem-twigs
24	Coconut	*Cocos nucifera*	Arecaceae	Agbon	Aku-beke	Mosara	The water neutralizes poison/drug	Nut
25	Pepper fruit	*Dennettia tripetala*	Annonaceae	Igbere,	Nmi-mi	—	Insect repellant, fever, cough, toothache, stimulant	Fruit, leaves, stem, twigs
	False thistle, leopard's tongue Bear's breech, white's ginger,	*Acanthus montanus*	Acanthaceae	Ahon-ekun, irunmuarugbo,	Nyin-yiog-wu		Syphilis, cough, emetic, urethral discharge, purgative, boils, anaemia, anthelmintics	Stem-twig, leaves, roots
26	Resurrection plant, life plant	*Bryophyllum pinnatum*	Crassulaceae	Eru-odundun,	—	Abomoda	Cough, diarrhoea, dysentery, wounds, fever, sedatives, diuretic, absc-esses, antifungal, epilepsy, antimicrobial, anticancer	Leaves, roots, leaf sap
27	Fertility tree, tree of life	*Newbouldia laevis*	Bignoniaceae	Akoko	Ogirisi	Aduruku	Round worms, elephantiasis, dysentery, malaria, convulsions, migraine, cough, yellow fever, stomachache, hernia, infertility, earache	Bark, leaves, root
28	African tulip	*Spathodea campanulata* (P. Beauv.)	Bignoniaceae	Akoko	Ogili-si, ogirisi	Aduruku	Asthma	Leaves
29	Pineapple	*Ananas comosus*	Bromeliaceae	Ope oyinbo	Nkw-aba	—	Antihypertension constipation	Unripe fruit

TABLE 1: Continued.

S/no.	Common name	Botanical name	Family name	Local names Yoruba	Local names Ibo	Local names Hausa	Use(s)	Part(s) used
30	Cock's comb, Heliotrope	*Heliotropium indicum* L	Boraginaceae	Agogo-igun, Apari-Igun,	Ogb-eria-kuko	Kalkashin korama	Convulsions, cancer, worms, rectal enema, mouth-wash	Whole plant
31	African cucumber, bitter gourd, balsam pear	*Momordica charantia* L.	Cucurbitaceae	Ejinrinw	Alo-ose	Kakayi	Diabetes, piles, convulsions, jaundice, sore, nervous disorders, diabetic recipe, emetic, night blindness, aphrodisiac, dysmenorrhoea, anthelmintic, antimicrobials	Whole plant, seeds, fruit root
32	Fluted pumpkin	*Telfaria occidentalis* Hook. F.	Cucurbitaceae	Ugu			Antianaemic, blood tonic	Leaf
33	Colocynth, wild gourd.	*Citrullus colocynthis* (L.)	Cucurbitaceae	Egunsi	Elili/egwusi	Egbsi/guna	Laxative digestion	Fruit
34	Paw paw	*Carica papaya* (Linn.)	Caricaceae	Ibepe	Okworo-gbogbo	Gwanda	Boil, purgative	Latex fruit
35	Velvet, black tamarid, tumble tree	*Dialium guineense* (Wild)	Leguminosae	Awin,	Icheku,	Tsamiyar kurmi	Fever, coughs, bronchitis, toothache, astringent, diuretic	Leaves, fruit, bark, twigs
36	White yam	*Dioscoreae cayennensis*	Dioscoreaceae danzaria	Ako isu	Ji-ocha	Doya	Antidiarrhea	Tubers
37	Physic nut	*Jatropha curcas* L.	Euphorbiaceae	Botuje,	Olulu-idu	lapalapa, Zugu, Ol	Ringworm, eczema, scabies, fever, guinea worms, herpes, rectal enema, black tongue, whitlow, impotence, irregular menses, convulsion, smallpox	Seed, leaves, stem, roots, sap
38	Caper bush	*Capparis thonningii*	Capparaceae	Eka-nawodi		ewon ekiri	Fever, headache, mental disorder, aphrodisiac, cough.	Root
39	African Walnut	*Tetracarpidium conophorum*	Euophorbiaceae	Awusa, asal	Ukpa	Hawuusa	Aphrodisiac	Fruit
40	Soya bean	*Glycine max*	Leguminosae	Ewa			Laxative	Seeds
41		*Grewia* sp	Tiliaceae	Ila-oko, lakolako			Religious purpose, mystic, soup with okra-like taste	Fruits
42	African/Native/Bush mango	*Irvingia gabonensis*	Irvingiaceae	Oro mopa	Ogb-ono	Mamujigoro	Condiment	Seed
43	Garlic	*Allium sativum* L.	Alliaceae	Aayu	Ayo-ishi	Tafarunua	Antibiotic antidiabetic Anti-hypertension	Bulb

TABLE 1: Continued.

S/no.	Common name	Botanical name	Family name	Local names — Yoruba	Local names — Ibo	Local names — Hausa	Use(s)	Part(s) used
44	Onion	Allium cepa Alabasa	Alliaceae	Alubosa	Alu-bosa	Yabasi	Antidiabetic	Bulb
45	Aloe, West African aloe	Aloe barteri	Liliaceae	Aloe	Aloe		Ringworm, anthelmintics, aphrodisiac, amenorrhoea, cough, skin infections, astringent, antitumour, pile, fruits for preventing snake bite	Leaves
46	Scent leaf, mint	Ocimum gratissimum	Lamiaceae	Efinrin nla	Nchanwu	Dadoya	Stomach ache	Leaf
47	Azadirachta indica (A. Juss.)	Neem tree	Meliaceae	Dongoyaro	Og-wu akom	Maina	Boils, antimalaria	Leaf, bark
48	Moringa oleifera Lam.	Horse radish tree, Moringa tree, "Never Die," drumstick tree	Moringaceae	Ewe-ile, Eweigbale,	Okweoyeibe,	zogale, Bagaaruwar makka	Inflammatory diseases, asthma, antipyretic, cough, earache, liver and pancreas diseases, venereal diseases, anthelmintic, hysteria, diarrhoea, diuretic diseases	Leaves, roots, stem bark, fruit
49	Persea americana (Mill)	Avocado pear	Lauraceae	Igba/apoka	Ube-beke		Antihypertension stomach ulcer	Fruit
50	Plantain	Musa paradisiaca L.	Musaceae	Ogede agagba	Abrika/Okirima	Okamu/ayaba	Potent astringent high iron	Unripe fruit
51	Guava	Psidium guajava L	Myrtaceae	Gurofa	Gova	Gwaabaa	Malaria	Leaf
52	Groundnut, peanut	Arachis hypogaea	Fabaceae	Egpa		Gedda	Oil as solvent, antimicrobials, insomnia	Nuts
53	Crab's eye	Abrus precatorius	Papilionaceae	Iwereje/ojuologbo	Anya nnunu	Da marzaya	Cough	Leaf
54	Cam wood	Baphia nitida	Sterculiaceae		Uri		Decoration	Latex
55	Bamboo	Bambusa vulgaris L.	Poaceae	Oparun	Atosi		Gonorrhoea, abortifacient, anthelmintics, emmenagogue, skin rashes of HIV/AIDS	Leaves, young shoots
56	Water willow	Deinbollia pinnata	Sapindaceae	Ogiri-egba,		Ekusi-Oloko	Cough, bronchial asthma, aphrodisiac	Leaves, root
57	Lemon grass	Cymbopogon citratus	Poaceae	Koriko-oba	Nche awula		Malaria	Leaf
58	Water leaf	Talinum triangulare	Portulacaceae	Gbure	Nte-oka/inene	Alenyruwa	Rat poison, vegetable	Root, Leaf
59	Pepper, chili	Capsicum annuum L.	Solanaceae	Ata wewe	Ose/totashi	Barkono	Stimulant	Fruit
60	Soursop,	Annona muricata	Annonaceae	Sawamsop			Relaxing nerves	Leaf
61	Wild Cassava	Jatropha gossypifolia	Euphorbiaceae	Botuje pupa, Binidi zugu	Ake mbogho		Ringworm, ascaris, antitumour, malaria, dysentery, dysmenorrhoea	Stem latex

TABLE 1: Continued.

S/no.	Common name	Botanical name	Family name	Local names Yoruba	Ibo	Hausa	Use(s)	Part(s) used
62	Black Mangrove	*Avicennia africana*	Avicenniaceae	Ogbun,	Ofun		Abortifacient, detergent	Leaves, stem, twigs
63	Baobab	*Adansonia digitata*	Bombacaceae	Ose,	Igiose,	kukaa, kulambali	Malaria, asthma, diarrhoea, kidney and bladder diseases, demulcent, prophylactic, antihistaminic, skin diseases, caries, antimicrobial	Leaves, fruit pulp, bark
64	Morning glory	*Ipomoea mauritiana*	Convolvulaceae	Atewogba, Tanpopo			Rheumatism, asthma, dropsy.	Whole plant
65	Bhadram, cherula	*Aerva lanata*	Amaranthaceae	Aje, Efunile	Ewowo,	Alhaji, Furfurata, fatumi	Ulcers, wounds, snake bite, diuretic, purgative, anthelmintic, sore throat, kidney and bladder stones	Whole plant
66	Acalypha	*Acalypha fimbriata*	Euphorbiaceae	Jinwinini,		kandiri	Syphilis, asthma, anthelmintics, ulcers, rheumatism, antimicrobial and antifungal	Leaves
67	Acalypha	*Acalypha godseffiana*	Euphorbiaceae	Jinwinini			Skin infection, Antimicrobials	Leaves, twigs
68	Hennaplant	*Lawsonia inermis*	Lythraceae	Lali,	Laali.	Lallee	Spermatorrhoea, jaundice, gonorrhoea, leucorrhoea, ulcers, menorrhagia, astringent, skin diseases, malaria	Leaves, flowers, bark
69	Wild lettuce	*Launaea taraxacifolia*	Compositae	Yanrin,		Yamurin, Nonanbarya	Yaws, fracture management	Leaves
70	Stinging bean	*Mucuna sloanei*	Leguminosae	Ewe-ina,	yerepe werepe	Kakara, osese	Haemorrhoids, diuretics, micturition problems in children, skin diseases	Seeds, roots
71	Cow-hage, cow-itch plant, velvet bean	*Mucuna pruriens*	Leguminosae	Esisi,	Werepe	Abbala, Kakara	Intestinal worms, genitourinary diseases.	Hairs on the pods
72	Devil's gut, parasitic vine	*Cassytha filiformis*	Lauraceae	Omoniginigini	omonigelegele,	sulunwahi.	Anthelmintics, antimicrobials, antifungal	Stem twigs
73	Morinda	*Morinda morindoides*	Rubiaceae		Oju-Ologbo		Fever, jaundice, asthma, dysentery, colic, emmenagogue, vermifuge, constipation	Root, bark, leaves, fruit
74	Millet	*Millettia thonningii*	Leguminosae	Ito,	okeokpa	Tuburku, Ajukwu	Fever, cough, respiratory ailment, anthelmintic, ophthalmia	Roots, bark
75	African Linden	*Mitragyna inermis*	Rubiaceae		Okobo,	Giyeya	Dysentery, leprosy, antipyretic, diuretic, gonorrhoea	Bark

TABLE 1: Continued.

S/no.	Common name	Botanical name	Family name	Local names Yoruba	Local names Ibo	Local names Hausa	Use(s)	Part(s) used
76	Sensitive plant	*Mimosa pudica*	Mimosaceae	aluro		Patanmo,	Guinea worms piles, kidney disease, fistula, boils	Leaves
77	Bullet wood	*Mimusops kummel*	Sapotaceae		Uku	Emido	Antipyretic, astringent, mouth wash, stomachic	Stem bark, seeds
78	Sword bean, horse bean	*Canavalia ensiformis*	Leguminosae		Ponpondo,	sese-nla	Antibiotic, antiseptic	Seed
79	Celosia	*Celosia laxa*	Amaranthaceae	Marugbo sanyantan,	Ajemawofo,	Mannafaa, sanyantan,	Antiscorbutic, purgative	Leaves
80	Indian chrysanthemum	*Chrysanthellum indicum*	Compositae			Abilere, Oyigi	Boils, fever, gonorrhoea, jaundice, heart-trouble, insecticide	Whole plant
81	Rattle box, rattle pea	*Crotalaria retusa*	Leguminosae	Koropo	Akidimuo	saworo, Yara,	Fever, cirrhosis, liver lesions, dysentery, colic, vermifuge	Root, seeds, juice of pods, leaves
82	Melon-pumpkin	*Cucurbita maxima*	Cucurbitaceae	Elegede,	Apala,	Kabeewaa	Tapeworm, diuretic, taenicide, otitis, utensils	Seeds, fruits
83	Flame of the forest	*Delonix regia*	Leguminosae	Seke	seke,	ayin.	Diuretic, anthelmintics, astringent, leucorrhoea	Leaves, bark, seeds, flower

tablespoon of the powder is mixed with four tablespoons of honey and licked three times daily for two weeks for intestinal ulcer. Some quantity of the root and fresh leaves are boiled separately; 1/2 and a full glass is drank daily for 1 week to intestinal ulcer and anaemia, respectively. Eating of unripe fruit either roasted, boiled, pounded, or processed into flour is a good treatment for diabetes.

3.1.5. Cough, Malaria Fever, and Repellant. Decoction from leaves of *Cymbopogon citrate* with onion and honey is used to cure cough, taken 3 times daily for 3 days. The leaf is boiled along with other herbs to treat malaria fever by bathing with it every night for 2 days. The leaf is burnt in homes to serve as repellant for mosquitoes.

3.1.6. Fever and Lactation. The bark, root, and leaf of *Morinda lucida* are used in infusion or decoction for the treatment of yellow fever and other forms of fever to be taken 1/2 glass, 2 times daily, and also bathing, for 3 days. The very bitter leaf decoction is applied to the breast of women at weaning of their infants to improve lactation. Twigs are used as chewing stick.

3.1.7. Malaria, Diabetes, Dysentery, Mouth Thrush, Toothache, and Sore Gums. The twig of *Anacardium occidentale* is used as chewing stick for mouth thrush, tooth ache, and sore gum. Decoction of the bark is a remedy for malaria fever, by drinking 1/2 glass 3 times daily. The bark and leaves are boiled, and a glass is taken twice daily for dysentery.

3.1.8. Fibroids, Cataract, Gonorrhea, Aphrodisiac, Cough, Inflammatory Symptoms, Toothache, and Sore Throat. Seed of *Spondias mombin* is boiled together with immature palm-nuts and 1/2 glass is taken thrice daily for 2 months for fibroid. Fresh leaves are ground and the juice is squeezed and mixed with one teaspoonful of lime juice and applied as eye drop twice daily for cataract. Fresh leaves are boiled and one glassful is drank thrice daily for gonorrhea. Decoction of leaves is used as an aphrodisiac. Decoction of the bark is taken for severe cough, toothache, and sore throat.

3.1.9. Abdominal Pains, Ulcers, Skin Disease, Dressing of Wound, and Prophylactic. Decoction of the whole plant of *Ageratum conyzoides* is a remedy for abdominal pains. Leaf juice is used for dressing wounds, ulcers, and other skin diseases. Leaves are used as tonic to aid fertility, because it prevents early miscarriage. It is also used as prophylactic and cure for trachoma in cattle.

3.1.10. Threatened Abortion, Convulsion, Epilepsy, Skin Infections, Conjunctivitis, Migraine, and Earache. A medium size-pot is filled with the fresh bark of *Newbouldia laevis* and boiled water for a long time. The preparation is then used to wash face and head every morning and night; oral taking of 1/2 glass of preparation twice daily for 6 days cures migraine and also stops vaginal bleeding in threatened abortion. Leaves and roots are boiled together and administered for fever, convulsion, and epilepsy. Stem bark is used for treating skin

infections. Decoction of leaves is used as an eye wash in conjunctivitis. Boiled leaves extract is used to treat general malaria.

3.1.11. Typhoid Fever, Menstrual Flow, Healthy Skin, Purgative, Diuretic, Anthelmintic, Expectorant, and Abortifacient. The fruit of *Ananas comosus* is cut, cooked, and drank for typhoid fever. The unripe fruit can be used as a purgative, diuretic, antihelmintic, expectorant, and abortifacient and is also taken to regulate and enhance menstrual flow. Fruit peel is used topically for healthy skin. The ripped fruit is taken regularly to recover from typhoid fever.

3.1.12. Sexually Transmitted Diseases, Stomach Troubles, Purgative, and Fungal Infection. Fruits of *Citrullus colocynthis* are recommended for the treatment of stomach troubles and sexually transmitted diseases. Fruit and leaf decoction is used as a purgative in man and animal. Seed shell powdered and mixed with palm oil is rubbed on skin to treat fungal infections.

3.1.13. Ringworm, Scabies, Eczema, Sexually Transmitted Diseases, Thrush Bleeding, Wounds, Toothache, and Skin Disease. The latex of *Jatropha curcas* is used to treat skin disease such as ring worm, scabies, and eczema. Twigs are used as chewing stick to prevent tooth decay, oral thrush, bleeding, wounds, and tooth ache. Roots are used to treat sexually transmitted diseases. Leaves are added to hasten fermentation of cassava. Decoction of leaves is used to sterilize umbilicus of new born babies.

3.1.14. Fibroid Improves Sperm Count, Fertility, and Menstrual Flow. Seed of *Tetracarpidium conophorum* is used in the treatment of fibroid. Boiled seeds are eaten to improve sperm count in men. Leaf juice is used to improve fertility in women and to regulate menstrual flow.

3.1.15. As Food. Leaves of *Telfairia occidentalis* are of highly nutritive value as vegetable for soup and other local dishes. Leaves are washed and the juice squeeze is mixed with milk and taken as a blood tonic. The boiled seeds are eaten as delicacy and source of oil. Some of the plants are used as herbs food and other uses.

The knowledge of the indigenous people about contraceptives was one of the informal innovative discoveries in this work. In this context, *Ageratum conyzoides, Tetracarpidium conophorum, Rhaphiostylis beninensis, Lonchocarpus cyanescens, carpolobia alba,* and *Chrysophyllum albidum* are used to invoke sterility, while *Moranthodoa leucantha* increases sexual vigor and *Mucuna soloanei* and *Senna occidentalis* are used by indigenous ladies as contraceptives. Few plant species known to be "poisonous," for example, *Ricinus communis* and *Scleria verrucosa* were reported to be very potent. We also observed that some tuberous plant species like *Colocasia esculenta* and *Dioscorea rotundata* and *Jatropha gossypifolia* and *Musa paradisiaca* are used to cure sexually transmitted diseases, to regularize menstruation, and to increase fertility. We further discovered that those plants with high nutritive value like *Colocasia esculenta, Basella alba, Telfairia occidentalis, Glycine soja, Gnetum africana, Arachis hypogea,* and *Solanum lycopersicum* are cultivated mainly for commercial purposes, as they are sold in nearby markets. Others like *Hibiscus senensis, Moringa oleifera* and *Sida acuta* in addition to food and medicinal values, have become beautiful ornamental plants. *Dracaena arborea, Anacardium occidentale, Basella alba, Spathodea campanulata, Allium sativum, Mucuna sloanei, Ocimum basilicum, Sida acuta, Laportea aestuans,* and *Trema orientalis* are used to treat constipation, indigestion, abdominal pain, and dysentery. Our result also shows that the people use *Xanthosoma* spp, *Calotropis procera, Vernonia amygdalina, Ageratum conyzoides, Chromolaena odorata, Newbouldia laevis, Spathodea campanulata,* and *Adenopus breviflorus* for skin diseases like wound, tumor, boils, burns, and cuts. *Dioscorea rotundata, Jatropha curcas, Ricinus communis, Irvingia gabonensis, Aloe barteri, Ocimum basilicum, Azadirachta indica, Baphia nitida, Mitracarpus scabrum, Glyphaea brevis,* and *Trema orientalis* are also used for the above purposes.

Respiratory disorders like cough, cold, tuberculosis, and asthma are cured using single herb or mixture of herbs like *Calotropis procera, Dennettia tripetala, Carica papaya, Allium sativum, Cymbopogon citratus, Chrysophyllum albidum,* and *Zingiber officinale.* Our study shows that most herbs are known to cure malaria and typhoid fever which are endemic in Port Harcourt metropolis: *Anacardium spondias, Dennettia tripetala, Ananas comosus, Adenopus breviflorus, Ipomea involucrate, Carica papaya, Securinega virosa, Hyptis pectinata, Sida acuta, Azarachta indica, Psidium guajava, Bambusa vulgaris, Cymbopogon citratus, Morinda lucida Citrus sinensis, Murraya koenigii, Capsicum annum Chrysophyllum albidum, Glyphaea brevis, Corchorus olitorius,* and *Trema orientalis.*

4. Discussion

This is the first ethnobotanical study of Port Harcourt metropolis in the Niger Delta region of Nigeria. Our study shows that the 83 plant species identified were useful as food and in the treatment of different human ailments, showing that traditional medical practice is an important component of our everyday life. Our findings are similar to [9]. Reference [1] suggested the need to institutionalize the traditional medicine in concert with orthodox medicine to achieve an effective national health care system in Nigeria. The authors maintained that an effective health cannot be achieved in Africa by orthodox medicine alone unless it has been complemented with traditional medicine, in support of [10].

The questionnaire and interviews gathered indicate that most people in the Metropolis depend on traditional medicine for their health needs because of their poor economic conditions. This is one of the major reasons why traditional medicine has continued to thrive in both rural and urban areas in Nigeria. The utilization of medicinal plants in traditional medicine was found to be effective, cheap, and practical. References [11, 12] noted the growing interest on the medicinal properties of a number of common

plants. The practice is fast developing due to poor economic situation, expensiveness, and inadequate availability of drugs. Reference [13] stated that the use of plants and products in health care is, even much higher particularly in those areas with little or no access to modern health services. These medicinal plants have been underutilized in the orthodox medicine but have now been recognized in ethnomedicinal preparation.

Gender and age influenced the traditional knowledge of our respondents. Males within 45–70 years have medicinal knowledge than females. This may be due to their involvement in trade or personal experience of using these plants for a very long time. In addition, the younger generation does not seem to have much trust in the traditional medicine system which may be attributed to increasing use of allelopathic medicines which are readily available and potent. Our findings also show that the indigenous people value some of these herbs for medicinal purposes than as food condiments. For instance garlic is more useful in treating fever, cough, constipation, asthma, nervous disorder, hypertension, ulcer, and antihelmentic than in mere seasoning of food. The same applies to onion, curry leaf, ginger, and scent leaf.

Traditional medicinal practices are known to still be an important component of everyday life in many regions of the world [14–16]. The use of plants in healthcare is even much higher particularly in areas with little or no access to modern health services [13]. Reference [1] gave a comprehensive treatise on the need to institutionalize traditional medicine into the health scheme according to WHO guidelines [17]. Most of the plants were used to treat malaria fever, underlying the importance of this disease in the region.

Conservation of indigenous plant species of medicinal importance is necessary as they remain source of health and wealth. There is need for closer collaboration between herbal medical practitioners, medical doctors, and other stakeholders in medical practice to bring traditional healers closer by engaging them in laboratory work, training as well as getting information on traditional prescriptions for specific diseases. Both traditional and orthodox medicines should complement each other, and their integration or harmonization is necessary for quality healthcare delivery, especially in the rural communities. The ethnobotany of Port Harcourt metropolis has been documented. Various plants have dual significance first as food, secondly as medicinal plants and can have some active constituents for future pharmaceutical analysis.

Conflict of Interests

There is no conflict of interests, as the information therein is purely for research purposes. The authors do not support self-medication and further suggest getting advice from medical practitioners before taking any of these herbs.

Acknowledgments

The authors are grateful to the local informants and herbal healers who shared their knowledge with them.

References

[1] R. N. Okigbo and E. C. Mmeka, "An appraisal of phytomedicine in Africa," *KMITL Science and Technology Journal*, vol. 6, no. 2, pp. 83–94, 2006.

[2] D. C. Mohana, S. Satish, and K. A. Raveesha, "Antibacterial evaluation of some plant extracts against some human pathogenic bacteria," *Advances in Biological Research*, vol. 2, no. 3-4, pp. 49–55, 2008.

[3] G. M. Adwan, B. A. Abu-shanab, and K. M. Adwan, "In vitro activity of certain drugs in combination with plant extracts against *Staphylococcus aureus* infections," *African Journal of Biotechnology*, vol. 8, no. 17, pp. 4239–4241, 2009.

[4] A. O. Ajayi and T. A. Akintola, "Evaluation of antibacterial activity of some medicinal plants on common enteric food-borne pathogens," *African Journal of Microbiology Research*, vol. 4, no. 4, pp. 314–316, 2010.

[5] F. C. Akharaiyi and B. Boboye, "Antibacterial and phytochemical evaluation of three medicinal plants," *Journal of Natural Products*, vol. 3, pp. 27–34, 2010.

[6] J. Hutchinson and J. M. Dalziel, *Flora of West Tropical Africa*, vol. 1, The Whitefriars Press, 1954.

[7] J. M. Dalziel, *The Useful Plants of West Tropical Africa*, Appendix to Flora of West Tropical Africa, Crown Agents for Overseas Government and Administration, 1937.

[8] L. S. Gill, *Ethnomedical Uses of Plants in Nigeria*, Ibadan University Press, 1988.

[9] A. A. Aiyeloja and O. A. Bello, "Ethnobotanical potentials of common herbs in Nigeria: a case study of Enugu state," *Educational Research and Review*, vol. 1, no. 1, pp. 16–22, 2006.

[10] A. A. Elujoba, O. M. Odeleye, and C. M. Ogunyemi, "Traditional medical development for medical and dental primary Health care delivery system in Africa," *African Journal of Traditional, Complementary and Alternative Medicine*, vol. 2, no. 1, pp. 46–61, 2005.

[11] M. A. Belewu, O. A. Olatunde, and T. A. Giwa, "Underutilized medicinal plants and spices: chemical composition and phytochemical properties," *Journal of Medicinal Plant Research*, vol. 3, no. 12, pp. 1099–1103, 2009.

[12] I. I. Ijeh, O. I. U. Njokwu, and E. C. Ekenze, "Medicinal evaluation of extracts of *Xylopia aethiopica* and *Ocimum gratissium*," *Journal of Medicinal and Aromatic Plant Sciences*, vol. 26, pp. 44–47, 2004.

[13] M. Saeed, M. Arshad, E. Ahmad, E. Ahmed, and M. Ishaque, "Ethnophytotherapies for the treatment of various diseases by the local people of selected areas of N.W.F.P (Pakistan)," *Pakistan Journal of Biological Sciences*, vol. 7, no. 7, pp. 1104–1108, 2004.

[14] R. W. Bussmann, "Manteniendo el balance de naturaleza y hombre La diversidad floritica Andina y su impotanicia porla diversidad cultural-ejemplos del norte de Peru y Sur de Ecuador," *Amaldoa*, vol. 13, no. 1-2, pp. 382–397, 2006.

[15] R. W. Bussmann and D. Sharon, "Traditional medicinal plant use in Northern Peru: tracking two thousand years of healing culture," *Journal of Ethnobiology and Ethnomedicine*, vol. 2, article 47, 2006.

[16] V. de Feo, "Medicinal and magical plants in the Northern Peruvian Andes," *Fitoterapia*, vol. 63, no. 5, pp. 417–440, 1992.

[17] World Health Organisation, *The Promotion and Development of Traditional Medicine*, Technical Report Series 622, World Health Organisation, Geneva, Switzerland, 1978.

7

An Effective Procedure for In Vitro Culture of *Eleusine coracana* (L.) and Its Application

Alla I. Yemets, Galina Ya. Bayer, and Yaroslav B. Blume

Institute of Food Biotechnology and Genomics, National Academy of Sciences of Ukraine, Osipovskogo Street 2a, Kiev 04123, Ukraine

Correspondence should be addressed to Alla I. Yemets; yemets.alla@gmail.com

Academic Editors: T. Berberich, K. P. Martin, and S. Ogita

Efficient protocols for callus production, plantlet regeneration, protoplast isolation, and micronucleation of finger millet (*Eleusine coracana* (L.) Gaertn.) were developed. White nodulated calli were formed on medium with N_6 macrosalts, MS microsalts, 2.4-dichlorophenoxyacetic acid ($2 \, \text{mg} \, \text{L}^{-1}$), kinetin ($0.4 \, \text{mg} \, \text{L}^{-1}$), 1-naphthalene acetic acid ($2 \, \text{mg} \, \text{L}^{-1}$), and certain additives. It was found that appropriate supplementation leads to formation of numerous shoots. Healthy rooted plantlets formed on hormone-free media. Although different tested additives had no significant effect on percentage of callus formation, it affected callus quality that further dictated plant-forming capacities. Seedlings were better source tissues for protoplasts isolation compared to callus cultures. About 5×10^6 protoplasts were isolated from one gram of seedling coleoptyles. Microcolonies were visible after 20–25 days' incubation on KM8p medium supplemented with glutamine ($100 \, \text{mg} \, \text{L}^{-1}$) and proline ($500 \, \text{mg} \, \text{L}^{-1}$). Here we also present a procedure of an efficient induction of micronuclei after chlorpropham ($10 \, \mu\text{M}$) and cytochalasin-B ($20 \, \mu\text{M}$) seedlings treatment with subsequent microprotoplasts isolation. This technique is discussed for the transfer of alien chromosomes and genes from finger millet by microprotoplast-mediated chromosome transfer.

1. Introduction

Techniques of plant biotechnology have emerged as an important aid to the traditional breeding methods for rapid genetic improvement and for integration of new genes into existing crop varieties. Success with the cell culture establishment in vitro and plant regeneration for most of the cereal and grass species has given impetus for further work with less researched crop species, especially millets which are of immense importance for solving the food and forage problems for many countries. Millets are important because they are grown in poor soils with limited inputs and they constitute a major source of food for resource-poor farmers of the areas of their cultivation [1]. The projected food demand for 2025 [2] will require the yield of millets to rise from 2.5 to 4.5 t ha^{-1}. Such yield increase could be largely achieved from improved varieties transgenically modified for biotic and abiotic stress resistance using different biotechnological methods [1].

Among various millets, finger millet *Eleusine coracana* (L.) Gaerth. is one of the well-known species for its outstanding properties as a subsistence food and forage crop [1]. Finger millet is a major crop in the arid and semiarid regions of developing countries of Asia and Africa; more small areas of it are also present in the Americas, Oceania, and Europe. Moreover, recently the new cultivars of *E. coracana* adapted to temperate climate zone have been bred in Ukraine [3–5]. It is planned to widely use this new perspective crop as additional very productive and economically advantageous source for forage and seed production, as well as alternative grass source for bioethanol production in countries with temperate climate of Eastern Europe.

The elaboration of effective protocols for in vitro cell culture production, plant regeneration, and protoplasts isolation will be the first step towards the biotechnology for improvement of this species by genetic transformation. The previous reports on *E. coracana* dealt only with regeneration of plants through organogenesis, multiple shoot production, and somatic embryogenesis [6–9], and these data are summarized in several reviews [1, 10, 11]. The attempts also have been made to establish gene transfer system for finger millet [12–17]. Moreover, up till now there was no obvious information about isolation and regeneration of protoplasts from finger millet that can be used for different cell and genetic engineering manipulations, including effective methods of microprotoplast-mediated chromosome transfer (MMCT) in interspecific or intergeneric breeding [18].

The present communication proposes an advanced and rapid method of plant regeneration through seedling callus cultures of *E. coracana* (L.) Gaerth. Here we also present an effective procedure of protoplast isolation and an efficient induction of micronuclei with subsequent microprotoplast isolation of finger millet. This technique is discussed for the transfer of alien chromosomes and genes from finger millet by MMCT.

2. Materials and Methods

Seeds of finger millet *Eleusine coracana* (L.) Gaerth. were surface-sterilized with 70% (v/v) alcohol for 10 min followed by 5% (v/v) sodium hypochlorite for 15 min twice. After rinsed for four times with sterile distilled water, they were used for callus induction and seedling production.

2.1. Establishment of Callus Cultures. Callus induction medium was stepwise determined in the following way: (1) basal medium selection, (2) hormonal optimization, and (3) supplements refining. Five different basal media were tested in step 1: media for callus initiation from *E. coracana* (MS macro- and microsalts, MS vitamins, 1 mg L^{-1} 2.4-dichlorphenoxy acetic acid, and 2.4-D, 3% sucrose) [19]; medium for *Setaria italica* (N$_6$ macro- and microsalts, N$_6$ vitamins, 200 mg L^{-1} casein hydrolysate, 2 mg L^{-1} 2.4-D, and 3% sucrose) [20]; medium for *Sorghum bicolor* (MS macro- and microsalts, MS vitamins, 200 mg L^{-1} glutamine, 2 mg L^{-1} 2.4-D, and 2% sucrose) [21]; and media B and E for *E. indica* [22].

For callus induction several combinations of phytohormones of 2.4-D (1; 2; 3 mg L^{-1}), kinetin, KIN (0.2; 0.4; 0.5; 0.8; 1 mg L^{-1}), 1-naphthalene acetic acid, NAA (1; 2 mg L^{-1}), and 6-benzylaminopurine, BAP (0.2; 0.4 mg L^{-1}) supplemented to the best basal medium resulting from step 1 were tested in step 2. In step 3, glutamine (100 mg L^{-1}), proline (500 mg L^{-1}), thiamine HCl (10 mg L^{-1}), tryptophan (200 mg L^{-1}), glycine (3 mg L^{-1}), and metal salts AgNO$_3$ (10 mg L^{-1}) and 1 μM CuSO$_4 \cdot$5H$_2$O (0.249 mg L^{-1}), as the commonly used ones for cereals, in five different combinations were tested as additional additives.

All media were solidified with 0.6% agar and were adjusted to pH 5.7–5.8 prior to autoclaving. Three dishes with 30 seeds each were used in each treatment. Each experiment was repeated for three times. Cultures were incubated for three weeks in darkness at 26 ± 2°C. The effectiveness of each tested medium was evaluated based on the size and morphology (color, surface structure, water content, and tissue density) of the resultant calli. Only calli exceeded 2-3 mm and more in size were scored.

2.2. Plant Regeneration. Organogenesis from callus was induced via alteration in medium hormonal concentrations.

TABLE 1: Compositions of enzyme solutions used for *E. coracana* protoplasts isolation.

Enzymes, in %	Variants of solutions						
	1	2	3	4	5	6	7
Cellulase "Onozuka" R-10	10	2	1	1	2	2	1.5
Drisellase	—	—	—	0.5	—	0.5	0.5
Pectolyase	1	0.2	1	1	0.5	1	1
Macerosyme R-10	—	—	1	0.6	1	0.5	—
Hemicellulase	—	—	0.5	0.5	0.5	0.2	0.2

However, for each callus, the supplement combination was unchanged from the step 3 of callus induction. Calli were incubated for three weeks on an MS [23] medium with 1.5 mg L^{-1} KIN and 0.2 mg L^{-1} NAA in light with 14 h photoperiod. The resultant shoots were grown into plantlets in a basal (hormone-free) MS medium into 100 mL tubes with perlite under light incubation. Plants with well-developed root were transferred into small pots containing sterile soil and covered with glass caps, and then they were transferred to the field after two-three weeks of growth.

2.3. Protoplast Isolation. Both callus tissues and explants of seedlings (coleoptiles and mesocotyles) from 6-day-old seedlings were used in protoplast isolation. To select the most optimal conditions the tissues were first cut and digested in seven different enzyme combination solutions (Table 1) dissolved in four different osmotic media: (1) CPW [24]; (2) medium for of isolation protoplasts of *E. indica* [22]; (3) 10x N$_6$ medium [25] + 80 mM KCl, 100 mM mannitol (pH 5.5); (4) 0.6 mM mannitol + 80 mM CaCl$_2$ (pH 5.5). Tissues were digested at 25°C for 18–24 hrs. Protoplasts were purified using the following steps.

(1) Filter the enzyme solution into a centrifuge tube through a 60–70 μm sterile nylon mesh. The debris on the mesh was washed with 3-4 mL of washing solution (155 mM NaCl; 30 mM KCl; 5 mM CaCl$_2$; and 15 mM glucose) to flush down trapped protoplasts.

(2) Protoplasts were pelleted by centrifuging at 1000 rpm for 5 min. After removing the supernatant, 2-3 mL of washing solution was added.

(3) Carefully layer the protoplast suspension onto the top of a 3-4 mL of 20% sucrose solution in a fresh centrifuge tube.

(4) A white band of protoplast was formed at the interface of the sucrose and washing solution layers after the tube was centrifuged at 1000 rpm for 5 min.

(5) The protoplast band was pipetted out and diluted to 5–10 mL in volume with the washing solution for cell counting.

The purified protoplasts were cultured for 20–25 days in darkness at 25°C in a KM8p [26] medium supplemented with glutamine (100 mg L^{-1}), proline (500 mg L^{-1}), 2.4-D (0.5 mg L^{-1}), and KIN (0.2 mg L^{-1}). The resultant cell aggregates were embedded in an agar medium made of 1:1

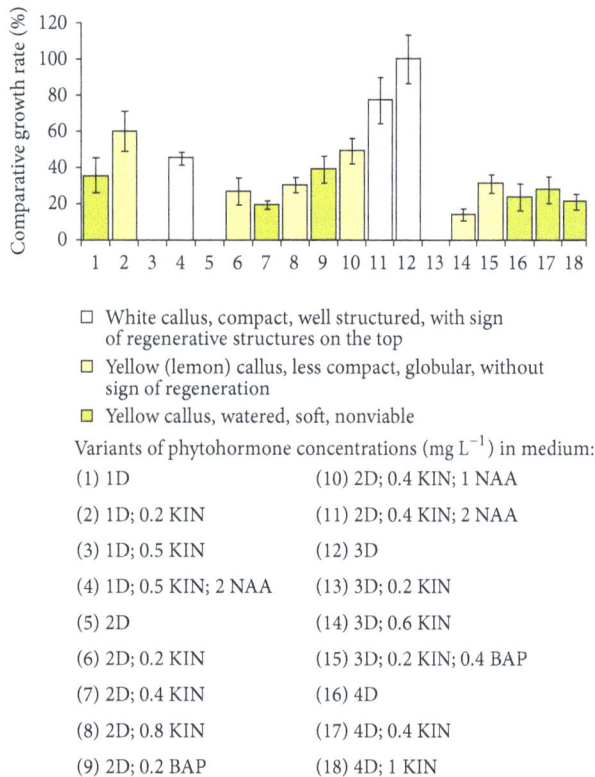

FIGURE 1: Influence of different concentrations of auxins and cytokinins on *E. coracana* callus induction.

(v/v) ratio of a KM8p medium with the above supplements to a KM8p medium supplemented with glutamine ($200 \, mg \, L^{-1}$), proline ($1000 \, mg \, L^{-1}$), tryptophan ($200 \, mg \, L^{-1}$), 2.4-D ($2 \, mg \, L^{-1}$), zeatin ($0.4 \, mg \, L^{-1}$), and agar (1.2%). Cultures were incubated in darkness for 3-4 days before being transferred into light for further incubation and regeneration.

2.4. Micronucleation.

Micronuclei were induced by the addition of 10–150 μM isopropyl N-(3-chlorophenyl) carbamate, CIPC (chlorpropham) (Sigma, USA), and 20 μM cytochalasin B (Sigma, USA) to 50 mL of liquid MS medium with $2 \, mg \, L^{-1}$ 2.4-D, $0.4 \, mg \, L^{-1}$ KIN, and $2 \, mg \, L^{-1}$ NAA where 3-day-old seedlings were placed for subculture for 24 h at 25°C in the dark on a rotary shaker (120 rpm). For microprotoplasts isolation 10 μM CIPC and 20 μM cytochalasin B were added also into enzyme solution 2 (Table 1) containing 80 mM $CaCl_2$ and 0.6 mM mannitol as osmoticum to prevent the fusion of micronuclei and to disrupt microfilaments during protoplast isolation. The microprotoplasts were purified using the following steps: (i) we filtered the enzyme solution into a centrifuge tube through nylon mesh and washed with washing solution, as described earlier for protoplast isolation; (ii) pelleted microprotoplasts by centrifuging at 1000 rpm for 5 min; after removing the supernatant, 2-3 mL of washing solution was added; (iii) centrifuged at 1000 rpm for 5 min to collect microprotoplasts then carefully removed the supernatant and added 3 mL KM8p medium supplemented with glutamine ($100 \, mg \, L^{-1}$), proline ($500 \, mg \, L^{-1}$), 2.4-D ($0.5 \, mg \, L^{-1}$), and KIN ($0.2 \, mg \, L^{-1}$).

For cytological analysis of microprotoplasts, $3 \, \mu g \, mL^{-1}$ 4, 6-diamodino-2-phenylindole (DAPI, Sigma, USA) was added to the suspension for 10 min. The micronuclei were observed under luminescent microscope Axiostar plus (Carl Zeiss, Jena, Germany).

3. Results and Discussion

3.1. Establishment of Callus Cultures.

Among the five culture media tested in step 1 of callus induction, only the Medium E supported healthy callus induction. The remaining media either failed to produce callus (Medium B) or resulted in small calli which soon became colorless and nonviable. Therefore, the basal components of Medium E [22], with N_6 macrosalts, MS microsalts, and B5 organics [27], were chosen as the basal medium in the subsequent callus induction medium optimization tests.

Since 2.4-D is commonly considered as the most powerful callus induction hormone, it was included in all step 2 callus induction media tested. Figure 1 clearly showed that, after combined with other hormones, $2 \, mg \, L^{-1}$ 2.4-D resulted in the highest, up to 100%, callus formation percentages. While lower or higher 2.4-D concentrations resulted in significantly lower percentage of callus formation.

It is known that the addition of a small amount of cytokinin to the auxin-containing medium frequently inserts positive effects on callus induction. In this work, we found that the addition of an $0.2–0.6 \, mg \, L^{-1}$ KIN always leads to a higher callus formation percentage over 2.4-D alone. Moreover KIN was also reported to have promotional effects on somatic embryogenesis of finger millet [9, 19]. To compare the effectiveness of KIN versus BAP, $0.2 \, mg \, L^{-1}$ KIN was replaced by the same concentration of BAP in one case, and in another case $0.6 \, mg \, L^{-1}$ KIN was substituted by $0.2 \, mg \, L^{-1}$ KIN + $0.4 \, mg \, L^{-1}$ BAP. The percentages of callus formation in both cases were clearly reduced. This suggested that, for *E. coracana*, KIN is a more effective cytokinin than BAP for callus induction. Similar results were reported by Pius et al. [19] for finger millet, although numerous reports indicated that BAP was preferred over KIN in induction of embryogenic calli of numerous cereal species [28].

There were three types of calli been observed in our works: (1) white, compact, well-structured callus; (2) yellow or lemon, less compact, globular callus; (3) yellow, watery, soft callus. It was found that the first type of callus (Figure 2(b)) was produced only when NAA was added to the 2.4D + KIN combinations. This type of callus was clearly superior in appearance compared to the remaining two types. It was also found that only the first type of callus stayed viable in the subsequent regeneration induction stage. Thus, despite that the NAA is not commonly used in cereal callus induction, we suggest that it is useful in light of callus regeneration capacity. This suggestion coincides with data of Kothari et al. [29] where NAA was added for plant regeneration of *E. coracana* from seeds used as initial explants.

We found also that there was no significant change in callus formation percentage regardless of the types of additional additive combinations tested supplemented to the basal media.

FIGURE 2: (a) Callus induction from mesocotyl of *E. coracana* from seed callus culture. (b) Morphology of compact, well-structured callus. (c) Plant regeneration of *E. coracana* from seed callus culture. (d) Rooted *E. coracana* plants in vitro.

3.2. Plant Regeneration. Although the percentage of callus formation was not affected by types of additives to the basal medium, different additive combinations greatly affected the quality of resultant tissues and resulted in significant differences in the final numbers of plantlets produced. Thus, we are presenting our results of this section in relation to the additives supplemented to the basal medium. It was found that the basal medium alone showed relatively low regeneration percentage and relatively poor tissue quality. The addition of glutamine and proline significantly improved such regeneration rate (Table 2) but failed to improve tissue quality. However, further supplementing of AgNO$_3$ or CuSO$_4$ cancelled out the advantage provided by glutamine and proline supplementation. This indicated possible toxic effects of such heavy metals.

Among the six supplementing treatments (Table 2), the supplementing of tryptophan only or supplementing of thiamine HCl and glycine in addition to glutamine and proline led to the highest percentages of regenerative calli. The resulting tissues of these two treatments appeared to have also the highest qualities and to be the most healthy. From healthy-appearing regenerated shoots (Figure 2(c)), roots developed upon transferring to hormone-free medium. Regenerating

TABLE 2: Influence of different supplements on plant regeneration of *E. coracana*.

Number	Supplements	Number of regenerants per 100 explants
1	—	0
2	100 mg L^{-1} glutamine 500 mg L^{-1} proline	30 ± 9.16
3	100 mg L^{-1} glutamine 500 mg L^{-1} proline 10 mg L^{-1} AgNO$_3$	25 ± 8.66
4	100 mg L^{-1} glutamine 500 mg L^{-1} proline 0,249 mg L^{-1} CuSO$_4 \cdot$ 5H$_2$O	0
5	100 mg L^{-1} glutamine 500 mg L^{-1} proline 10 mg L^{-1} thiamine HCl 3 mg L^{-1} glycine	56 ± 9.92
6	200 mg L^{-1} tryptophan	45 ± 9.94

$^*P = 0.05$.

shoots from the nonsupplemented control treatment were of poor quality and failed to produce viable plants. The

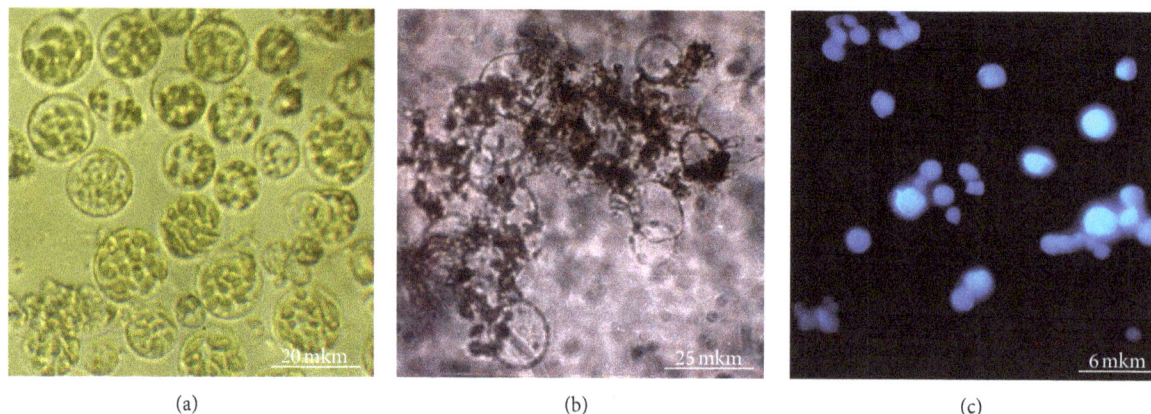

(a)   (b)   (c)

FIGURE 3: (a) *E. coracana* protoplasts isolated from seedlings. (b) Microcolonies formation in liquid KM8p medium. (c) Micronuclei in *E. coracana* cells.

supplementation of $CuSO_4$ inhibited root formation and did not result in plant production either. Although the supplementing of only glutamine and proline or with $AgNO_3$ in addition had high regeneration percentages, due to their poor tissue qualities (yellow calli with little compact structure), the number of viable plants formed was not high. In spite of their lowest regeneration percentages, treatments that led to good tissue qualities (white, well-structured calli) resulted in the highest number of viable plant production. These were treatments supplemented either with tryptophan alone or with thiamine HCl and glycine in addition to glutamine and proline.

Green and Philips [30] demonstrated that the supplementation of proline enhances somatic embryogenesis in maize, and Sirivardana and Nabros [31] showed that tryptophan favored somatic embryogenesis in some rice cultivars. Eapen and George [8], on the other hand, experienced decreased embryo germination frequency in finger millet when glutamine, proline, and tryptophan were supplemented. Thiamine was used by Chu et al. [25] to induce large quantity of cereal plant regenerants.

At the same time the supplementing of $CuSO_4$ did not lead to positive effect in plant regeneration. These data agreed with that of Purnhauser and Gyulai [32] that $CuSO_4$ supplementation inhibited rape regeneration. In contrast, Cu^{2+} enhanced rhizogenesis in barley cultures (up to 93.7%) [33], wheat, and triticale [32]. Dahleen [34] also used $CuSO_4$ to increase barley regeneration. The reason that led to the above discrepancy is not clear, and the role of copper in regeneration is not well understood. Since Cu^{2+} is the cofactor of many important enzymes in electron transport and protein/carbohydrate biosynthesis, Purnhauser and Gyulai [32] suggested the possibility that some of these enzymes might play certain roles in plant regeneration.

In conclusion, our results suggested that the number of plants (Figure 2(d)) eventually produced did not correlate directly with the regeneration percentage of the tissues; instead, it was closely associated with the quality of the tissues which were dictated by the hormones and supplements in the basal medium. From the obtained results, for future experiments we propose to use the following additional composition of media for callus induction and plant regeneration of finger millet: $100\,mg\,L^{-1}$ Glu, $500\,mg\,L^{-1}$ Pro, $10\,mg\,L^{-1}$ B_1, and $3\,mg\,L^{-1}$ Gly.

3.3. Protoplast Isolation and Culture.

Among the four protoplast osmotic media, only Medium 4 supported viable protoplast isolation. Despite the CPW (Medium 1) being successful in supporting protoplast isolation for several cereal species, *Poa pratensis* [35], rice [36], and *S. bicolor* [37], its osmolarity was too low for *E. coracana* and led to the swelling of the isolated protoplasts that burst at the washing step. The same was true for Medium 3. Medium 2, on the other hand, had too high osmolarity and resulted in the condensation of all plastids at the center of the protoplasts.

Among the seven tested enzyme mixtures, Solution 2 supported the best protoplast isolation. Protoplasts resulting from this enzyme solution were spherical in shape, rich in cytoplasm, and void of large vacuoles. It was possible to reduce the time of digestion to 4 hrs.

Protoplast yield from callus was 1×10^4 protoplasts per gram of tissue; while the per-gram yield from seedling coleoptyle tissues was 5×10^6 protoplasts (Figure 3(a)). We feel that, for *E. coracana*, seedling explants resulted in sufficiently high protoplast yield upon enzyme digestion. Protoplasts started regenerating cell walls within 4-5 days in the supplemented KM8p liquid medium. Cell division took place 1-2 days later. Microcolonies (Figure 3(b)) were visible in 20–25-day-old cultures. Then they were embedded into solid KM8p medium for further plant regeneration. Thus, we described here for the first time the elaborated method of effective protoplast isolation from *E. coracana*, which could be used in future the in different programs on cellular engineering of millets, for example, through methods of symmetric or asymmetric somatic hybridization.

3.4. Micronucleation.

Polygenic traits or traits with unknown biochemical or molecular mechanisms (e.g., resistance to certain diseases or stresses and other economically important traits) are still recalcitrant to transfer using methods of genetic engineering. Recently, an alternative asymmetric somatic hybridization method using microprotoplasts (microprotoplast fusion) has been developed [18]. Since microprotoplasts contain only one or a few intact chromosomes, a limited number of chromosomes can be transferred via microprotoplast fusion, resulting in the production of chromosome addition lines with even a single and specific intact chromosome between sexually incompatible species [18, 38]. In order to apply this technique for improvement of monocot plant species, we established an efficient system for mass-preparation of microprotoplasts of *E. coracana*. Here, we present an effective method of the induction by antimitotic agents and isolation of microprotoplast from finger millet somatic cells. As antimitotic drugs 10–150 μM CIPC and 20 μM cytochalasin B were used for micronuclei induction in cells of finger millet seedling. Different concentration of CIPC and 20 μM cytochalasin B were added to liquid MS medium supplemented with 2 mg L^{-1} 2.4-D, 0.4 mg L^{-1} KIN, and 2 mg L^{-1} NAA where 3-day-old seedlings were placed for subcultivation for 24 h at 25°C in the dark on a rotary shaker. For microprotoplasts isolation both drugs were added also into enzyme solution to prevent the fusion of micronuclei and to disrupt microfilaments during protoplast isolation. It has been found that the most effective concentration of CIPC was 10 μM, which in combination with 20 μM cytochalasin B led to the highest percentage of micronuclei formation (Figure 3(c)). As a rule the cells with 3-4 micronuclei were observed after such treatment of finger millet cells. It is corresponding to the previously obtained data where CIPC in this concentration was successfully used to obtain microprotoplasts from developing microspores of lily species [38]. The system described here can be used for the transfer of one or a few chromosomes via microprotoplast fusion from finger millet to different economically important grasses or cereals. Because it is known that finger millet is little affected by disease and insects [39]. Chromosome addition lines produced via such technique may contribute to genetic improvement of the above-mentioned plant species.

Conflict of Interests

None of the authors has a conflict of interests with any mentioned commercial identity in the paper.

Acknowledgment

The authors wish to express their thanks to Professor Emeritus C. Hu (Wm. Paterson University of New Jersey, USA) for critical evaluation of the paper and improvement of its English version.

References

[1] S. L. Kothari, S. Kumar, R. K. Vishnoi, A. Kothari, and K. N. Watanabe, "Applications of biotechnology for improvement of millet crops: review of progress and future prospects," *Plant Biotechnology*, vol. 22, no. 2, pp. 81–88, 2005.

[2] N. E. Borlaug, "Feeding a world of 10 billion people: the miracle ahead," *In Vitro Cellular and Developmental Biology—Plant*, vol. 38, no. 2, pp. 221–228, 2002.

[3] N. O. Stadnichuk and A. A. Abramov, "Finger millet (*Eleusine coracana* (L.) Gaertn.), variety 'Tropikanka,'" Ukrainian plant variety certificate No. 1252, State Service for Plant Variety Protection (invention No. 9950001 from 12. 12. 1999), 2001.

[4] N. O. Stadnichuk, G. Y. Bayer, A. I. Yemets, D. B. Rakhmetov, and B. Y. Blume, "Finger millet (*Eleusine coracana* (L.) Gaertn.), variety 'Yaroslav-8,'" Ukrainian plant variety certificate No. 09551, State Service for Plant Variety Protection (invention No. 03.11.2008), 2009.

[5] N. O. Stadnichuk, A. I. Yemets, D. B. Rakhmetov, and B. Y. Blume, "Finger millet (*Eleusine coracana* (L.) Gaertn.), variety 'Yevgeniya,'" Ukrainian plant variety certificate No. 120109, State Service for Plant Variety Protection (invention No. 10344001 from 03.11.2010), 2011.

[6] T. Wakizuka and T. Yamaguchi, "The induction of enlarged apical domes in vitro and multi-shoot formation from finger millet (*Eleusine coracana*)," *Annals of Botany*, vol. 60, no. 3, pp. 331–336, 1987.

[7] S. Eapen and L. George, "High frequency plant regeneration through somatic embryogenesis in finger millet (*Eleusine coracana* Gaertn.)," *Plant Science*, vol. 61, no. 1, pp. 127–130, 1989.

[8] S. Eapen and L. George, "Influence of phytohormones, carbohydrates, aminoacids, growth supplements and antibiotics on somatic embryogenesis and plant differentiation in finger millet," *Plant Cell, Tissue and Organ Culture*, vol. 22, no. 2, pp. 87–93, 1990.

[9] P. Sivadas, S. L. Kothari, and N. Chandra, "High frequency embryoid and plantlet formation from tissue cultures of the Finger millet-*Eleusine coracana* (L.) Gaertn.," *Plant Cell Reports*, vol. 9, no. 2, pp. 93–96, 1990.

[10] V. D. Reddy, K. V. Rao, T. P. Reddy, and P. B. K. Kishor, "Finger millet," in *Compendium of Transgenic Crop Plants: Transgenic Cereals and Forage Grasses*, C. Kole and T. C. Hall, Eds., pp. 191–198, Blackwell, London, UK, 2008.

[11] S. A. Ceasar and S. Ignacimuthu, "Genetic engineering of millets: current status and future prospects," *Biotechnology Letters*, vol. 31, no. 6, pp. 779–788, 2009.

[12] P. Gupta, S. Raghuvanshi, and A. K. Tyagi, "Assessment of the efficiency of various gene promoters *via* biolistics in leaf and regenerating seed callus of millets, *Eleusine coracana* and *Echinochloa crusgalli*," *Plant Biotechnology*, vol. 18, no. 4, pp. 275–282, 2001.

[13] A. M. Latha, K. V. Rao, and V. D. Reddy, "Production of transgenic plants resistant to leaf blast disease in finger millet (*Eleusine coracana* (L.) Gaertn.)," *Plant Science*, vol. 169, no. 4, pp. 657–667, 2005.

[14] S. Mahalakshmi, G. S. B. Christopher, T. P. Reddy, K. V. Rao, and V. D. Reddy, "Isolation of a cDNA clone (*PcSrp*) encoding serine-rich-protein from *Porteresia coarctata* T. and its expression in yeast and finger millet (*Eleusine coracana* L.) affording salt tolerance," *Planta*, vol. 224, no. 2, pp. 347–359, 2006.

[15] A. Yemets, V. Radchuk, O. Bayer et al., "Development of transformation vectors based upon a modified plant α-tubulin gene as the selectable marker," *Cell Biology International*, vol. 32, no. 5, pp. 566–570, 2008.

[16] S. A. Ceasar and S. Ignacimuthu, "*Agrobacterium*-mediated transformation of finger millet (*Eleusine coracana* (L.) Gaertn.)

using shoot apex explants," *Plant Cell Reports*, vol. 30, no. 9, pp. 1759–1770, 2011.

[17] M. Sharma, A. Kothari-Chajer, S. Jagga-Chugh, and S. L. Kothari, "Factors influencing *Agrobacterium tumefaciens*-mediated genetic transformation of *Eleusine coracana* (L.) Gaertn," *Plant Cell, Tissue and Organ Culture*, vol. 105, no. 1, pp. 93–104, 2011.

[18] A. I. Yemets and B. Y. Blume, "Antimitotic drugs for microprotoplast-mediated chromosome transfer in plant genomics, cell engineering and breeding," in *The Plant Cytoskeleton: Genomic and Bioinformatic Tools for Biotechnology and Agriculture*, Y. B. Blume, W. V. Baird, A. I. Yemets, and D. Breviario, Eds., pp. 435–454, Springer, Berlin, Germany, 2008.

[19] J. Pius, L. George, S. Eapen, and P. S. Rao, "Influence of genotype and phytohormones on somatic embryogenesis and plant regeneration in finger millet," *Proceedings of the Indian National Science Academy*, vol. 60, pp. 53–56, 1994.

[20] T. G. Lu and C. S. Sun, "Cryopreservation of foxtail millet (*Setaria italica* L.)," *Biotechnology in Agriculture and Forestry*, vol. 98, pp. 1099–1103, 1995.

[21] A. M. Rao, K. P. Sree, and P. B. K. Kishor, "Enhanced plant regeneration in grain and sweet sorghum by asparagine, proline and cefotaxime," *Plant Cell Reports*, vol. 15, no. 1-2, pp. 72–75, 1995.

[22] A. I. Yemets, L. A. Klimkina, L. V. Tarassenko, and Y. B. Blume, "Efficient callus formation and plant regeneration of goosegrass [*Eleusine indica* (L.) Gaertn.]," *Plant Cell Reports*, vol. 21, no. 6, pp. 503–510, 2003.

[23] T. Murashige and F. Skoog, "A revised medium for rapid growth and bioassays with tobacco tissue culture," *Physiologia Plantarum*, vol. 15, pp. 473–497, 1962.

[24] E. C. Cocking and J. F. Peberdy, *The Use Protoplasts from Fungi and Higher Plants as Genetic Systems*, Department of Botany, University of Nottingham, Nottingham, UK, 1974.

[25] C. C. Chu, C. C. Wang, C. S. Sun et al., "Establishment of an efficient medium for anther culture of rice through comparative experiments on the nitrogen sources," *Scientia Sinica*, vol. 18, pp. 659–668, 1975.

[26] K. N. Kao and M. R. Michayluk, "Nutritional requirements for growth of *Vicia hajastana* cells and protoplasts at a very low population density in liquid media," *Planta*, vol. 126, no. 2, pp. 105–110, 1975.

[27] O. L. Gamborg, R. A. Miller, and K. Ojima, "Nutrient requirements of suspension cultures of soybean root cells," *Experimental Cell Research*, vol. 50, no. 1, pp. 151–158, 1968.

[28] J. Patnaik, S. Sahoo, and B. K. Debata, "Somatic embryogenesis and plantlet regeneration from cell suspension cultures of palmarosa grass (*Cymbopogon martinii*)," *Plant Cell Reports*, vol. 16, no. 6, pp. 430–434, 1997.

[29] S. L. Kothari, K. Agarwal, and S. Kumar, "Inorganic nutrient manipulation for highly improved in vitro plant regeneration in finger millet—*Eleusine coracana* (L.) Gaertn," *In Vitro Cellular and Developmental Biology—Plant*, vol. 40, no. 5, pp. 515–519, 2004.

[30] C. E. Green and R. L. Phillips, "Plant regeneration from tissue cultures of maize," *Crop Science*, vol. 15, pp. 417–421, 1998.

[31] S. Siriwardana and M. W. Nabors, "Tryptophan enhancement of somatic embryogenesis in rice," *Plant Physiology*, vol. 73, pp. 143–146, 1983.

[32] L. Purnhauser and G. Gyulai, "Effect of copper on shoot and root regeneration in wheat, triticale, rape and tobacco tissue cultures," *Plant Cell, Tissue and Organ Culture*, vol. 35, no. 2, pp. 131–139, 1993.

[33] A. M. Castillo, B. Egaña, J. M. Sanz, and L. Cistué, "Somatic embryogenesis and plant regeneration from barley cultivars grown in Spain," *Plant Cell Reports*, vol. 17, no. 11, pp. 902–906, 1998.

[34] L. S. Dahleen, "Improved plant regeneration from barley callus cultures by increased copper levels," *Plant Cell, Tissue and Organ Culture*, vol. 43, no. 3, pp. 267–269, 1995.

[35] K. A. Nielsen, E. Larsen, and E. Knudsen, "Regeneration of protoplast-derived green plants of Kentucky blue grass (*Poa pratensis* L.)," *Plant Cell Reports*, vol. 12, no. 10, pp. 537–540, 1993.

[36] Z. S. Zhang Shiping, "Efficient plant regeneration from Indica (group 1) rice protoplasts of one advanced breeding line and three varieties," *Plant Cell Reports*, vol. 15, no. 1-2, pp. 68–71, 1995.

[37] R. V. Sairam, N. Seetharama, P. S. Devi, A. Verma, U. R. Murthy, and I. Potrykus, "Culture and regeneration of mesophyll-derived protoplasts of sorghum [*Sorghum bicolor* (L.) Moench]," *Plant Cell Reports*, vol. 18, no. 12, pp. 972–977, 1999.

[38] H. Saito and M. Nakano, "Preparation of microprotoplasts for partial genome transfer via microprotoplast fusion in *Liliaceous* ornamental plants," *Japan Agricultural Research Quarterly*, vol. 36, no. 3, pp. 129–135, 2002.

[39] S. Neves, "*Eleusine*," in *Wild Crop Relatives: Genomic and Breeding Resources, Millets and Grasses*, C. Kole, Ed., pp. 113–133, Springer, Berlin, Germany, 2011.

8

Modulating Plant Calcium for Better Nutrition and Stress Tolerance

Dominique (Niki) Robertson

Department of Plant Biology, North Carolina State University, P.O. Box 7612, Raleigh, NC 27695, USA

Correspondence should be addressed to Dominique (Niki) Robertson; niki@ncsu.edu

Academic Editors: M. Adrian, E. Collakova, G. T. Maatooq, I. Paponov, and K. Takeno

External Ca^{2+} supplementation helps plants to recover from stress. This paper considers genetic methods for increasing Ca^{2+} to augment stress tolerance in plants and to increase their nutritional value. The transport of Ca^{2+} must be carefully controlled to minimize fluctuations in the cytosol while providing both structural support to new cell walls and membranes, and intracellular stores of Ca^{2+} for signaling. It is not clear how this is accomplished in meristems, which are remote from active transpiration—the driving force for Ca^{2+} movement into shoots. Meristems have high levels of calreticulin (CRT), which bind a 50-fold excess of Ca^{2+} and may facilitate Ca^{2+} transport between cells across plasmodesmatal ER. Transgenes based on the high-capacity Ca^{2+}-binding C-domain of *CRT1* have increased the total plant Ca^{2+} by 15%–25% and also increased the abiotic stress tolerance. These results are compared to the overexpression of *sCAX1*, which not only increased total Ca^{2+} up to 3-fold but also caused Ca^{2+} deficiency symptoms. Coexpression of *sCAX1* and *CRT1* resolved the symptoms and led to high levels of Ca^{2+} without Ca^{2+} supplementation. These results imply an important role for ER Ca^{2+} in stress tolerance and signaling and demonstrate the feasibility of using Ca^{2+}-modulating proteins to enhance both agronomic and nutritional properties.

1. Introduction

Plants sense and respond to environmental stimuli using networks of sensors, second messengers, kinases, and transcription factors to regulate gene expression and adapt to the new conditions. Ca^{2+} is perhaps the best-known second messenger but is also required for proper cell wall structure and membrane integrity [1]. Although Ca^{2+} is present at relatively high concentrations (0.1–80 mM) in cell walls and organelles, cytoplasmic levels of Ca^{2+} are maintained at ~100 nM [2–4]. Signal transduction in plants requires the ability to mobilize and sequester Ca^{2+} from both internal and external Ca^{2+} stores. Because both deficiency and high concentrations of Ca^{2+} cause localized cell death, the transport of Ca^{2+} throughout the plant must be tightly regulated [5, 6].

Plants grown under Ca^{2+} deficient conditions are more susceptible to plant pathogens and show reduced growth of apical meristems, chlorotic leaves, and cell wall breakdown leading to softening of tissues [2]. But adding Ca^{2+} does more than just alleviate these symptoms, it bolsters plant growth by increasing root length and helps them to withstand or recover from stress [7–14]. Supplemental Ca^{2+} is also used to improve fruit characteristics and can function to delay ethylene-induced senescence [15]. This information is not new, a report in *Science* published over 40 years ago described the effect of 1 mM Ca^{2+} in preventing severe NaCl toxicity in beans [16].

The precise effects of extracellular Ca^{2+} on a plant system is likely to be complex, because Ca^{2+} has multiple roles, and because different plants show different responses to supplemental Ca^{2+}. For example, in most plants extracellular Ca^{2+} reduces Na^+ accumulation, which alleviates salt stress. But in some plants (such as maize), Na^+ levels remain constant, but a beneficial effect on plant growth is still apparent [17]. Supplemental Ca^{2+} in a few plants, such as rice, has no apparent effect on salt tolerance [18] (see [19]). Supplementation with K^+ has either no effect or is detrimental [20].

In a recent report, a solution of Ca^{2+} was found to be beneficial when sprayed directly onto the leaves of drought-stressed tea plants [21]. How does simply spraying Ca^{2+} onto leaves benefit plants? Why have not plants figured out how to increase their own stores, since Ca^{2+} is readily available in most environments? Alternatively, is this a part of what makes some plants "weedy"? Is there a barrier to the effective long-distance transport of Ca^{2+}? Can we engineer a "work-around", or alternative mechanism for Ca^{2+} transport, to help them recover from stress or even to prevent damage in the first place?

To begin to understand how extracellular Ca^{2+} benefits plants, it is necessary to understand more about the function and mobility of Ca^{2+} at both the cellular and the whole plant level. Once we understand the different roles of Ca^{2+}, how Ca^{2+} is sequestered and transported within the plant, released for cellular signaling—and then rapidly sequestered away from detrimental interactions—then we can begin to think about revising or tailoring some of its pathways. This paper will provide a brief, whole-plant overview of Ca^{2+}-regulated pathways and functions with the goal of identifying potential strategies for engineering additional Ca^{2+} ions into soluble plant reserves, so that they are readily available for signaling and growth. It is hoped that this approach can be part of a strategy to design more nutritional crop plants that are also more resilient to stress.

2. Ca^{2+} Stores and Signaling

It is commonly believed that Ca^{2+}, one of the most abundant minerals in the earth, evolved as a signaling molecule because of the dual needs of the cell for soluble phosphate and Ca^{2+} and the propensity for the two to precipitate out as an insoluble salt [22]. Phosphate also plays a critical role in signal transduction, but its role as an energy intermediate requires a presence in the cytoplasm [23]. In plants, metabolic pathways that use ATP are found largely in the cytoplasm and are kept separate from Ca^{2+} stores, which are found primarily in the apoplast, vacuole, and endoplasmic reticulum (ER) and to a lesser extent in mitochondria, chloroplasts, and the nucleus [4]. In animal cells and early in the plant lineage, the ER was the major source of Ca^{2+}, and its release was controlled by another second messenger, inositol (1,4,5) triphosphate (IP$_3$), through activation of ER-localized IP$_3$ receptors [24, 25]. Similar IP$_3$ receptors have not been found in plants; however, the phosphoinositide pathway is conserved in plants [26–29]. Members of the phosphoinositide signaling pathway show transcriptional regulation by environmental and developmental stimuli in Arabidopsis [30], and Ca^{2+} release by IP$_3$ is conserved [31, 32].

In addition to the apoplast, the ER and vacuoles are the major and metabolically relevant sources of cellular Ca^{2+} [33–35]. Cytosolic Ca^{2+} levels fluctuate and are controlled by a system of membrane-localized Ca^{2+} pumps and Ca^{2+} channels located in the plasmalemma, vacuole, and ER [4, 5, 36]. The electrochemical potential for Ca^{2+} to enter the cytoplasm, across the plasma membrane, was calculated by Spalding and Harper to be about −52 kJ/mol [22]. Therefore, Ca^{2+} can enter cells passively through ion channels but

requires energy to be pumped out of the cytoplasm. Although energetically unfavorable, removal of Ca^{2+} is rapid and efficient, resulting in 1000-fold and higher [Ca^{2+}] differences between the cytosol and surrounding organelles and apoplast [37].

Unlike animal systems, mutations in Ca^{2+} transport proteins often do not produce dramatic phenotypes [22], suggesting that plants are more tolerant of cytosolic Ca^{2+} or that they have overlapping and redundant systems. This has made it difficult to correlate electrophysiological experiments with genetics to identify exactly which Ca^{2+} channels function in signaling (or storage) and when. Ca^{2+} was shown to be released from the ER and possibly other membranes by cADP-ribose, an NAD$^+$ metabolite, similar to what happens in animal cells, over a decade ago [34], however, it now seems clear that cyclic nucleotide-gated channels (CNGC) are found in the plasmalemma [48]. One of the few proteins that do have a phenotype, the phenotype of cngc2, is similar to cax1/cax3 (see Section 3) suggesting that it plays a major role in allowing nonsignaling Ca^{2+} entry into leaf cells [49]. There are 20 CNGC genes in Arabidopsis and an additional 20 genes that encode glutamate receptor-like channels (GLR), another type of Ca^{2+} channel found in the plasmalemma [48]. A third type of channel, the two-pore Ca^{2+} channel (TPC1), was first identified as a plasmalemma protein but is now known to be localized to the tonoplast membrane.

There are two major groups of proteins that function in Ca^{2+} removal from the cytoplasm [50]. Autoinhibitory Ca^{2+} ATPase (ACA) uses the energy of ATP to pump Ca^{2+} out of the cytoplasm and into organelles such as the vacuole and ER. The second group of proteins function as antiporters and are called CAtion eXchange proteins (CAX), found on the tonoplast membrane. CAX exchanges two protons for one Ca^{2+}, using the energy of the proton gradient to dampen cytoplasmic Ca^{2+} signals [51].

2.1. Calcium Signatures. Cytoplasmic increases in Ca^{2+} in response to high concentrations of salt were noted at least 25 years ago in plants [52], but the specificity of Ca^{2+} signaling is still not well understood. There are two nonexclusive models for how Ca^{2+} functions as a second messenger. The Ca^{2+} signature model posits that information is encoded in the shape, duration, and frequency of Ca^{2+} transients and the diversity of cellular Ca^{2+} stores, all of which may facilitate the formation of microdomains that support and respond to localized Ca^{2+} changes [4, 53]. These localized changes are specific to the inducing stimulus and result in specific changes to Ca^{2+}-modulated proteins and their targets [5, 39, 54–56]. A second model suggests that Ca^{2+} transients function as a simple binary switch, either on or off, and it is the Ca^{2+} sensor (a Ca^{2+}-modulated protein) that links different stimuli to the adaptive response [22].

The best-studied examples of Ca^{2+}-mediated signal transduction include guard cell opening, nodulation, and tip growth of polarized structures such as pollen tubes [57–66]. Specific Ca^{2+} signatures have also been reported, for example, in response to different chemicals in the root (aluminum, glutamic acid, and ATP [67]) and, at the whole plant level, in response to ozone [68]. Examples of other stimuli that cause transient increases in cytosolic Ca^{2+} concentrations include touch, cold shock, heat shock, oxidative stress, anoxia, hypo-osmotic shock, salinity, wounding, gravity, and pathogen infection [37, 56, 69–81]. Developmental signals including fertilization, senescence, abscission, and ripening also involve Ca^{2+}-regulated proteins [82–88].

There is evidence for tissue-specific differences in Ca^{2+} flux in response to the same stimulus, for example, salt stress. Salt tolerance is a complex trait involving responses to cellular osmotic and ionic stresses and their consequent secondary stresses (e.g., oxidative stress) [89, 90]. Roots show a biphasic transient increase in cytosolic Ca^{2+} following exposure to acute salt stress [73]. In contrast to cold shock, which is restricted to areas near the root meristem, salt shock increases cytosolic Ca^{2+} along the entire root [91]. To distinguish tissue-specific differences in Ca^{2+} flux, different transgenic plants transformed with a gene encoding aequorin (a reporter gene for Ca^{2+}) targeted to the cytoplasm of the epidermis, endodermis, or pericycle of Arabidopsis roots were used [73]. Prolonged oscillations in aequorin luminescence in the endodermis and pericycle occurred that were distinct from the epidermis [73]. This demonstrated that the same stimulus was transduced differently depending on the cell type, which could be due in part to the evolution of multiple family members in genes that transport Ca^{2+} (Section 2).

2.2. Calcium Sensors.
Understanding the transduction of Ca^{2+} signatures has increased in the past decade due to rapid progress in deciphering the cellular network of Ca^{2+}-responsive proteins. There are several families of Ca^{2+}-binding proteins in plants [92–95]. Proteins such as calmodulin, calcineurin B-like proteins (CBL), and Ca^{2+}-dependent protein kinases (CDPK) "sense" Ca^{2+}, having one or more EF-hand domains that bind Ca^{2+} with high affinity. The Arabidopsis genome encodes ~250 EF-hand containing proteins [96], although it should be noted that the presence of an EF-hand domain does not necessarily mean that a protein is activated by Ca^{2+} [97]. Calmodulins can interact with transcription factors, directly transducing Ca^{2+} signals into changes in gene expression [98–103]. There is also evidence of Ca^{2+} signals within the nucleus, where CDPKs can phosphorylate and activate transcription factors [104, 105], and in the chloroplast [4, 106]. It is becoming clear that the cellular location of all parts of the signal transduction pathway plays an important role in proper signal transduction [105]. Sensors "relay" information from Ca^{2+} signatures (or the binary switch) into downstream events that include phosphorylation, changes in gene expression and protein-protein interactions [107]. The variety of Ca^{2+} binding proteins in plants suggests that intracellular Ca^{2+} levels, transport, release, and uptake are interdependent and tightly regulated [92].

2.3. CIPK/CBL Network.
Batistic and Kudla [23] argue that a new system of Ca^{2+}-regulated proteins has evolved to replace the IP_3 receptor network as plants adapted to life on land. In Arabidopsis this system comprises 10 calcineurin B-like proteins (CBLs), which function as Ca^{2+} sensors, and 26 CBL-interacting protein kinases (CIPKs) [23]. Elegant experiments combining microscopy and biochemistry have been used to decipher the logistics of this pathway [108]. In addition to Ca^{2+} sensing, variations in both the cellular distribution and the interaction partners of members in this pathway contribute to an elaborate system capable of interpreting information from a variety of different stimuli [109]. To date, CBL/CIPK complexes have been shown to participate in the transduction of signals caused by the abiotic stress response, abscisic acid, potassium and nitrate uptake mechanisms, anaerobic response, cold, salt, sugar, cytokinin, and light [44, 47, 74, 110–122].

Kudla's group has demonstrated that CIPK6/CBL4 interactions can lead to relocation of the K^+ channel, AKT2, from the ER membrane to the plasmalemma [113]. Two lipid modifications of CBL4, myristoylation and palmitoylation, are required for it to associate with the ER to begin the relocalization. CIPK6 serves as a scaffold in this process as phosphorylation is not required [113]. Lipid modifications are also required for CBL1 association with the plasmalemma, where it interacts with CIPK23 to activate a second K^+ channel, AKT1 [124]. This interaction results in K^+ uptake under low K^+ conditions [124] while the CIPK6/CBL4 interaction is needed for normal growth [113].

There is indirect evidence for the role of the CBL/CIPK network in biotic stress as members of this family respond to salicylic acid [125]. CIPK6L was induced by Ca^{2+} in apples, and exogenous Ca^{2+} also induced both *CIPK* and *CBL* from pea [45, 125] and a *CIPK* from rice [122].

The overexpression of different CIPK/CBL proteins involved in abiotic stress has been shown to confer increased drought tolerance (Table 1). In addition to nutrient deprivation and abiotic stress, some CIPK/CBL members target particular developmental pathways during abiotic stress including root growth, pollination, and germination [47, 112, 126]. The impact of ectopic *CIPK6* expression on root growth was shown to be mediated through auxin [44, 126]. Although *CIPK6* expression was shown to confer tolerance to salt, the positive impact of its overexpression in Arabidopsis and tobacco on root growth suggests that those plants may also do well under water-limiting conditions. This is discussed in more details in Section 6.

2.4. Ca^{2+} Binding Proteins and Modulation of Ca^{2+} Stores.
Suberization of the cell walls in the endodermis might prevent apoplastic Ca^{2+} from participating in cytosolic signaling events, because the deposition of the wax onto the cell walls would inhibit Ca^{2+} mobility. White and Knight used this insight to demonstrate that different stimuli do result in the cell accessing different stores of Ca^{2+} [91]. Transgenic plants that expressed apoaequorin only in the endodermis were used, and the root tips, which had different levels of suberization, were examined for luminescence in the presence of luciferin, which is directly proportional to the concentration of Ca^{2+}. While salt stress resulted in the production of

TABLE 1: Ectopic expression of CIPK/CBL members; the effect on abiotic stress.

Gene	Source of gene	Target organism	Impact	Reference
AtCBL1	Arabidopsis	Arabidopsis	Reduces transpiration, increases abiotic stress tolerance	[38]
AtCBL1	Arabidopsis	Arabidopsis	Increased salt and drought tolerance, reduced freezing tolerance	[39]
AtCBL2	Arabidopsis	Arabidopsis	Enhanced susceptibility to low K^+	[40]
AtCBL3	Arabidopsis	Arabidopsis	Enhanced susceptibility to low K^+	[40]
ZmCBL4	Zea mays	Arabidopsis	Increased salt tolerance	[41]
AtCBL5	Arabidopsis	Arabidopsis	Increased drought tolerance	[42]
OsCBL8	Rice	Rice	Increased salt tolerance	[43]
CaCIPK6	Chickpea	Tobacco	Increased salt tolerance, enhanced root development	[44]
MdCIPK6L	Apple	Apple, Arabidopsis, tomato	Enhanced tolerance to salt, osmotic, drought and chilling stress; no effect on root growth	[45]
OsCIPK03	Rice	Rice	Enhanced tolerance to cold by increased proline and soluble sugars	[46]
AtCIPK9	Arabidopsis	Arabidopsis	Enhanced susceptibility to low K^+	[40]
OsCIPK12	Rice	Rice	Enhanced tolerance to drought by increased proline and soluble sugars	[46]
OsCIPK15	Rice	Rice	Enhanced tolerance to salt	[46]
OsCIPK23	Rice	Rice	Increased drought tolerance	[47]

a continuous luminescent Ca^{2+} signal along the endodermis, cooling the roots produced a signal that was confined to a terminal 4-mm region of the root tip, where suberization was incomplete or lacking [91]. This was an elegant demonstration that signal propagation from salt and cooling require access to different Ca^{2+} stores. Moore et al. concluded that cytoplasmic signaling in response to salt stress utilized intracellular stores of Ca^{2+}, although it is still not clear what part of the cell contained the store [91].

2.4.1. The Vacuole as a Ca^{2+} Store. Although a considerable amount of Ca^{2+} is present in the apoplast, the vacuole is the main storage organelle for Ca^{2+} within the plant cell. However, there is little direct evidence for the vacuole as a source of Ca^{2+} for signaling [4, 127, 128], although the identification of Ca^{2+} channels in the tonoplast membrane is not complete either. Furthermore, most of the Ca^{2+} in the vacuole is complexed with chelators such as malate, isocitrate, and citrate and is, therefore, not readily available for signaling [4].

There is evidence for an important role for the vacuole in depleting cytosolic Ca^{2+}, which is critical for preventing association with phosphate and for shaping putative Ca^{2+} signatures. Using mathematical modeling, Bose et al. suggest that the activity of the known major Ca^{2+} efflux proteins (two members, each of the ACA and CAX gene families) is sufficient to describe a wide variety of Ca^{2+} signatures, including all of the current experimental results, without having to take into consideration how Ca^{2+} enters the cytosol [50]. Figure 1 shows a diagram of the major Ca^{2+} efflux proteins in a leaf cell.

Two vacuolar Ca^{2+} ATPases, ACA4 and ACA11, have been shown experimentally to be important for removing excess cytoplasmic Ca^{2+} [129]. When genes for both of these

pumps were mutated, groups of cells in the mesophyll began undergoing programmed cell death (PCD). This phenotype requires salicylic acid, suggesting that the increased cytoplasmic Ca^{2+} by itself was not toxic [129]. It could be that PCD has the lowest threshold for sensing an activating cytoplasmic Ca^{2+} signal. While many stimuli could activate the release of Ca^{2+} into the cytoplasm (light, gravity, etc.), without appropriate dampening by ACAs the signal could spread to other parts of the cell to trigger unintended responses. It will be interesting to know if the propensity for cell death is an indirect effect of altered cytosolic Ca^{2+} on a PCD-related Ca^{2+} sensor, or if ACA4 and ACA11 are specifically involved in PCD.

2.4.2. The ER as a Ca^{2+} Store. The ER also contains high levels of Ca^{2+} and is an attractive candidate for storing signaling Ca^{2+} [130]. Calreticulin (CRT) is an ER luminal chaperone that has two Ca^{2+} binding domains. The P-domain contains a high affinity, EF hand-like structure that binds 1-2 moles of Ca^{2+} per mole protein [131]. The C-domain is the least conserved among organisms but contains a disproportionately high number of acidic amino acid residues that function to bind large amounts of Ca^{2+} with weak affinity. The C-domain has been estimated to bind 30–50 moles of Ca^{2+} per mole of protein [131]. Because of its low affinity, C-domain binding requires a relatively high concentration of Ca^{2+}, such as in the ER. Although estimates are scarce, the concentration in the ER of pollen tubes has been estimated to be ~100–500 μM; about 1000-fold higher than in the cytoplasm [130]. The ER of animal cells contains ~1 mM Ca^{2+} but the concentration is nonuniform [132]. This is also likely to be true in plants due to the conservation of ER pumps and Ca^{2+} binding proteins such as CRT and Calnexin (CXN) [133]. CXN is a membrane-bound ER protein that functions with CRT and

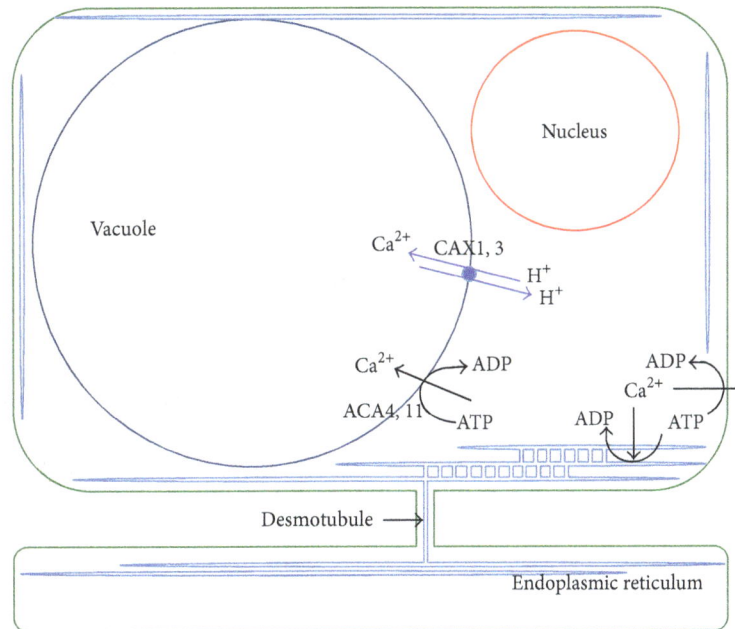

FIGURE 1: Major Ca^{2+} efflux systems in a leaf cell, and structure of the ER spanning two cells. CAX1 is the major cation exchanger in leaf cells, but CAX3 can compensate if CAX1 activity is compromised. Not shown is a vacuolar proton ATPase that uses ATP to pump protons into the vacuole. The energy from the proton gradient is used to pump Ca^{2+} into the vacuole. There are also two Ca^{2+} pumps on the tonoplast membrane, ACA4 and ACA11. The ER and plasma membrane also have Ca^{2+} pumps (lower right). Ca^{2+} pumps are also found on the nuclear envelope and chloroplast (not shown). The reticulate nature of the ER is modeled next to the plasma membrane but reticulation (and the ER) is found throughout the cell. Cortical ER is found near the cell wall and is less dynamic than ER in the interior. A desmotubule spans a single plasmodesma between the upper cell and a partial cell on the bottom, but of course there are multiple plasmodesmatal connections between most cells (except guard cells and between the epidermis and mesophyll). Both the cytosol and the ER lumen are continuous across the plasmodesmata.

BiP (another chaperone) in glycosylation and quality control of ER proteins [133, 134].

Most plants have two forms of CRT [135, 136]. In Arabidopsis, CRT1a and CRT1b (also called CRT2) have the highest homology and form the first group, while CRT3, which is specifically needed for viral cell-to-cell movement [137], is in the second group. All three CRTs function as chaperones and play an important role in protein folding and glycosylation [136, 138–141]. The C-domain of CRT3 is reduced in size compared to CRT1a and CRT1b, but was specifically required for proper folding of the brassinosteroid receptor, BRI1 [142]. All three CRTs have been implicated in innate immunity for proper folding of different receptor proteins [143–148], and CRT1 appears to participate in signaling [149]. CRT has also been associated with increased tolerance to abiotic stress [147, 150].

CRT is highly expressed in meristematic and reproductive tissues. It shows lower expression associated with vascular tissue. CRT1 and CRT2 are largely coexpressed, except that CRT2 is high in senescing leaves, perhaps as a mechanism for retrieving Ca^{2+}. CRT2 also shows guard cell -specific expression.

The ER also contains at least one Ca^{2+} ATPase, ACA2, that is activated by calmodulin and inhibited by a CDPK [151, 152]. Inhibition of an ER-type Ca^{2+}ATPase (ECA1) in pollen tubes decreased ER Ca^{2+} and inhibited pollen tube growth

suggesting that the ER serves as a Ca^{2+} store for signaling [130]. In addition, mutants with 4-fold lower ECA1 activity showed poor growth on medium with low Ca^{2+} (0.2 mM versus 1.5 mM, normal) [153]. It is not clear why ECA1 is needed to pump Ca^{2+} into the ER under low Ca^{2+} conditions.

In animal cells, Ca^{2+} is constantly leaking out of the ER and constantly being pumped back in by SERCAs, membrane pumps that are similar to ECA [132]. But the major mechanism for ensuring adequate ER levels of Ca^{2+} is a specialized plasma membrane pump that responds only to low ER Ca^{2+}. In a mechanism called store-operated Ca^{2+} entry [154], the pump (Orai1) forms a structure adjacent to an ER protein (STIM1) that contains an EF hand to sense ER Ca^{2+} levels. Together, they allow ER Ca^{2+} levels to be refilled [132]. It is not known if a similar mechanism could function in plants.

2.5. Ca^{2+} Transduction and Regulation of Gene Expression. How many genes and proteins are associated with Ca^{2+} regulation? In addition to the ~250 EF-hand containing proteins, ~700 are thought to be involved with Ca^{2+} signaling for Arabidopsis, according to proteomic data [155]. These proteins generate Ca^{2+} signatures and transduce the signal into changes in protein phosphorylation, protein localization, protein-protein interactions, and changes in gene expression. It is the latter that is most difficult to identify due to the difficulty in testing Ca^{2+} without other secondary effects that result when stimuli such as NaCl are used that also cause chemical and ionic perturbations of the system. Knight's group addressed this by using an applied voltage to alter membrane permeability in combination with transgenic aequorin to monitor changes in cytoplasmic Ca^{2+} levels [156]. Conditions for a transient increase in cytoplasmic Ca^{2+} from less than 100 nM to almost 600 nM were established, and microarrays were used to profile genetic changes. A combination of transient and oscillating Ca^{2+} fluxes produced the greatest number of genes (269) with increased expression levels, while a single long increase in Ca^{2+} to 200 nM produced only 10 genes with increased expression.

Analysis of the promoter regions of the Ca^{2+}-upregulated genes revealed a surprising bias for genes that respond to abiotic stress. Three out of the four Ca^{2+}-regulated promoter motifs were previously identified as being important for abiotic stress responses and included the ABA-response element and the drought-responsive element [156]. This bias could be due to the nature of the Ca^{2+} flux, which may have resembled signatures produced from an apoplastic source of Ca^{2+}, or could be a feature of Ca^{2+} regulation.

In addition to the cytoplasm, transient Ca^{2+} fluctuations have also been reported in the nucleus, chloroplast, mitochondrion, and peroxisome [4]. Ca^{2+} oscillations in the cytosol and chloroplast have been linked to circadian rhythms [32, 157]. It is not known whether these fluctuations also lead to changes in gene expression.

We used a genetic method to specifically increase Ca^{2+} in the ER by taking advantage of the high capacity, low affinity Ca^{2+} binding activity of the C-domain from CRT. A green fluorescent protein-calcium binding peptide (GFP-CBP) fusion protein consisting of the C-domain from *Zea mays CRT1* was fused to the C-terminal region of GFP [158]. The GFP-CBP construct included a signal protein for ER-targeting and the C-terminal region of *CRT1*, which contains an HDEL sequence for ER retention. Total Ca^{2+} in seedling shoots was increased by ~25%, when GFP-CBP was expressed in Arabidopsis using a constitutive promoter. Microarray analysis of seedlings expressing GFP-CBP compared to seedlings expressing GFP showed that 31 genes were upregulated by >3.5-fold. As expected, none of these genes included the cytosolic Ca^{2+}-regulated genes identified by Whalley et al. Only one of the genes was involved in Ca^{2+} regulation—*CIPK6* [158]. Whalley et al. also identified a single CIPK, *CIPK9* [156]. The other genes we found were enriched for microsome-associated proteins and glycine-rich proteins, which are often targeted to the cell wall [158]. One of the proteins encoded a subunit of the anaphase-promoting complex [158]. This expression pattern could indicate a regulatory role for ER Ca^{2+} levels in mitosis. We will come back to this in Section 5.2.

Of course steady-state modulation of Ca^{2+} levels in an organelle is quite different from generating a cytosolic Ca^{2+} signal. According to the eFP browser [159], *CIPK6* is induced by salt, drought, and abscisic acid and is expressed at a low level in guard cells, leaves, flowers, and developing fruit and seed. Although some of the genes coexpressed with *CIPK6* in the GFP-CBP plants showed similar expression profiles to *CIPK6*, there is nothing to suggest a connection with ER Ca^{2+}.

2.6. Summary of Cellular Ca^{2+} Dynamics. Cells contain stores of Ca^{2+} in the apoplast and in various compartments within the cell. Cytoplasmic Ca^{2+} is kept low to prevent interference with phosphate-containing pathways. Signal transduction uses discrete Ca^{2+} fluxes to connect stimuli with adaptive responses. Different stores of Ca^{2+} are used in the generation of these fluxes and the location, magnitude, and duration of the fluxes appear to contain information for the appropriate response. Vacuolar pumps and antiporters participate in removing Ca^{2+} from the cytoplasm before deleterious interactions occur. It has been difficult to determine which intracellular stores participate in different kinds of signaling, but the ER is an attractive candidate because of its distribution throughout the cell, and the ability of CRT to bind large quantities of Ca^{2+} with low affinity.

We still need more information on the plant's ability to generate stimulus-specific Ca^{2+} signatures. What is the source of the Ca^{2+} used for different signals? What dampens the signature? How is information about the signal (magnitude, oscillations, and duration) transduced into specific responses? With respect to the original question—what, exactly, could the presence of supplemental Ca^{2+} contribute to increase stress tolerance? Are certain Ca^{2+} stores normally limited, or does spraying Ca^{2+} onto a plant trigger oscillations as Ca^{2+} is assimilated? Understanding Ca^{2+}-regulated networks is plagued by the ubiquity of the molecule, and dissecting pathways in different cells and tissues is still tedious and difficult. However, the combination of biochemistry, Ca^{2+} reporter genes, and genetics is providing tremendous information that is building a solid foundation for understanding Ca^{2+} regulation.

The next section begins to discuss tissue-specific differences in Ca^{2+} levels to better understand how exogenous Ca^{2+} is assimilated.

3. Calcium Distribution within the Leaf

Eating roots and leaves is the best way for vegans (people who do not eat meat, fish, or dairy products) to increase Ca^{2+} intake [160]. This makes sense because Ca^{2+} is transported from roots to shoots through transpiration, and leaves carry out the bulk of transpiration. But not all cells within a leaf have equivalent Ca^{2+} levels. In grasses, Ca^{2+} is found mainly in the upper epidermis [161]. In dicots, Ca^{2+} levels are low in both upper and lower epidermis, but are higher in mesophyll, a distribution that facilitates Ca^{2+} control over stomatal aperture [161, 162].

A landmark study looked at the distribution of Ca^{2+} in different cell types of the leaf and found that mesophyll cells have ~6-fold more Ca^{2+} than epidermal cells, due largely to the differential expression of *CAX1* in those cells [162]. CAX1 is located on the tonoplast membrane and couples proton export with Ca^{2+} transport into the vacuole.

caxl/3 double mutants not only had reduced growth, reduced photosynthesis, and thicker cell walls, but also had higher apoplastic levels of Ca^{2+} [162]. This resulted in reduced stomatal apertures, which led to reduced growth due to a lack of carbon assimilation compared to nonmutant lines [162]. Although the cell walls were thicker, they were also more brittle and contained more pectin. Supplementation with low Ca^{2+} media reduced free apoplastic Ca^{2+} levels and suppressed the phenotype, while returning the plants to normal Ca^{2+} caused the phenotype to return. Free Ca^{2+} (sorbitol-exchangeable) was ~3-fold higher in the apoplast of caxl/3 double mutants compared to the nonmutant line. In fact, CAX1, CAX3, CAX4, and ACA4 (encoding a Ca^{2+} ATPase) and ACA11 are coregulated to make sure total Ca^{2+} levels are constant [162].

Why was high apoplastic Ca^{2+} a problem? Guard cells use Ca^{2+} to signal downstream components to close or open stomata. In the presence of excess Ca^{2+}, stomata remain closed even under conditions favorable for gas exchange and carbon fixation. The exact mechanism for how extracellular Ca^{2+} interferes with guard cell signaling is not known. As mentioned in Section 2, the electrochemical gradient for Ca^{2+} across the cell membrane strongly favors passive Ca^{2+} entry—it is the removal of Ca^{2+} from the cytoplasm that requires energy. Thus, the presence of high levels of free Ca^{2+} on the other side of the plasmalemma may either make it difficult to remove Ca^{2+} from the cytoplasm or make it too easy for Ca^{2+} to enter it. Extracellular Ca^{2+} has been shown to cause guard cells to close by generating H_2O_2 and NO, which generate an intracellular Ca^{2+} spike, leading to stomatal closure [63].

Thus, keeping free Ca^{2+} out of the apoplast enables proper guard cell function and allows normal plant growth. CAX1 keeps apoplastic Ca^{2+} low by storing it in the vacuole [162]. Rather than viewing the apoplast as a separate entity that protected plant cells from extracellular threats, it now seems important to acknowledge that unbound extracellular Ca^{2+} must be maintained in equilibrium across the apoplast/symplast boundary. At least in leaves, it is the vacuole, a membrane-bound organelle on the symplastic side of the divide, not the cell wall, that serves as the reservoir for excess accumulation of Ca^{2+}.

Where does Ca^{2+} come from? In leaf cells, Ca^{2+} is transported through the xylem by transpiration [2]. Ca^{2+} is one of the most immobile ions in the plant, with Mg^{2+} and Mn^{2+} not far behind [2]. In the leaf, Ca^{2+} is thought to diffuse through the apoplast up to about 15 cells away from the xylem. Transpiration would seem to direct Ca^{2+} to guard cells, which are mostly on the lower side of leaves, but the pattern of veins, anatomy of the leaf, and presence of air spaces all help to dissipate the pattern of water flow [163].

The pattern of Ca^{2+} transport is thought to vary with the developmental stage of the leaf, the species, and environmental conditions [163]. In eudicots, Ca^{2+} is trapped in the vacuoles of mesophyll cells by CAX1 [163], while in monocots higher relative levels of Ca^{2+} are found in the epidermis [2,123]. Root pressure can contribute to the transport of Ca^{2+},

especially when humidity is high and transpiration low [164]. Ca^{2+} deficiency is first noticed as tip burn, and diseases such as blossom end rot in tomato are a visual demonstration of the limited mobility of Ca^{2+}. Since leaves develop acropetally, the apex is the last to differentiate. This suggests that dividing cells may be particularly vulnerable to Ca^{2+} depletion. We will come back to this in Section 5.

4. Ca^{2+} Is Transported from the Roots to the Shoot by Transpiration through the Xylem

There could be three points of control for transpiration—uptake in the root apoplast, entry into the xylem across the endodermis, and exit through guard cells. The apoplast shows very little electrical resistance and allows the free exchange of most ions. Ca^{2+} is absorbed from the soil by the apoplast and by cation channels in the root epidermis [165]. The extent of symplastic transport of Ca^{2+} between cells is not known, although a cadmium resistant channel was recently identified that facilitates radial movement of Ca^{2+} in roots [166].

Two pathways for Ca^{2+} transport to the shoot can be experimentally tested, a symplastic or cell-to-cell pathway and an apoplastic pathway. The symplastic pathway involves passage through at least one membrane. The Casparian strip of the endodermis, which contains suberin, restricts solute passage through the apoplast, and promotes passage through the symplastic pathway. Studies with radio-labeled Ca^{2+} suggest that this pathway predominates in onion [6]. Identification of enhanced suberin (esb) mutants in Arabidopsis allowed the role of the endodermis to be directly tested [167]. Shoot Ca^{2+} levels decreased ~50% compared to wild type. If there was no change, it could be concluded that transport was entirely apoplastic or entirely symplastic. So the reduction in Ca^{2+} transport suggests that restriction by the Casparian strip of the endodermis is incomplete—some apoplastic flow is permitted through the Casparian strip in its wild type state. There was no change in Mg^{2+} in the esb mutants, which is also transported through the phloem, but Zn^{2+} and Mn^{2+} also decreased [167]. Surprisingly, accumulation of the monovalent ions Na^+, S^+, and K^+ increased. Transpiration was also decreased and the plants were less susceptible to wilting.

The existence of the apoplastic pathway was demonstrated from experiments that showed that the ratios of Ca^{2+}, Br^{2+}, and Sr^{2+} do not change after they are applied to roots, although channels and pumps have a clear preference for Ca^{2+} [168]. In many plants, the amount of Ca^{2+} transported depends on the rate of transpiration, which is consistent with solvent drag, not symplastic processes [168]. In some plants under certain conditions, Ca^{2+} transport may be almost entirely apoplastic with channels at the destination cell controlling cellular Ca^{2+} entry, followed by rapid assimilation into different organelles by pumps and antiporters. Ca^{2+} transport through the endodermal cytosol in the symplastic pathway is thought to be achieved using Ca^{2+} channels and pumps, but must be carefully regulated to avoid interfering

TABLE 2: Concentration of Ca^{2+} and Mg^{2+} in shoots of different angiosperm families (data taken from [123]).

family	Ca^{2+}	Mg^{2+}
Plantaginaceae (48)*	17.38 (3.59)*	1.69 (0.13)
Polygonaceae (6)	5.90 (1.23)	2.87 (0.41)
Poaceae (6)	3.33 (0.25)	1.33 (0.07)

*Number in parenthesis is n.
**Data are the average value of the mineral in mg/g dry weight, with SE in parenthesis.

with signaling pathways. According to White, apoplastic transport may be necessary to meet the demand for adequate Ca^{2+} in the shoot [168]. Breaks in the endodermis, for example where lateral roots emerge, allow Ca^{2+} transport without an intervening symplastic step.

Transpiration is considered to be the driving force for Ca^{2+} transport into shoots and leaves, and Ca^{2+} travels with the bulk water flow [2, 6, 163, 169, 170]. The pattern of Ca^{2+} deficiency symptoms can be explained by a combination between demand for Ca^{2+} and variation in transpiration. Tip burn, which affects the leaf margin and the undeveloped distal region of the leaf, is thought to result from a lack of well-developed veins in the undifferentiated part of the leaf and high rates of cell wall deposition.

Recent experiments actually compared the shoot accumulation of several minerals in members of 7 different plant families grown together under different fertilizer regimes [123]. The correlation with phylogeny (versus fertilizer treatment or residual) was the strongest for Ca^{2+} (70%) and total Ca^{2+} varied over 5-fold (Table 2). In contrast, Mg (with a 32.8% correlation with phylogeny) showed little more than a 2-fold variation. Dicotyledonous plants are known to accumulate more Ca^{2+} than monocots, partly as a function of the structure of their cell walls, and there only was ~3-fold variation in Ca^{2+} in different dicot families (Table 2). To put this in perspective, there was a ~2-fold variation in Ca^{2+} among Arabidopsis ecotypes, which are all members of the same species [171]. The molecular basis for the difference in Ca^{2+} levels between different families is not known, but the data suggest that factors are at play that ultimately limit the amount of Ca^{2+} absorbed from the soil.

The endodermis clearly has a role in regulating water transport, and likely helps the plant to conserve water by preventing unrestricted transpiration. Gilliham et al. argue that Ca^{2+} transport and transpiration are linked—Ca^{2+} regulates both stomatal activity in leaves and aquaporin (water channel) density and function in roots [163]. Thus, Ca^{2+} could increase its own transport by affecting aquaporin function [14]. Global mechanisms such as this may also play a role in limiting the amount of Ca^{2+} that ultimately reaches the shoot. In support of this, the overexpression of an aquaporin in Arabidopsis increased Ca^{2+} levels by ~33% under normal conditions and almost doubled Ca^{2+} under 100 mM NaCl [172]. The regulation of hydraulic conductivity (aquaporin function) under stress is reviewed by Aroca et al. [173].

A second mechanism for Ca^{2+} regulation of Ca^{2+} leaf concentration has been proposed [174]. A plasma membrane-localized CAlcium Sensing receptor, CAS, is upregulated in guard cells. High levels of apoplastic Ca^{2+} cause stomata to close, a process that requires CAS. When transpiration levels are high, Ca^{2+} has the potential to be too high. CAS mutants grown in soil had ~40% more Ca^{2+} than wild type plants [174]. Together with the aquaporin overexpression [172], this suggests that global regulation of Ca^{2+} levels occurs primarily through mechanisms found in the shoot, not through the endodermis in the root.

5. An ER Ca^{2+} Network for Meristems

Meristems are critically important for plant growth and reproduction. Meristems require high amounts of Ca^{2+} because of cell wall deposition and organelle biogenesis, but it is not clear how Ca^{2+} moves from areas with high rates of transpiration (leaves) into the protected region of the meristem (Figure 2). An alternative mechanism for Ca^{2+} transport is through the endoplasmic reticulum (ER). The ER is contiguous with the nuclear envelope and forms a symplastic continuum throughout the plant by spanning cell walls through plasmodesmata. Consistent with the idea of CRT as a Ca^{2+} transporter/regulator, high levels of CRT are found in plasmodesmata [175, 176] and in meristems [177]. This may be especially important in meristems, where the need for Ca^{2+} is high due to the formation of new cell walls, but the ability to transpire Ca^{2+} is limited by the lack of differentiated xylem. Transport through the ER would avoid the problem of cytoplasmic transit disrupting signaling pathways and could either augment apoplastic transport to ensure the protection of developing areas of the plant or bypass it, depending on where Ca^{2+} enters the ER.

If the ER functions in Ca^{2+} transport, why has not this been detected in leaves? A key aspect of the proposed Ca^{2+} network in meristems is the presence of CRT, whose gene shows high expression in meristematic tissues [177]. As described in Section 2.4.2, CRT has three conserved domains, one of which binds 30–50 Ca^{2+} ions with low affinity (the C-domain). CRT may function in intercellular Ca^{2+} distribution by acting as a buffer, partly neutralizing the charge. CRT is further proposed here to act as a sort of matrix to facilitate Ca^{2+} absorption and movement by the cell and to provide a gradient for additional Ca^{2+} to be transported cell-to-cell from mature tissues. But because CRT is not expressed at high levels in mature leaves, Ca^{2+} transport appears to follow a bulk flow pattern of distribution with the rate of transpiration dictating where it accumulates.

5.1. Desmotubules Allow Movement through the Plasmodesmata. Cytoplasmic Ca^{2+} transients have been demonstrated to result in rapid closure of plasmodesmata [178]. The biggest obstacle to Ca^{2+} transport through an ER network is the plasmodesmata. Plasmodesmata consist of a central desmotubule (see Figure 1), which is derived from the compaction of the two sides of the ER tubule that traverses the cell wall. A thin

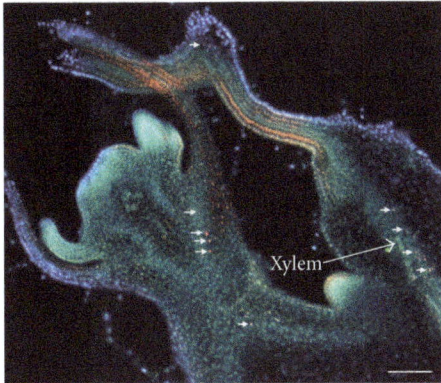

FIGURE 2: How can transpiration, which requires differentiated xylem, deliver sufficient Ca^{2+} to sustain meristematic growth? Longitudinal section of *Nicotiana benthamiana* stained with DAPI (blue) to detect DNA and hybridized to fluorescent oligos complementary to tomato golden mosaic virus DNA (pink, short arrows). Viral DNA is transported through phloem and makes a nice marker for developing vascular tissue. Image was visualized with a triple-fluorescence cube for DAPI, FITC, and Texas Red. Orange and green colors are the result of autofluorescence. Most of the cells in this section are undifferentiated. Long arrow points to a tracheary element that has differentiated, but most of the leaf is still developing (section through this leaf is oblique). Bar = 50 microns.

cytoplasmic sleeve that lies between the desmotubule and the plasma membrane serves as the conduit for cytoplasmic proteins and solutes that show intercellular trafficking [175].

CRT has been localized to plasmodesmata [175, 176] and could serve as a Ca^{2+} donor to maintain an internal network of stored Ca^{2+}. High concentrations of CRT on either side of the plasmodesmata may result in a Ca^{2+} gradient, which could facilitate the distribution of Ca^{2+} to adjacent cells. Any cytoplasmic Ca^{2+} transients would occur independently of luminal concentrations [178].

Despite the narrow aperture of the desmotubule, transit of fluorescent molecules across the desmotubule appeared to be rapid. Microinjection studies were used to study the spread of the small molecular weight fluorescent tracers carboxyfluorescein and FITC-conjugated triglutamic acid in epidermal cells of tobacco and *Torenia* [179]. About 10% of the injections resulted in a punctate pattern of label that corresponded to the pattern obtained with $DiOC_6$, a fluorescent dye that labels ER. This was explained by insertion of the needle into the lumen of the ER. In each case, the fluorescent molecules rapidly spread into adjacent cells through the desmotubule of the plasmodesmata. Spread of the fluorescence was more rapid through the desmotubule than through the cytoplasmic sleeve of the plasmodesmata and occurred more readily (100% of the cases versus ~88% for injections into the cytoplasm) [179]. Fluorescent dextrans corresponding to 10 kDa showed luminal transport in Torenia in 3 out of 3 injections. This demonstrates that sufficient space exists within the desmotubule for cell-to-cell Ca^{2+} transport. Although movement of ER-targeted GFP through the desmotubule was not demonstrated, Martens et al. discuss

the possibility of the desmotubule functioning both as a conduit for cell-to-cell Ca^{2+} transport and as a mechanism for whole-plant signaling [180].

GFP fusions have also been used to study intercellular trafficking in leaf epidermal cells following microinjection. A CRT-GFP fusion protein in the ER lumen did not traffic into adjacent cells, but calnexin-GFP, an ER membrane-localized protein, did spread cell to cell [181]. CXN also binds Ca^{2+} and functions with CRT as a protein chaperone. It contains an N-terminal Ca^{2+}-binding domain on the luminal side and an acidic tail of ~90 amino acids. These characteristics could enable it to transport Ca^{2+} across the plasmodesmata.

Why would transport through the desmotubule be needed? Plasmodesmata are regulated by Ca^{2+}. When a cold shock was used to increase cytoplasmic Ca^{2+} from 100 to 200 mM, there was a 4-fold increase in resistance, but the resistance returned to normal within 10 sec [182]. Thus, cytoplasmic Ca^{2+} transients would be expected to close plasmodesmata. By compartmentalizing Ca^{2+} away from the cytoplasm, it could equilibrate between cells at levels that would interfere with plasmodesmata function if it were on the cytosolic side of the plasmodesmata.

5.2. Ca^{2+} and Cell Division. Vascular tissue forms de-novo and differentiates acropetally (phloem) and basipetally (xylem) in developing leaves after they have begun to expand and differentiate. The leaf midvein does not connect to the stem until after xylem and phloem have differentiated, and the leaves have begun to actively photosynthesize. The high rates of cell division in developing leaf primordia require significant amounts of Ca^{2+} to bind to cell wall pectin, stabilize the plasma membrane, and ensure completion of mitosis.

Ca^{2+} plays a major role in mitosis at anaphase, where it concentrates at the spindle poles at levels that cause microtubule depolymerization [183]. Interestingly, two proteins, one of them a *CRT*-like protein, have been identified in plants that could facilitate this process. Tonsoku (TSK) localized to the nucleoplasm while tonsoku-associated protein (TSA) has a signal peptide and was found in cytoplasmic vesicles derived from the ER [184]. During anaphase, the two proteins colocalized and appeared to interact. TSA has 10 repeats of an EFE motif consisting of acidic amino acids and was shown to bind Ca^{2+} in vitro. Although there was no homology with *CRT*, it may have a very similar function—to provide a matrix for storing Ca^{2+} until it is needed. Although a function has not been reported for these proteins, other than to bind Ca^{2+} and colocalize, it seems possible that they would be needed, along with kinesins [185], for depolymerization of microtubules during anaphase.

When plant cells enter prophase, the nuclear envelope (which is contiguous with the ER) disintegrates into vesicles. Following anaphase, a new cell wall is deposited, which requires vesicle secretion and membrane fusion. The ER is well positioned to provide Ca^{2+} during this process, which would be needed for stabilizing the developing cell wall by binding to pectate. The dynamic nature of cell-to-cell movement through desmotubules could ensure that the ER has a ready supply of Ca^{2+} available for the new cell wall that could be delivered through the vesicles.

CRT is known to be expressed at high levels in dividing cells, but the reason for this has not been obvious. Clearly, there is a higher need for glycosylated proteins as new cells are formed, but it appears to play more than a structural role, as overexpression of *CRT* has been shown to increase regeneration [186]. One possibility is that cell division, and possibly regeneration, may have become linked to the expression of *CRT*, such that if ER Ca^{2+} levels were not adequate to support new growth, the process of cell division would arrest. Our microarray results provide some tantalizing evidence in favor of this hypothesis. We found that GFP-CBP caused a 3.7-fold increase in the expression of At5g26635, which encodes one subunit of a putative anaphase-promoting complex. However, this is probably too late in the cell cycle to arrest development. A more likely explanation is that ER Ca^{2+} levels need to be high enough to facilitate the depolymerization of microtubules.

Nevertheless, the relationship between ER Ca^{2+}, *CRT* expression, and mitotic activity would be interesting to study—to determine why "meristem burn" is not a problem, for example. It would also be interesting to examine *CRT* expression in tomato, since it suffers from the occurrence of blossom end rot, discussed in Section 6. Blossom end rot is a Ca^{2+}-related disorder that results in tissue softening and necrosis at the distal end of the tomato fruit, which contains the highest proportion of dividing cells. Tomato is not the only fruit to undergo extensive cell division during fruit development (papayas, watermelons, and jack fruit are also quite large); what makes it more susceptible?

5.3. Summary. In summary, a gradient of Ca^{2+} ions in the ER is proposed to help the plant guard its vulnerable meristem from fluctuations in the transpiration of Ca^{2+}. *CRT* networks in the ER could provide a conduit for Ca^{2+} transport to ensure that adequate levels of Ca^{2+} reach the meristem to support growth. *CRT* could serve as a buffer to help neutralize charge and to draw Ca^{2+} towards the meristem. Cell division may be coupled to *CRT* expression to ensure that adequate levels of Ca^{2+} are present when the cell divides. Transport through the ER would avoid competition with the vacuole and protect the cytoplasm while ensuring that enough Ca^{2+} is transported to meet the demands of the cell wall and organelles.

6. Genetic Manipulation of Ca^{2+} Stores

Many postmenopausal women take supplemental Ca^{2+} to help prevent osteoporosis, a crippling disease related to aging. With the demographics of most developed countries showing a rise in the aging population, the impact of nutrient deficiencies on human health is likely to increase. Many people do not like to take Ca^{2+} in the form of a pill because of its large size, which is needed due to the relatively poor absorption of chemical Ca^{2+}. The best way to obtain more nutrients is to consume more fruits and vegetables, especially roots and leaves for Ca^{2+}, Mg^{2+}, and K^+ [160, 187]. Unfortunately, almost 10% of the adult population of the

USA and UK are deficient for those three elements [160], due, in part, to consumption of cereal grains rather than vegetables (although breakfast cereals are often sprayed with supplemental Ca^{2+}). Although other countries who rely on rice as a major staple face a similar problem, they are much more likely to combine it with vegetables, if they have the money. Thus, Ca^{2+} deficiencies are more of a problem in developed countries, which are also more likely to have an aging population. Since fortifying plants with supplemental Ca^{2+} increases their tolerance to stress, it would be prudent to consider the genetic alteration of Ca^{2+} stores with transgenes that benefit consumers as well as farmers.

Ca^{2+} levels show a high degree of heritability but vary from species to species. Ca^{2+} distribution was shown to vary ~2-fold among Arabidopsis ecotypes and was correlated with Mg^{2+} in all tissues except seeds, [171]. In general, there is more Ca^{2+} in shoots than in roots, and the distribution within leaves is nonuniform. In grasses, Ca^{2+} is found only in the upper epidermis. In dicots, Ca^{2+} levels are low in both upper and lower epidermis, but are higher in mesophyll, a distribution that facilitates Ca^{2+} control over stomatal aperture. Much of the variation in Ca^{2+} levels could be traced to the expression of *CAX1* [162]. So far, Ca^{2+} levels have been altered by mutation or overexpression in the vacuole, ER, and apoplast using two proteins—*CAX1* and a derivative of *CRT*. This section will describe these results and examine their collateral impact on abiotic stress responses and make some recommendations for future experiments.

6.1. Transgenic Expression of CAX Family Members. The protein family with the best potential for increasing bioavailable Ca^{2+} in plants is *CAX*, located on the tonoplast membrane [188]. As previously mentioned, *CAX1* expression levels are the primary determinant for Ca^{2+} levels in Arabidopsis [162, 189]. In leaves, *CAX1* functions to clear free Ca^{2+} from the apoplast, so that guard cell signaling, which requires extracellular Ca^{2+}, can be regulated properly (Section 3). In *sCAX1*, the N-terminal autoinhibitory loop has been removed so that it can transport increased amounts of Ca^{2+} into the vacuole [190].

Ectopic *sCAX1* expression increased Ca^{2+} in potato tubers by 2-3 folds with no change in morphology or yield when supplemented with 2 mM $CaCl_2$ during the first 3 months [191]. However, CAX1 transports other cations in addition to Ca^{2+}, which are not as beneficial from a nutritional standpoint. Hirschi's group, therefore, modified the *CAX2* gene, which shows a greater specificity for Ca^{2+}, to eliminate its Mn^{2+} transport function and then showed ~50%–60% increase in Ca^{2+} in transgenic potatoes [192].

Tomato was transformed with *CAX4*, which is more specific for Ca^{2+} than *CAX1* [193]. This resulted in a 40% increase in total Ca^{2+} and was not associated with Ca^{2+} deficiency symptoms even in the absence of $CaCl_2$ supplementation. *CAX4* increased fruit firmness (and, therefore, postharvest life), but did not impact ethylene production or sugar content [193]. In addition, root growth was enhanced [193]. Later experiments in Arabidopsis demonstrated that *CAX4* expression, which is uniquely confined to roots, is needed for normal root growth and that *cax4* mutants had reduced DR5 : GUS expression [194]. DR5 is a synthetic promoter that responds to auxin. The authors postulate that cytosolic Ca^{2+} levels may have increased, due to altered CAX4-mediated efflux into the vacuole. This may have affected polar transport of auxin, which is regulated by CDPKs [194]. The impact of *CAX* genes on root growth is important and deserves further study. Although CAX1 and CAX4 are thought to act primarily by depleting cytosolic Ca^{2+}, roots of *CAX1* transformants were less sensitive to inhibition by applied auxin than the wild type [195], while roots of *CAX4* transformants were more sensitive [194]. It would be very interesting to know how these alterations in root phenotype affect tolerance to abiotic stress.

There are no deleterious effects of *sCAX1* expression under normal conditions if supplemental Ca^{2+} is added. Otherwise, Ca^{2+} deficiency symptoms result, which include increased sensitivity to salt and cold stress. The yeast 2-hybrid experiments demonstrated that CAX1 interacts with SOS2, a CIPK that usually requires SOS3 (a CBL) for activity, through its N-terminal domain [196]. This may help to deplete the cytosol of excess Ca^{2+} following salt stress, which is known to produce a transient increase in Ca^{2+}. Overexpression of *sCAX1* increased the plant's sensitivity to salt, perhaps by being too efficient in the removal of excess Ca^{2+}, leading to store depletion [196]. The impact of drought and osmotic stress on *sCAX1* overexpression has not been studied, but would be expected to show similar responses (enhanced sensitivity). In contrast to *sCAX1* transformants, mutants of *CAX3* show increased salt sensitivity [197]. Both decreased Ca^{2+} transport into the vacuole during salt stress and decreased H^+ ATPase activities at the plasma membrane were associated with the *cax3* mutation.

Overexpression of *CAX1* also resulted in increased sensitivity to salt while mutations in this gene produce salt and drought tolerant plants [195]. Interestingly, exogenous Ca^{2+} can reverse salt sensitivity in *CAX1* transgenic plants and can also reverse the salt tolerance of *cax1* mutants [195]. CAX1 may be involved in sequestering Ca^{2+} to the vacuole following release into the cytoplasm. If Ca^{2+} signals cannot be dampened by transport into the vacuole, cytosolic levels may remain high, activating salt tolerance pathways. Conversely, if Ca^{2+} is sequestered into the vacuole at a faster rate than normal (as in the *CAX1* over-expressors), cytosolic levels may never reach the threshold required to activate pathways for salt tolerance. *cax1* mutants showed developmental abnormalities including reduced root growth and delayed flowering [195].

Ectopic expression of *sCAX1* in tobacco was also associated with increased sensitivity to cold shock [188]. This correlated with the positive impact of mutations in *cax1* on cold tolerance [51]. The negative impact of *CAX1* on cold tolerance was shown to be due to decreased upregulation (relative to wild type) of *DREB1* and a subset of cold-responsive genes induced by DREB1 [51]. These results are interesting because they are the first to demonstrate altered gene expression by *CAX1*, although the signal transduction pathway has yet to be demonstrated. Although *DREB1* was upregulated by cold in the *cax1* mutants, there were no changes in gene expression associated with exposure to dehydration or salt [51].

There are also beneficial effects of ectopic *CAX* expression. Both *CAX1* and *CAX4* expressions have been associated with enhanced tolerance to heavy metals [194, 198–201]. The potential impacts on other traits are difficult to assess. *CAX1* and *CAX3* have been shown to regulate phosphate homeostasis by repressing phosphate starvation-associated genes [202]. A *cax1/3* double mutant resulted in increased shoot phosphorous accumulation [202]. Grafting experiments suggested that CAX1 and CAX3 could be involved in the generation of a shoot to root signal that represses phosphate transport [202], but the impact of *sCAX1* over-expressing plants on phosphate transport has not been determined.

Unfortunately, *sCAX1* expression has not contributed towards mitigation of Ca^{2+} deficiency diseases. Massive cell death is associated with Ca^{2+} deficiency resulting, for example, in fruit that is not suitable for consumption. Tomato fruit development is especially susceptible to cell death (blossom end rot) caused by Ca^{2+} deficiency [203] a situation aggravated by increased salinity [204]. Blossom end rot in tomato is known to be related to Ca^{2+} deficiency [205]. Instead of helping to prevent blossom end rot, *sCAX1* expression resulted in 100% of the tomato fruits developing the disorder [206]. This may have been due to reduced free Ca^{2+} in the apoplast, where it likely helps to stabilize membrane structure, among other things. Ca^{2+} deficiency near the plasma membrane causes destabilization, which could precipitate the disorder [206]. Although *sCAX1* expression may make Ca^{2+} more bioavailable to humans, it does not appear to have the same effect in plants.

In contrast to Arabidopsis, over expression of soybean *CAX1* homolog in Arabidopsis increased salt tolerance [207]. *GmCAX1* has an N-terminal autoinhibitory loop, also found in *AtCAX1*, but shows only 65% homology to it and 68% homology to *CAX2* [207]. In contrast to Arabidopsis, *GmCAX1* was not induced by cold suggesting that the regulation and function of different *CAX* homologs may show considerable variation across species [207]. It is not clear what this means for predicting the impact of overexpression of *sCAX1* in other species. As acknowledged [197], it may be difficult to predict the effects of overexpression of a major transporter on the phenotype of any plant.

6.2. CRT and CBP. CRT is a multifunctional protein that is highly conserved in eukaryotic cells [208–210]. It has at least three functional domains: a globular N-domain, a proline rich, high affinity ($K_d = 1.6 \mu M$), low capacity ($B_{max} = 1$ mol/mol of protein) Ca^{2+}-binding domain (the P-domain), and a highly acidic, low affinity ($K_d = 0.3$–2 mM), high capacity ($B_{max} = 20$–50 mol/mol of protein) Ca^{2+}-binding domain (the C-domain) [211]. In animals, *CRT* has been suggested to be involved in Ca^{2+} signaling [212, 213], chaperone activity [211], cell adhesion [214], gene expression [215], apoptosis [216], and in controlling store-operated fluxes through the plasma membrane [217–219]. Overexpression of *CRT* in both plants [220] and animals [221] increases total ER Ca^{2+} stores.

We found that ectopic expression of the maize *CRT1* or a Ca^{2+}-Binding Peptide (CBP) consisting of only the *CRT* C-domain can not only increase Ca^{2+} stores, but also enhance the survival of Arabidopsis plants grown in low Ca^{2+} medium [222, 223], suggesting that the extra Ca^{2+} could be used by the plant in times of stress. The hypothesis guiding this research is that the CBP sequesters Ca^{2+} in the ER in a manner similar to CRT. However, Ca^{2+} may bind the CBP protein in the ER, but then travel as a complex through the secretory system to the vacuole, cytoplasm, or even the nucleus [224]. It is highly unlikely that Ca^{2+} will be bound by ER-CBP in the cytoplasm, because of its low affinity. It is, therefore, reasonable to use the ER-CBP as a tool for altering intracellular stores of Ca^{2+}.

Our previous work demonstrated that intracellular Ca^{2+} levels could be manipulated in Arabidopsis by heat shock induction of an ER-targeted GFP-CBP peptide constructed by translationally fusing the green fluorescent protein gene to a sequence corresponding to 126 amino acids derived from the maize calreticulin C-domain [223]. ER-CBP plants induced on Ca^{2+} containing medium survived longer than similarly heat-shocked ER-GFP control plants when transferred to Ca^{2+} depleted medium [223]. This work suggested that the ER capacity for Ca^{2+} could be directly related to a physiological response, early senescence in the absence of Ca^{2+}. Importantly, ER Ca^{2+} could be modulated without the addition of external Ca^{2+} and deleterious effects due to Ca^{2+} depletion were not apparent. To further examine physiological differences in these plants and to avoid the complications of heat shock induction, we transformed Arabidopsis with the same GFP-CBP construct (or CBP without GFP, for indo-1 experiments) but under the control of the constitutive 35S cauliflower mosaic virus promoter.

Why not over-express *CRT* to increase Ca^{2+}? Overexpression of *ZmCRT1* in tobacco cells increased Ca^{2+} by 2-fold, and transformation of Arabidopsis with *ZmCRT1* reduced the rate of senescence following transfer to low Ca^{2+} media [222]. There are two potential problems with over-expressing full-length *CRT*, silencing of the endogenous gene, and deleterious effects under some conditions. Overexpression of *CRT2* resulted in the production of dwarfed plants, caused by high levels of salicylic acid [145]. Although overexpression of Chinese cabbage *CRT1* enhanced shoot and root regeneration in tobacco, the subsequent growth of tobacco plants was retarded [225]. *CRT1* overexpression was also shown to be deleterious in rice [186].

My group initially used a soybean heat shock promoter to drive the expression of a maize *CRT1 C-domain*, which we called CBP for Ca^{2+} binding peptide, fused to GFP to stabilize it. This turned out to be unnecessary although it was very useful for detecting gene silencing. Nevertheless, we were able to increase Ca^{2+} in heat-shocked plants by ~15%. Now we know that total Ca^{2+} levels can be increased by ~25% using constitutively expressed ER-localized *CBP* [158]. Arabidopsis plants transformed with 35S:CBP showed better salt and drought tolerance and had longer roots, even in the absence of stress [158]. There were no detectable differences in GFP-CBP plants compared to GFP or control plants under

normal conditions except that seed production was slightly higher and seedling root growth was increased [158].

Preliminary experiments using both cytoplasmic aequorin-expressing plants and indole-1 ratio imaging suggested that there were no significant differences in [Ca^{2+}]$_{cyt}$ concentrations between 35S:CBP-expressing Arabidopsis and wild type or 35S:GFP control plants [226]. However, after 4-5 days growth in Ca^{2+}-deficient media, the peak [Ca^{2+}]$_{cyt}$ in control plants was significantly lower than in CBP-expressing plants in response to a 150–300 mM NaCl challenge [226]. This suggested that expression of *CBP* allowed plants to respond to stimuli over a longer period of time due to the excess ER-localized reserves of Ca^{2+}. This was a very interesting result that could provide a mechanism for how CBP benefits plants with respect to stress tolerance.

Microarray results of 35S:GFP-CBP compared to 35S:GFP plants showed that genes for endomembrane and cell wall-associated proteins were upregulated [158]. One Ca^{2+}-regulated gene was strongly upregulated (greater than 3.5-fold), *CIPK6*. As described in Section 2.3, CIPK6 is a protein kinase that interacts with a Ca^{2+} sensor protein, CBL. Mutants in *AtCIPK6* are sensitive to salt [44], and overexpression of a constitutively active mutant of *AtCIPK6* in tobacco confers salt tolerance and also increases root length [44, 126], which are both found in *CBP*-expressing plants. We, therefore, asked if the enhanced salt tolerance was due to co-expression of *CIPK6*. When *CBP* was crossed with a *cipk6* knockdown mutant (50% reduction in mRNA) and then challenged with NaCl, it showed the same response as wild type plants. This was somewhat disappointing, as we believed that CBP would enhance stress tolerance by providing a Ca^{2+} reserve. Of course the induction of *CIPK6* may have been caused by the presence of additional ER Ca^{2+}; but the eradication of the response by a single mutation was surprising. It remains possible that there is an extra advantage of *CBP* expression in drought tolerance or under different conditions. The *cipk6* mutant has been complemented with a *CIPK6* transgene (D. Chattopadhyay, pers. Comm.).

CBP-expressing plants also downregulate *CIPK23*, which is also involved in salt tolerance, by 2-fold [158]. We believe this is why the *CBPxcipk6* plants showed a similar response to NaCl as the controls, despite the presence of ~50% CIPK6 in the knockdown mutant.

How does CIPK6 enhance salt tolerance? Recent experiments from Kudla's group have shown that CIPK6 interacts with AKT2, a K$^+$ channel [113]. Interaction occurs on the ER membrane, although both proteins are translated in the cytosol. CIPK6 interacts specifically with CBL4, which was originally identified as an SOS (salt overly sensitive) mutant [227–229]. When CBL4 is modified by both myristoylation and palmitoylation, the AKT2/CIPK6/CBL4 complex moves from the ER membrane to the plasma membrane, where AKT2 participates as a K$^+$ channel. Mutations in *CIPK6*, *AKT2*, and *CBL4* confer similar phenotypes when grown under short days, reduced leaf number and size and delayed flowering [113]. K$^+$ is needed for phloem transport, and the reduced size of the mutant plants is restored under long day conditions [113]. This phenotype is consistent with a reduction in phloem transport, but does not provide an explanation for the altered response by *cipk6* to NaCl.

The role of AKT1, which is modulated by CIPK23/ CBL1/CBL9 in a similar manner as AKT2, was recently called into question. Mutants defective in *akt1* or *cipk23* showed better drought tolerance than wild type plants, suggesting that CIPK23/CBL1/CBL9 regulation of AKT1 may actually decrease abiotic stress tolerance [230]. However, overexpression of CBL1 and CIPK23 has been shown to increase tolerance to abiotic stress [39, 47]. Clearly, more experiments are needed to understand the relationship between K^+ and abiotic stress.

In addition to Arabidopsis, CBP has been transformed into potato and rice ([231], S. Y. Lee, R. Qu, and D. Robertson, in preparation). The goal for CBP expression in potato was to prevent internal heat necrosis (INH), a disorder affecting the quality of potato tubers [232]. There is strong but indirect evidence for an involvement of Ca^{2+} in this disorder. The application of antitranspirants to potato leaves reduced total Ca^{2+} levels and increased Ca^{2+} in tubers. This led to a decreased incidence of the disorder. However, when 3 independent transgenic potato lines (cv. Atlantic) expressing a 35S : CBP gene were grown under greenhouse conditions, the incidence of INH correlated positively with expression of 35S : CBP, which also increased potato tuber yield and total Ca^{2+} in leaves [231]. It was not possible to measure Ca^{2+} in tubers. There were also increased levels of Mg^{2+} and Mn^{2+} in the CBP-expressing plants, and reduced levels of K^+ [231]. Although the increased yield was statistically significant, the experiment would need to be repeated. It is not known if it was the increased yield that was responsible for greater incidence of INH, but it is unlikely that it could be separated from the expression of CBP.

It would be interesting to know if *CBP* expression in other plants (besides Arabidopsis) causes an increase in *CIPK6* orthologs, and these experiments are currently in progress for rice (Lee, Qu, and Robertson, unpublished). Does the induction of *CIPK6* depend on a flux or an increase of ER Ca^{2+}? Could this result from ACA and ECA activity in removing Ca^{2+} from the cytosol? Confocal microscopy of the GFP-CBP fusion protein showed ER and, to a lesser extent, nuclear activity [158]. Although CBP would not be expected to bind Ca^{2+} in the cytosol, it could bind Ca^{2+} in the nucleus. Acidic domains can act as transcriptional coactivators [233], providing a possible mechanism for CBP action. These results illustrate the difficulty of using genetic methods to modulate specific stores of Ca^{2+}. Although targeting of CBP to the nucleus could be used as a control, the molecular weight of CBP is estimated to be ~5 kDa so it should enter the nucleus without a targeting sequence.

6.3. Coexpression of sCAX1 and CRT1. 100% of tomato plants expressing *sCAX1* developed blossom end rot, a Ca^{2+} deficiency related disorder that leads to necrosis in the distal, developing end of the fruit [206]. These plants were grown in a greenhouse under conditions where none of the nontransgenic control plants developed the syndrome.

The expression of *sCAX1* was shown to reduce apoplastic Ca^{2+} levels, which increased membrane leakiness [234]. Coexpression of *CRT* resulted in a significant decrease in Ca^{2+} deficiency symptoms in both tomato and tobacco without the addition of supplemental Ca^{2+} [235]. This is very interesting and, if it can be repeated in other species, may suggest several things about the ER and vacuole with respect to signaling. Questions that this observation raises include the following:

(1) How is the Ca^{2+} level in the shoot increased, without an increase in transpiration? (This is relevant to all *sCAX*-expressing plants.)

(2) Does the ER form a symplastic Ca^{2+} network distinct from the apoplast and vacuole?

(3) Is *CRT* needed to keep a bioavailable pool of Ca^{2+} inside the ER for signaling? If so, then extra *CRT* may have successfully competed with *CAX1* for the limited pool of free Ca^{2+} in the apoplast in the dual transgenic plants.

(4) Can the vacuole serve as the source of Ca^{2+} for some stimuli?

(5) What would *CRT* overexpression in a *cax1/3* mutant do? Could it help to bind excess apoplastic Ca^{2+}?

6.4. Other Transgenes for Manipulating Ca^{2+} Stores in Plants. Several Ca^{2+}-related proteins have the potential to serve as a mechanism for altering Ca^{2+} stores in plants. Theoretically, any part of the cell except the cytoplasm could sustain increased levels of Ca^{2+} without deleterious consequences, although this needs to be experimentally verified. As a group, plants vary in Ca^{2+} content and show differential sensitivity to Ca^{2+} as a nutrient [236]. Since we know there is variation in Ca^{2+} levels between plants, even between ecotypes of Arabidopsis (and that variation correlates with *CAX1* expression [189]), we should be able to genetically manipulate it.

One of the benefits of large-scale scientific experiments ("omics") is the availability of data for gene expression and ion concentrations for a variety of closely related plants. Arabidopsis ecotypes have been collected from around the world, and there are hundreds of accessions, each of which shows less genetic variation than would be found between two species, but together there is a large pool of variation that can be correlated with a variety of different phenotypes. The leaf ionome of 31 of these accessions has now been completed, and Conn and his colleagues have outlined methods for using this data to identify candidate genes controlling elemental accumulation [237]. This promises to be a very productive avenue of research, especially if some of the candidates can be correlated with positive agronomic properties.

In addition to proteins found in Arabidopsis, there are other Ca^{2+} binding proteins that have been identified in various species. Examples include a celery vacuole-associated dehydrin-like protein [238] and a radish vacuolar Ca^{2+} binding protein [239] that is induced by lack of Ca^{2+}. Neither of these proteins has mutants nor has been overexpressed, so it is not clear how much Ca^{2+} can be increased by using them.

Recently, TPC1, the slow vacuolar channel found in all plants, has been shown to contain a novel Ca^{2+} binding site that senses Ca^{2+} and alters its activity. Mutants have been created that are insensitive to feedback inhibition by luminal Ca^{2+}, which leads to an increase in the store of vacuolar Ca^{2+} [240].

Simply adding Ca^{2+} to fertilizers can increase leaf Ca^{2+} levels by up to 3-fold [241], and there is an argument that transgene manipulation may be unnecessary as breeding for increased Ca^{2+} levels should be sufficient to meet nutritional requirements for Ca^{2+}. There are two arguments against this notion: adding Ca^{2+} to the right compartment has the potential to boost the resiliency of plants to stress and providing Ca^{2+} loosely complexed to protein might result in enhanced nutritional absorption. Since overexpression of CRT can be detrimental to plant growth [145, 225], transgenic approaches that separate out the C-domain are the most straightforward approach to boosting ER Ca^{2+}.

6.5. Biofortification Studies. The potential role for CAX in biofortification has been demonstrated in carrots expressing sCAX1 [242]. Human consumption of the genetically engineered carrots resulted in a 41% increase in Ca^{2+} absorption compared to controls, demonstrating the bioavailability of vacuolar Ca^{2+} in this system [242]. Lettuce was also transformed with sCAX1 and contained 25%–32% more Ca^{2+} than controls [243]. The response of a human panel to the engineered lettuce was positive for its sensory characteristics [243]. As long as sCAX1-expressing plants have good agronomic properties and can be grown in the presence of excess Ca^{2+} or cotransformed with CRT (or, better, CBP), this is a very promising method for biofortification.

The absorption of Ca^{2+} from vegetables can be complicated by the presence of "antinutrients" such as oxalic acid, which forms insoluble Ca^{2+} oxalate crystals. As long as the diet is varied, it should not have a significant impact. Antinutrients are more important when choosing a plant for transgenic modification. These requirements are fulfilled in carrots, a good choice for one of the first plants to be transformed for increased Ca^{2+} absorption [242].

It has never been tested in clinical trials, but the delivery of Ca^{2+} ions complexed with protein, such as found in the ER in the form of the C-domain of CRT (CBP), could increase the absorption rate of Ca^{2+}. Although the use of sCAX1 to increase vacuolar Ca^{2+} has achieved remarkable increases in Ca^{2+} absorption on a per gram basis [242], the overall efficiency of Ca^{2+} absorption was 10% less than for controls. The reason for this is not clear, unless the level of antinutrients increased (which would be important to know). Comparing the efficiency of absorption between CBP transgenic and sCAX1 transgenic carrots could help to determine if Ca^{2+} absorption efficiency decreases as its concentration increases, or whether the cellular context of the extra Ca^{2+} plays a role in absorption. In the long run, it will be important to be able to use Ca^{2+} as efficiently as possible. Since CBP and the combination of CBP and sCAX1 lead to higher total Ca^{2+} levels without external supplementation, the added nutritional benefit may not require supplemental Ca^{2+} to be added.

Because the CRT C-domain is not highly conserved, it should be possible to choose sequences that retain a high number of acidic amino acids, which are known to bind Ca^{2+}, without causing silencing of the endogenous CRT genes. The potential for CBP expression alone to increase Ca^{2+} absorption from food should be tested, because Ca^{2+} loosely bound to a protein may be even more bioavailable than Ca^{2+} salts in the vacuole. CBP has not been associated with Ca^{2+} deficiency symptoms under normal or stress conditions in the laboratory. It would be interesting to compare Ca^{2+} uptake from sCAX-expressing plants to those expressing CBP, along with a combination of the two transgenes. The long-term goal for sustainable agriculture should be to maximize the efficiency of Ca^{2+} supplementation in the human diet, so that the effective use of Ca^{2+} as a fertilizer can be maximized.

6.6. Summary. Transgenic expression of sCAX1 or CAX4 may be the best way to increase vegetative sources of Ca^{2+} but this can require supplementation with $CaCl_2$. When coexpressed with CRT1, the need for Ca^{2+} supplementation appears to be reduced, but more studies are needed to determine the effect of two Ca^{2+} binding transgenes on agronomic properties, because CRT1 overexpression by itself can have deleterious effects on plant growth under certain conditions.

Transgenic expression of CBP also increases total Ca^{2+} but not by as much as CAX1. This may be a better transgene to co-express with CAX1 than CRT1 because it retains Ca^{2+} binding but lacks most of the functions of CRT1. CBP expression by itself increases root growth under nonstress conditions and reduces the effects of drought and salt stress, perhaps in part by increasing root growth but we think also by providing a more extensive store of bioavailable Ca^{2+}.

7. Conclusion

As described in the beginning, many studies show that Ca^{2+} applied externally can benefit plants by increasing stress tolerance. Even postharvest fruit characteristics are improved following a $CaCl_2$ soak. It is still not known where in the plant this supplemental Ca^{2+} is absorbed and distributed, or how it is used to benefit the plant. How much is actually necessary for the enhanced growth and stress responses? Is it the change in Ca^{2+} concentration or the absolute amount of Ca^{2+} available to the plant that is relevant?

One explanation for the beneficial response is that Ca^{2+} induces genes involved in abiotic stress tolerance, such as members of the CIPK/CBL family, some of which are known to be induced by exogenous Ca^{2+} [45, 125]. But rather than overexpressing Ca^{2+}-regulated genes, it may be more beneficial to increase the Ca^{2+} stores that are used to cause their induction. Finding ways to genetically increase Ca^{2+} levels in plants may allow us to capture the Ca^{2+}-stimulated enhancement under normal conditions or with minimal Ca^{2+} supplementation. Additional research on targeting Ca^{2+}

binding proteins to various organelles may, therefore, be useful.

More robust signaling pathways and stress responses would seem to be a good thing in the face of global climate change. By increasing just the second messenger, one could conceivably preserve the ability of the plant to adapt to different stresses. By increasing the degree of stress response, but not the specific pathway, plants may be better able to deploy valuable reserves into tolerating a wide variety of different stresses. This would make the ubiquity of Ca^{2+} an asset rather than an impediment to research. The more we understand about Ca^{2+}-regulated pathways, the more we can optimize the response to adverse conditions. One thing is clear, more exploratory research on the ectopic expression of Ca^{2+} binding or exchange proteins could be very promising for plants, agronomists, and consumers.

Acknowledgments

The author would like to thank Dr. Sang Yoon Lee for his dedication and initiative in working on the CBP project and for many interesting discussions. Drs. Pei-Lan Tsou and Sarah Wyatt started this work, and it was Dr. Wendy Boss's idea to use the *CRT* C-domain as a transgene for manipulating ER Ca^{2+}. I would also like to thank Dr. George Allen for his critical comments and encouragement and Dr. Steven Nagar for Figure 2. The CBP project was originally funded by NASA.

References

[1] F. J. Maathuis, "Physiological functions of mineral macronutrients," *Current Opinion in Plant Biology*, vol. 12, no. 3, pp. 250–258, 2009.

[2] P. J. White and M. R. Broadley, "Calcium in plants," *Annals of Botany*, vol. 92, no. 4, pp. 487–511, 2003.

[3] K. D. Hirschi, "The calcium conundrum. Both versatile nutrient and specific signal," *Plant Physiologyogy*, vol. 136, no. 1, pp. 2348–2442, 2004.

[4] S. Stael, B. Wurzinger, A. Mair, N. Mehlmer, U. C. Vothknecht, and M. Teige, "Plant organellar calcium signalling: an emerging field," *Journal of Experimental Botany*, vol. 63, pp. 1525–1542, 2011.

[5] C. K. Y. Ng and M. R. Mcainsh, "Encoding specificity in plant calcium signalling: hot-spotting the ups and downs and waves," *Annals of Botany*, vol. 92, no. 4, pp. 477–485, 2003.

[6] E. Cholewa and C. A. Peterson, "Evidence for symplastic involvement in the radial movement of calcium in onion roots," *Plant Physiologyogy*, vol. 134, no. 4, pp. 1793–1802, 2004.

[7] H. Upadhyaya, B. K. Dutta, L. Sahoo, and S. K. Panda, "Comparative Effect of Ca, K, Mn and B on Post-Drought Stress Recovery in Tea (*Camellia sinensis* (L) O. *Kuntze*)," *American Journal of Plant Sciences*, vol. 3, no. 4, pp. 443–460, 2012.

[8] T. Jiang, X. Zhan, Y. Xu, L. Zhou, and L. Zong, "Roles of calcium in stress-tolerance of plants and its ecological significance," *Chinese Journal of Applied Ecology*, vol. 16, no. 5, pp. 971–976, 2005.

[9] C. A. Jaleel, P. Manivannan, B. Sankar et al., "Water deficit stress mitigation by calcium chloride in Catharanthus roseus: effects on oxidative stress, proline metabolism and indole alkaloid accumulation," *Colloids and Surfaces B*, vol. 60, no. 1, pp. 110–116, 2007.

[10] H. Nayyar and S. K. Kaushal, "Alleviation of negative effects of water stress in two contrasting wheat genotypes by calcium and abscisic acid," *Biologia Plantarum*, vol. 45, no. 1, pp. 65–70, 2002.

[11] S. M. Juice, T. J. Fahey, T. G. Siccama et al., "Response of sugar maple to calcium addition to northern hardwood forest," *Ecology*, vol. 87, no. 5, pp. 1267–1280, 2006.

[12] C. Sulochana and N. Sanithramma, "Effect of Calcium in Amelioration of PEG, (600) Induced Water Stress in Ground Nut (*Araclus hypogaea* L.) Cultivars during Seedling Growth," *Journal of Plant Biology*, vol. 38, 2001.

[13] Y. E. Kolupaev, G. E. Akinina, and A. V. Mokrousov, "Induction of heat tolerance in wheat coleoptiles by calcium ions and its relation to oxidative stress," *Russian Journal of Plant Physiologygyogy*, vol. 52, no. 2, pp. 199–204, 2005.

[14] Y. Wu, X. Liu, W. Wang, S. Zhang, and B. Xu, "Calcium regulates the cell-to-cell water flow pathway in maize roots during variable water conditions," *Plant Physiologyogy and Biochemistry*, vol. 58, pp. 212–219, 2012.

[15] M. S. Aghdam, M. B. Hassanpouraghdam, G. Paliyath, and B. Farmani, "The language of calcium in postharvest life of fruits, vegetables and flowers," *Scientia Horticulturae*, vol. 144, pp. 102–115, 2012.

[16] P. A. Lahaye and E. Epstein, "Salt toleration by plants: enhancement with calcium," *Science*, vol. 166, no. 3903, pp. 395–396, 1969.

[17] P. A. Essah, R. Davenport, and M. Tester, "Sodium influx and accumulation in *Arabidopsis*," *Plant Physiologyogy*, vol. 133, no. 1, pp. 307–318, 2003.

[18] C. M. Grieve and H. Fujiyama, "The response of two rice cultivars to external Na/Ca ratio," *Plant and Soil*, vol. 103, no. 2, pp. 245–250, 1987.

[19] L. Shaoyun, L. Yongchao, G. Zhenfei, L. Baosheng, and L. Mingqi, "Enhancement of drought resistance of rice seedlings by calcium," *Zhongguo Shuidao Kexue*, vol. 13, pp. 161–164, 1999.

[20] Z. Rengel, "The role of calcium in salt toxicity," *Plant, Cell and Environment*, vol. 15, pp. 625–632, 1992.

[21] H. Upadhyaya, S. K. Panda, and B. K. Dutta, "$CaCl_2$ improves post-drought recovery potential in *Camellia sinensis* (L) O. Kuntze," *Plant Cell Reports*, vol. 30, no. 4, pp. 495–503, 2011.

[22] E. P. Spalding and J. F. Harper, "The ins and outs of cellular Ca^{2+} transport," *Current Opinion in Plant Biology*, vol. 14, no. 6, pp. 715–720, 2011.

[23] O. Batistic and J. Kudla, "Analysis of calcium signaling pathways in plants," *Biochimica et Biophysica Acta*, vol. 8, pp. 1283–1293, 1820.

[24] A. M. Cameron, J. P. Steiner, A. J. Roskams, S. M. Ali, G. V. Ronnett, and S. H. Snyder, "Calcineurin associated with the inositol 1,4,5-trisphosphate receptor- FKBP12 complex modulates Ca^{2+} flux," *Cell*, vol. 83, no. 3, pp. 463–472, 1995.

[25] M. D. Sjaastad, R. S. Lewis, and W. J. Nelson, "Mechanisms of integrin-mediated calcium signaling in MDCK cells: regulation of adhesion by IP3- and store-independent calcium influx," *Molecular Biology of the Cell*, vol. 7, no. 7, pp. 1025–1041, 1996.

[26] I. Y. Perera, I. Heilmann, and W. F. Boss, "Transient and sustained increases in inositol 1,4,5-trisphosphate precede the differential growth response in gravistimulated maize pulvini," *Proceedings of the National Academy of Sciences of the United States of America*, vol. 96, no. 10, pp. 5838–5843, 1999.

[27] J. M. Stevenson, I. Y. Perera, I. Heilmann, S. Persson, and W. F. Boss, "Inositol signaling and plant growth," *Trends in Plant Science*, vol. 5, no. 6, pp. 252–258, 2000.

[28] R. Zhong, D. H. Burk, W. H. Morrison, and Z. H. Ye, "FRAGILE FIBER3, an *Arabidopsis* gene encoding a type ii inositol polyphosphate 5-phosphatase, is required for secondary wall synthesis and actin organization in fiber cells," *Plant Cell*, vol. 16, no. 12, pp. 3242–3259, 2004.

[29] F. M. Carland and T. Nelson, "Cotyledon Vascular Pattern$_2$-mediated inositol (1,4,5) triphosphate signal transduction is essential for closed venation patterns of *Arabidopsis* foliar organs," *Plant Cell*, vol. 16, no. 5, pp. 1263–1275, 2004.

[30] W. H. Lin, R. Ye, H. Ma, Z. H. Xu, and H. W. Xue, "DNA chip-based expression profile analysis indicates involvement of the phosphatidylinositol signaling pathway in multiple plant responses to hormone and abiotic treatments," *Cell Research*, vol. 14, no. 1, pp. 34–45, 2004.

[31] I. Y. Perera, C. Y. Hung, C. D. Moore, J. Stevenson-Paulik, and W. F. Boss, "Transgenic *Arabidopsis plants* expressing the type 1 inositol 5-phosphatase exhibit increased drought tolerance and altered abscisic acid signaling," *Plant Cell*, vol. 20, no. 10, pp. 2876–2893, 2008.

[32] R. H. Tang, S. Han, H. Zheng et al., "Coupling diurnal cytosolic Ca^{2+} oscillations to the CAS-IP 3 pathway in *Arabidopsis*," *Science*, vol. 315, no. 5817, pp. 1423–1426, 2007.

[33] J. Groenendyk, J. Lynch, and M. Michalak, "Calreticulin, Ca^{2+}, and calcineurin—signaling from the endoplasmic reticulum," *Molecules and Cells*, vol. 17, no. 3, pp. 383–389, 2004.

[34] L. Navazio, P. Mariani, and D. Sanders, "Mobilization of CA^{2+} by cyclic ADP-ribose from the endoplasmic reticulum of cauliflower florets," *Plant Physiologyogy*, vol. 125, no. 4, pp. 2129–2138, 2001.

[35] G. Mailhot, J. L. Petit, C. Demers, and M. Gascon-Barre, "Influence of the in vivo calcium status on cellular calcium homeostasis and the level of the calcium-binding protein calreticulin in rat hepatocytes," *Endocrinology*, vol. 141, pp. 891–900, 2000.

[36] T. Yang and B. W. Poovaiah, "Calcium/calmodulin-mediated signal network in plants," *Trends in Plant Science*, vol. 8, no. 10, pp. 505–512, 2003.

[37] D. Sanders, J. Pelloux, C. Brownlee, and J. F. Harper, "Calcium at the crossroads of signaling," *Plant Cell*, vol. 14, no. supplement 1, pp. S401–S417, 2002.

[38] V. Albrecht, S. Weinl, D. Blazevic et al., "The calcium sensor CBL1 integrates plant responses to abiotic stresses," *The Plant Journal*, vol. 36, no. 4, pp. 457–470, 2003.

[39] Y. H. Cheong, K. N. Kim, G. K. Pandey, R. Gupta, J. J. Grant, and S. Luan, "CBL1, a calcium sensor that differentially regulates salt, drought, and cold responses in *Arabidopsis*," *Plant Cell*, vol. 15, no. 8, pp. 1833–1845, 2003.

[40] L. L. Liu, H. M. Ren, L. Q. Chen, Y. Wang, and W. H. Wu, "A protein kinase CIPK9 interacts with calcium sensor CBL3 and regulates K$^+$ homeostasis under low-KK$^+$ stress in *Arabidopsis*," *Plant Physiology*, vol. 161, pp. 266–277, 2013.

[41] M. Wang, D. Gu, T. Liu et al., "Overexpression of a putative maize calcineurin B-like protein in *Arabidopsis* confers salt tolerance," *Plant Molecular Biology*, vol. 65, no. 6, pp. 733–746, 2007.

[42] Y. H. Cheong, S. J. Sung, B. G. Kim et al., "Constitutive overexpression of the calcium sensor CBL5 confers osmotic or drought stress tolerance in *Arabidopsis*," *Molecules and Cells*, vol. 29, no. 2, pp. 159–165, 2010.

[43] Z. Gu, B. Ma, Y. Jiang, Z. Chen, X. Su, and H. Zhang, "Expression analysis of the calcineurin B-like gene family in rice (Oryza sativa L.) under environmental stresses," *Gene*, vol. 415, no. 1-2, pp. 1–12, 2008.

[44] V. Tripathi, B. Parasuraman, A. Laxmi, and D. Chattopadhyay, "CIPK6, a CBL-interacting protein kinase is required for development and salt tolerance in plants," *The Plant Journal*, vol. 58, no. 5, pp. 778–790, 2009.

[45] R. K. Wang, L. L. Li, Z. H. Cao et al., "Molecular cloning and functional characterization of a novel apple MdCIPK6L gene reveals its involvement in multiple abiotic stress tolerance in transgenic plants," *Plant Molecular Biology*, vol. 79, pp. 123–135, 2012.

[46] Y. Xiang, Y. Huang, and L. Xiong, "Characterization of stress-responsive CIPK genes in rice for stress tolerance improvement," *Plant Physiologyogy*, vol. 144, no. 3, pp. 1416–1428, 2007.

[47] W. Yang, Z. Kong, E. Omo-Ikerodah, W. Xu, Q. Li, and Y. Xue, "Calcineurin B-like interacting protein kinase OsCIPK23 functions in pollination and drought stress responses in rice (*Oryza sativa* L.)," *Journal of Genetics and Genomics*, vol. 35, no. 9, pp. 531.S1–543.S2, 2008.

[48] K. R. Konrad, M. M. Wudick, and J. A. Feijo, "Calcium regulation of tip growth: new genes for old mechanisms," *Current Opinion in Plant Biology*, vol. 14, pp. 721–730, 2011.

[49] M. Gilliham, A. Athman, S. D. Tyerman, and S. J. Conn, "Cell-specific compartmentation of mineral nutrients is an essential mechanism for optimal plant productivity—another role for TPC1?" *Plant Signaling & Behavior*, vol. 6, pp. 1656–1661, 2011.

[50] J. Bose, I. I. Pottosin, S. S. Shabala, M. G. Palmgren, and S. Shabala, "Calcium efflux systems in stress signaling and adaptation in plants," *Frontiers in Plant Science*, vol. 2, p. 85, 2011.

[51] R. Catala, E. Santos, J. M. Alonso, J. R. Ecker, J. M. Martinez-Zapater, and J. Salinas, "Mutations in the Ca^{2+}/H$^+$ transporter CAX1 increase CBF/DREB1 expression and the cold-acclimation response in *Arabidopsis*," *Plant Cell*, vol. 15, pp. 2940–2951, 2003.

[52] J. Lynch, V. S. Polito, and A. Lauchli, "Salinity stress increases cytoplasmic Ca activity in maize root protoplasts," *Plant Physiologyogy*, vol. 90, pp. 1271–1274, 1989.

[53] A. J. Laude and A. W. M. Simpson, "Compartmentalized signalling: Ca^{2+} compartments, microdomains and the many facets of Ca^{2+} signalling," *FEBS Journal*, vol. 276, no. 7, pp. 1800–1816, 2009.

[54] A. Trewavas, "Le calcium, c'est la vie: calcium makes waves," *Plant Physiologyogy*, vol. 120, no. 1, pp. 1–6, 1999.

[55] S. Papp, E. Dziak, M. Michalak, and M. Opas, "Is all of the endoplasmic reticulum created equal? The effects of the heterogeneous distribution of endoplasmic reticulum Ca^{2+}-handling proteins," *Journal of Cell Biology*, vol. 160, no. 4, pp. 475–479, 2003.

[56] J. M. Fasano, G. D. Massa, and S. Gilroy, "Ionic signaling in plant responses to gravity and touch," *Journal of Plant Growth Regulation*, vol. 21, no. 2, pp. 71–88, 2002.

[57] I. C. Mori, Y. Murata, Y. Yang et al., "CDPKs CPK6 and CPK3 function in ABA regulation of guard cell S-type anion- and Ca^{2+}-permeable channels and stomatal closure," *PLoS Biology*, vol. 4, no. 10, p. e327, 2006.

[58] H. Marten, K. R. Konrad, P. Dietrich, M. R. G. Roelfsema, and R. Hedrich, "Ca^{2+}-dependent and -independent abscisic acid activation of plasma membrane anion channels in guard cells of Nicotiana tabacum," *Plant Physiologyogy*, vol. 143, no. 1, pp. 28–37, 2007.

[59] S. J. Su, Y. F. Wang, A. Frelet et al., "The ATP binding cassette transporter AtMRP5 modulates anion and calcium channel activities in *Arabidopsis* guard cells," *Journal of Biological Chemistry*, vol. 282, no. 3, pp. 1916–1924, 2007.

[60] L. Cárdenas, "New findings in the mechanisms regulating polar growth in root hair cells," *Plant Signaling and Behavior*, vol. 4, no. 1, pp. 4–8, 2009.

[61] D. Cho, S. A. Kim, Y. Murata et al., "De-regulated expression of the plant glutamate receptor homolog AtGLR3.1 impairs long-term Ca^{2+}-programmed stomatal closure," *The Plant Journal*, vol. 58, no. 3, pp. 437–449, 2009.

[62] R. S. Siegel, S. Xue, Y. Murata et al., "Calcium elevation-dependent and attenuated resting calcium-dependent abscisic acid induction of stomatal closure and abscisic acid-induced enhancement of calcium sensitivities of S-type anion and inward-rectifying K$^+$ channels in *Arabidopsis* guard cells," *The Plant Journal*, vol. 59, no. 2, pp. 207–220, 2009.

[63] W. H. Wang, X. Q. Yi, A. D. Han et al., "Calcium-sensing receptor regulates stomatal closure through hydrogen peroxide and nitric oxide in response to extracellular calcium in *Arabidopsis*," *J Exp Bot*, vol. 63, pp. 177–190, 2011.

[64] W. H. Wang and H. L. Zheng, "Mechanisms for calcium sensing receptor-regulated stomatal closure in response to the extracellular calcium signal," *Plant Signaling & Behavior*, vol. 7, pp. 289–291, 2012.

[65] W. Capoen, J. D. Herder, J. Sun et al., "Calcium spiking patterns and the role of the calcium/calmodulin-dependent kinase CCaMK in lateral root base nodulation of sesbania rostrata," *Plant Cell*, vol. 21, no. 5, pp. 1526–1540, 2009.

[66] P. K. Hepler, J. G. Kunkel, C. M. Rounds, and L. J. Winship, "Calcium entry into pollen tubes," *Trends in Plant Science*, vol. 17, pp. 32–38, 2011.

[67] M. Rincón-Zachary, N. D. Teaster, J. Alan Sparks, A. H. Valster, C. M. Motes, and E. B. Blancaflor, "Fluorescence resonance energy transfer-sensitized emission of yellow cameleon 3.60 reveals root zone-specific calcium signatures in *Arabidopsis* in response to aluminum and other trivalent cations," *Plant Physiologyogy*, vol. 152, no. 3, pp. 1442–1458, 2010.

[68] E. F. Short, K. A. North, M. R. Roberts, A. M. Hetherington, A. D. Shirras, and M. R. McAinsh, "A stress-specific calcium signature regulating an ozone-responsive gene expression network in *Arabidopsis*," *The Plant Journal*, vol. 71, no. 6, pp. 948–961, 2012.

[69] K. Takahashi, M. Isobe, M. R. Knight, A. J. Trewavas, and S. Muto, "Hypoosmotic shock induces increases in cytosolic Ca^{2+} in tobacco suspension-culture cells," *Plant Physiologyogy*, vol. 113, no. 2, pp. 587–594, 1997.

[70] J. C. Sedbrook, P. J. Kronebusch, G. G. Borisy, A. J. Trewavas, and P. H. Masson, "Transgenic AEQUORIN reveals organ-specific cytosolic Ca^{2+} responses to anoxia in *Arabidopsis thaliana* seedling," *Plant Physiologyogy*, vol. 111, no. 1, pp. 243–257, 1996.

[71] M. C. Rentel and M. R. Knight, "Oxidative stress-induced calcium signaling in *Arabidopsis*," *Plant Physiologyogy*, vol. 135, no. 3, pp. 1471–1479, 2004.

[72] H. Song, R. Zhao, P. Fan, X. Wang, X. Chen, and Y. Li, "Overexpression of *AtHsp90.2, AtHsp90.5* and *AtHsp90.7* in *Arabidopsis thaliana* enhances plant sensitivity to salt and drought stresses," *Planta*, vol. 229, no. 4, pp. 955–964, 2009.

[73] E. Kiegle, C. A. Moore, J. Haseloff, M. A. Tester, and M. R. Knight, "Cell-type-specific calcium responses to drought, salt and cold in the *Arabidopsis* root," *The Plant Journal*, vol. 23, no. 2, pp. 267–278, 2000.

[74] C. Huang, S. Ding, H. Zhang, H. Du, and L. An, "CIPK7 is involved in cold response by interacting with CBL1 in *Arabidopsis thaliana*," *Plant Science*, vol. 181, no. 1, pp. 57–64, 2011.

[75] J. Szczegielniak, L. Borkiewicz, B. Szurmak et al., "Maize calcium-dependent protein kinase (ZmCPK11): local and systemic response to wounding, regulation by touch and components of jasmonate signaling," *Plant Physiologyogy*, vol. 146, pp. 1–14, 2012.

[76] H. Nie, C. Zhao, G. Wu, Y. Wu, Y. Chen, and D. Tang, "SR1, a calmodulin-binding transcription factor, modulates plant defense and ethylene-induced senescence by directly regulating NDR1 and EIN3," *Plant Physiologyogy*, vol. 158, pp. 1847–1859, 2012.

[77] W. Urquhart, K. Chin, H. Ung, W. Moeder, and K. Yoshioka, "The cyclic nucleotide-gated channels AtCNGC11 and 12 are involved in multiple Ca^{2+}-dependent physiological responses and act in a synergistic manner," *Journal of Experimental Botany*, vol. 62, no. 10, pp. 3671–3682, 2011.

[78] I. C. Mori and J. I. Schroeder, "Reactive oxygen species activation of plant Ca^{2+} channels. A signaling mechanism in polar growth, hormone transduction, stress signaling, and hypothetically mechanotransduction," *Plant Physiologyogy*, vol. 135, no. 2, pp. 702–708, 2004.

[79] G. D. Massa, J. M. Fasano, and S. Gilroy, "Ionic signaling in plant gravity and touch responses," *Gravitational and Space Biology Bulletin*, vol. 16, no. 2, pp. 71–82, 2003.

[80] L. Xiong, K. S. Schumaker, and J. K. Zhu, "Cell signaling during cold, drought, and salt stress," *Plant Cell*, vol. 14, no. supplement 1, pp. S165–S183, 2002.

[81] M. Maffei, S. Bossi, D. Spiteller, A. Mithöfer, and W. Boland, "Effects of feeding *Spodoptera littoralis* on Lima bean leaves. I. Membrane potentials, intracellular calcium variations, oral secretions, and regurgitate components," *Plant Physiologyogy*, vol. 134, no. 4, pp. 1752–1762, 2004.

[82] U. Jongebloed, J. Szederkényi, K. Hartig, C. Schobert, and E. Komor, "Sequence of morphological and physiological events during natural ageing and senescence of a castor bean leaf: sieve tube occlusion and carbohydrate back-up precede chlorophyll degradation," *Physiologia Plantarum*, vol. 120, no. 2, pp. 338–346, 2004.

[83] W. Ma, A. Smigel, R. K. Walker, W. Moeder, K. Yoshioka, and G. A. Berkowitz, "Leaf senescence signaling: the Ca^{2+}-Conducting *Arabidopsis* cyclic nucleotide gated channel2 acts through nitric Oxide to repress senescence programming," *Plant Physiologyogy*, vol. 154, no. 2, pp. 733–743, 2010.

[84] W. Ma and G. A. Berkowitz, "Cyclic nucleotide gated channel and Ca^{2+}-mediated signal transduction during plant senescence signaling," *Plant Signaling and Behavior*, vol. 6, no. 3, pp. 413–415, 2011.

[85] S. Masuda, K. Mizusawa, T. Narisawa, Y. Tozawa, H. Ohta, and K. I. Takamiya, "The bacterial stringent response, conserved in chloroplasts, controls plant fertilization," *Plant and Cell Physiology*, vol. 49, no. 2, pp. 135–141, 2008.

[86] C. Dumas and T. Gaude, "Fertilization in plants: is calcium a key player?" *Seminars in Cell and Developmental Biology*, vol. 17, no. 2, pp. 244–253, 2006.

[87] M. Schiøtt, S. M. Romanowsky, L. Bækgaard, M. K. Jakobsen, M. G. Palmgren, and J. F. Harper, "A plant plasma membrane Ca^{2+} pump is required for normal pollen tube growth and fertilization," *Proceedings of the National Academy of Sciences*

of the United States of America, vol. 101, no. 25, pp. 9502–9507, 2004.

[88] M. A. M. Aboul-Soud, A. M. Aboul-Enein, and G. J. Loake, "Nitric oxide triggers specific and dose-dependent cytosolic calcium transients in *Arabidopsis*," *Plant Signaling and Behavior*, vol. 4, no. 3, pp. 191–196, 2009.

[89] J. K. Zhu, "Salt and drought stress signal transduction in plants," *Annual Review of Plant Biology*, vol. 53, pp. 247–273, 2002.

[90] J. K. Zhu, "Regulation of ion homeostasis under salt stress," *Current Opinion in Plant Biology*, vol. 6, no. 5, pp. 441–445, 2003.

[91] C. A. Moore, H. C. Bowen, S. Scrase-Field, M. R. Knight, and P. J. White, "The deposition of suberin lamellae determines the magnitude of cytosolic Ca^{2+} elevations in root endodermal cells subjected to cooling," *The Plant Journal*, vol. 30, no. 4, pp. 457–465, 2002.

[92] J. F. Harper, G. Breton, and A. Harmon, "Decoding Ca^{2+} signals through plant protein kinases," *Annual Review of Plant Biology*, vol. 55, pp. 263–288, 2004.

[93] A. A. Ludwig, T. Romeis, and J. D. G. Jones, "CDPK-mediated signalling pathways: specificity and cross-talk," *Journal of Experimental Botany*, vol. 55, no. 395, pp. 181–188, 2004.

[94] A. C. Harmon, "Calcium-regulated protein kinases of plants," *Gravitational and Space Biology Bulletin*, vol. 16, no. 2, pp. 83–90, 2003.

[95] E. M. Hrabak, C. W. M. Chan, M. Gribskov et al., "The *Arabidopsis* CDPK-SnRK superfamily of protein kinases," *Plant Physiologyogy*, vol. 132, no. 2, pp. 666–680, 2003.

[96] I. S. Day, V. S. Reddy, G. Shad Ali, and A. S. Reddy, "Analysis of EF-hand-containing proteins in *Arabidopsis*," *Genome Biology*, vol. 3, no. 10, 2002.

[97] M. Boudsocq, M. J. Droillard, L. Regad, and C. Lauriere, "Characterization of *Arabidopsis* calcium-dependent protein kinases: activated or not by calcium?" *Biochemical Journal*, vol. 447, no. 2, pp. 291–299, 2012.

[98] Y. Galon, R. Aloni, D. Nachmias et al., "Calmodulin-binding transcription activator 1 mediates auxin signaling and responds to stresses in *Arabidopsis*," *Planta*, vol. 232, no. 1, pp. 165–178, 2010.

[99] Y. Galon, O. Snir, and H. Fromm, "How calmodulin binding transcription activators (CAMTAs) mediate auxin responses," *Plant Signaling and Behavior*, vol. 5, no. 10, pp. 1311–1314, 2010.

[100] Y. Qiu, J. Xi, L. Du, J. C. Suttle, and B. W. Poovaiah, "Coupling calcium/calmodulin-mediated signaling and herbivore-induced plant response through calmodulin-binding transcription factor AtSR1/CAMTA3," *Plant Molecular Biology*, vol. 79, pp. 89–99, 2012.

[101] C. J. Doherty, H. A. Van Buskirk, S. J. Myers, and M. F. Thomashow, "Roles for *Arabidopsis* CAMTA transcription factors in cold-regulated gene expression and freezing tolerance," *Plant Cell*, vol. 21, no. 3, pp. 972–984, 2009.

[102] L. Du, G. S. Ali, K. A. Simons et al., "Ca^{2+}/calmodulin regulates salicylic-acid-mediated plant immunity," *Nature*, vol. 457, no. 7233, pp. 1154–1158, 2009.

[103] N. A. Eckardt, "CAMTA proteins: a direct link between calcium signals and cold acclimation?" *Plant Cell*, vol. 21, no. 3, p. 697, 2009.

[104] J. Vadassery, S. Ranf, C. Drzewiecki et al., "A cell wall extract from the endophytic fungus *Piriformospora indica* promotes growth of *Arabidopsis* seedlings and induces intracellular calcium elevation in roots," *The Plant Journal*, vol. 59, no. 2, pp. 193–206, 2009.

[105] J. Vadassery and R. Oelmüller, "Calcium signaling in pathogenic and beneficial plant microbe interactions: what can we learn from the interaction between Piriformospora indica and *Arabidopsis thaliana*," *Plant Signaling & Behavior*, vol. 4, no. 11, pp. 1024–1027, 2009.

[106] S. Stael, A. G. Rocha, T. Wimberger, D. Anrather, U. C. Vothknecht, and M. Teige, "Cross-talk between calcium signalling and protein phosphorylation at the thylakoid," *Journal of Experimental Botany*, vol. 63, pp. 1725–1733, 2011.

[107] K. Hashimoto and J. Kudla, "Calcium decoding mechanisms in plants," *Biochimie*, vol. 93, pp. 2054–2059, 2011.

[108] O. Batistič, R. Waadt, L. Steinhorst, K. Held, and J. Kudla, "CBL-mediated targeting of CIPKs facilitates the decoding of calcium signals emanating from distinct cellular stores," *The Plant Journal*, vol. 61, no. 2, pp. 211–222, 2010.

[109] O. Batistič and J. Kudla, "Plant calcineurin B-like proteins and their interacting protein kinases," *Biochimica et Biophysica Acta*, vol. 1793, no. 6, pp. 985–992, 2009.

[110] O. Batistic, M. Rehers, A. Akerman et al., "S-acylation-dependent association of the calcium sensor CBL2 with the vacuolar membrane is essential for proper abscisic acid responses," *Cell Research*, vol. 22, pp. 1155–1168, 2012.

[111] W. Z. Lan, S. C. Lee, Y. F. Che, Y. Q. Jiang, and S. Luan, "Mechanistic analysis of AKT1 regulation by the CBL-CIPK-PP2CA interactions," *Molecular Plant*, vol. 4, no. 3, pp. 527–536, 2011.

[112] H. L. Piao, Y. H. Xuan, S. H. Park et al., "OsCIPK31, a CBL-interacting protein kinase is involved in germination and seedling growth under abiotic stress conditions in rice plants," *Molecules and Cells*, vol. 30, no. 1, pp. 19–27, 2010.

[113] K. Held, F. Pascaud, C. Eckert et al., "Calcium-dependent modulation and plasma membrane targeting of the AKT2 potassium channel by the CBL4/CIPK6 calcium sensor/protein kinase complex," *Cell Research*, vol. 21, no. 7, pp. 1116–1130, 2011.

[114] L. Li, B. G. Kim, Y. H. Cheong, G. K. Pandey, and S. Luan, "A Ca^{2+} signaling pathway regulates a K^+ channel for low-K response in *Arabidopsis*," *Proceedings of the National Academy of Sciences of the United States of America*, vol. 103, no. 33, pp. 12625–12630, 2006.

[115] J. Xu, H. D. Li, L. Q. Chen et al., "A Protein Kinase, Interacting with Two Calcineurin B-like Proteins, Regulates K+ Transporter AKT1 in *Arabidopsis*," *Cell*, vol. 125, no. 7, pp. 1347–1360, 2006.

[116] Y. H. Cheong, G. K. Pandey, J. J. Grant et al., "Two calcineurin B-like calcium sensors, interacting with protein kinase CIPK23, regulate leaf transpiration and root potassium uptake in *Arabidopsis*," *The Plant Journal*, vol. 52, no. 2, pp. 223–239, 2007.

[117] B. G. Kim, R. Waadt, Y. H. Cheong et al., "The calcium sensor CBL10 mediates salt tolerance by regulating ion homeostasis in *Arabidopsis*," *The Plant Journal*, vol. 52, no. 3, pp. 473–484, 2007.

[118] S. C. Lee, W. Z. Lan, B. G. Kim et al., "A protein phosphorylation/dephosphorylation network regulates a plant potassium channel," *Proceedings of the National Academy of Sciences of the United States of America*, vol. 104, no. 40, pp. 15959–15964, 2007.

[119] G. K. Pandey, J. J. Grant, Y. H. Cheong, B. G. Kim, L. G. Li, and S. Luan, "Calcineurin-B-like protein CBL9 interacts with target kinase CIPK3 in the regulation of ABA response in seed germination," *Molecular Plant*, vol. 1, no. 2, pp. 238–248, 2008.

[120] H. C. Hu, Y. Y. Wang, and Y. F. Tsay, "AtCIPK8, a CBL-interacting protein kinase, regulates the low-affinity phase of the primary nitrate response," *The Plant Journal*, vol. 57, no. 2, pp. 264–278, 2009.

[121] S. Luan, W. Lan, and S. Chul Lee, "Potassium nutrition, sodium toxicity, and calcium signaling: connections through the CBL-CIPK network," *Current Opinion in Plant Biology*, vol. 12, no. 3, pp. 339–346, 2009.

[122] K. N. Kim, J. S. Lee, H. Han, S. A. Choi, S. J. Go, and I. S. Yoon, "Isolation and characterization of a novel rice Ca^{2+}-regulated protein kinase gene involved in responses to diverse signals including cold, light, cytokinins, sugars and salts," *Plant Molecular Biology*, vol. 52, no. 6, pp. 1191–1202, 2003.

[123] P. J. White, M. R. Broadley, J. A. Thompson et al., "Testing the distinctness of shoot ionomes of angiosperm families using the Rothamsted Park Grass Continuous Hay Experiment," *New Phytologist*, vol. 196, pp. 101–109, 2012.

[124] D. Geiger, D. Becker, D. Vosloh et al., "Heteromeric AtKC1·AKT1 channels in *Arabidopsis* roots facilitate growth under K+-limiting conditions," *Journal of Biological Chemistry*, vol. 284, no. 32, pp. 21288–21295, 2009.

[125] N. Tuteja and S. Mahajan, "Further characterization of calcineurin B-like protein and its interacting partner CBL-interacting protein kinase from Pisum sativum," *Plant Signaling and Behavior*, vol. 2, no. 5, pp. 358–361, 2007.

[126] V. Tripathi, N. Syed, A. Laxmi, and D. Chattopadhyay, "Role of CIPK6 in root growth and auxin transport," *Plant Signaling and Behavior*, vol. 4, no. 7, pp. 663–665, 2009.

[127] E. Peiter, "The plant vacuole: emitter and receiver of calcium signals," *Cell Calcium*, vol. 50, no. 2, pp. 120–128, 2011.

[128] R. Hedrich and I. Marten, "TPC1—SV channels gain shape," *Molecular Plant*, vol. 4, no. 3, pp. 428–441, 2011.

[129] Y. Boursiac, S. M. Lee, S. Romanowsky et al., "Disruption of the vacuolar calcium-ATPases in *Arabidopsis* results in the activation of a salicylic acid-dependent programmed cell death pathway," *Plant Physiologyogy*, vol. 154, no. 3, pp. 1158–1171, 2010.

[130] M. Iwano, T. Entani, H. Shiba et al., "Fine-Tuning of the cytoplasmic Ca^{2+} concentration is essential for pollen tube growth," *Plant Physiologyogy*, vol. 150, no. 3, pp. 1322–1334, 2009.

[131] M. Michalak, J. Groenendyk, E. Szabo, L. I. Gold, and M. Opas, "Calreticulin, a multi-process calcium-buffering chaperone of the endoplasmic reticulum," *Biochemical Journal*, vol. 417, no. 3, pp. 651–666, 2009.

[132] D. E. Clapham, "Calcium Signaling," *Cell*, vol. 131, no. 6, pp. 1047–1058, 2007.

[133] L. E. V. Del Bem, "The evolutionary history of calreticulin and calnexin genes in green plants," *Genetica*, vol. 139, no. 2, pp. 255–259, 2011.

[134] J. P. Lièvremont, R. Rizzuto, L. Hendershot, and J. Meldolesi, "BiP, a major chaperone protein of the endoplasmic reticulum lumen, plays a direct and important role in the storage of the rapidly exchanging pool of Ca^{2+}," *Journal of Biological Chemistry*, vol. 272, no. 49, pp. 30873–30879, 1997.

[135] S. Persson, M. Rosenquist, K. Svensson, R. Galvão, W. F. Boss, and M. Sommarin, "Phylogenetic analyses and expression studies reveal two distinct groups of calreticulin isoforms in higher plants," *Plant Physiologyogy*, vol. 133, no. 3, pp. 1385–1396, 2003.

[136] A. Christensen, K. Svensson, L. Thelin et al., "Higher plant calreticulins have acquired specialized functions in *Arabidopsis*," *PLoS ONE*, vol. 5, no. 6, p. e11342, 2010.

[137] M. H. Chen, G. W. Tian, Y. Gafni, and V. Citovsky, "Effects of calreticulin on viral cell-to-cell movement," *Plant Physiologyogy*, vol. 138, no. 4, pp. 1866–1876, 2005.

[138] Y. Saito, Y. Ihara, M. R. Leach, M. F. Cohen-Doyle, and D. B. Williams, "Calreticulin functions in vitro as a molecular chaperone for both glycosylated and non-glycosylated proteins," *The EMBO Journal*, vol. 18, no. 23, pp. 6718–6729, 1999.

[139] X. Y. Jia, L. H. He, R. L. Jing, and R. Z. Li, "Calreticulin: conserved protein and diverse functions in plants," *Physiologia Plantarum*, vol. 136, no. 2, pp. 127–138, 2009.

[140] I. L. Conte, N. Keith, C. Gutiérrez-González, A. J. Parodi, and J. J. Caramelo, "The interplay between calcium and the in vitro lectin and chaperone activities of calreticulin," *Biochemistry*, vol. 46, no. 15, pp. 4671–4680, 2007.

[141] A. Christensen, K. Svensson, S. Persson et al., "Functional characterization of *Arabidopsis* calreticulin1a: a key alleviator of endoplasmic reticulum stress," *Plant and Cell Physiology*, vol. 49, no. 6, pp. 912–924, 2008.

[142] H. Jin, Z. Hong, W. Su, and J. Li, "A plant-specific calreticulin is a key retention factor for a defective brassinosteroid receptor in the endoplasmic reticulum," *Proceedings of the National Academy of Sciences of the United States of America*, vol. 106, no. 32, pp. 13612–13617, 2009.

[143] Y. Saijo, N. Tintor, X. Lu et al., "Receptor quality control in the endoplasmic reticulum for plant innate immunity," *The EMBO Journal*, vol. 28, no. 21, pp. 3439–3449, 2009.

[144] Y. Qiu, J. Xi, L. Du, and B. W. Poovaiah, "The function of calreticulin in plant immunity: new discoveries for an old protein," *Plant Signaling & Behaviorv*, vol. 7, no. 8, pp. 907–910, 2012.

[145] Y. Qiu, J. Xi, L. Du, S. Roje, and B. W. Poovaiah, "A dual regulatory role of*Arabidopsis*calreticulin-2 in plant innate immunity," *The Plant Journal*, vol. 69, pp. 489–500, 2011.

[146] J. Li, Z. H. Chu, M. Batoux et al., "Specific ER quality control components required for biogenesis of the plant innate immune receptor EFR," *Proceedings of the National Academy of Sciences of the United States of America*, vol. 106, no. 37, pp. 15973–15978, 2009.

[147] Y. Q. An, R. M. Lin, F. T. Wang, J. Feng, Y. F. Xu, and S. C. Xu, "Molecular cloning of a new wheat calreticulin gene TaCRT1 and expression analysis in plant defense responses and abiotic stress resistance," *Genetics and Molecular Research*, vol. 10, pp. 3576–3585, 2011.

[148] J. L. Caplan, X. Zhu, P. Mamillapalli, R. Marathe, R. Anandalakshmi, and S. P. Dinesh-Kumar, "Induced ER chaperones regulate a receptor-like kinase to mediate antiviral innate immune response in plants," *Cell Host and Microbe*, vol. 6, no. 5, pp. 457–469, 2009.

[149] H. G. Kang, C. S. Oh, M. Sato et al., "Endosome-associated CRT1 functions early in Resistance gene-mediated defense signaling in *Arabidopsis* and tobacco," *Plant Cell*, vol. 22, no. 3, pp. 918–936, 2010.

[150] X. Y. Jia, C. Y. Xu, R. L. Jing et al., "Molecular cloning and characterization of wheat calreticulin (CRT) gene involved in drought-stressed responses," *Journal of Experimental Botany*, vol. 59, no. 4, pp. 739–751, 2008.

[151] I. Hwang, J. F. Harper, F. Liang, and H. Sze, "Calmodulin activation of an endoplasmic reticulum-located calcium pump involves an interaction with the N-terminal autoinhibitory domain," *Plant Physiologyogy*, vol. 122, no. 1, pp. 157–167, 2000.

[152] I. Hwang, H. Sze, and J. F. Harper, "A calcium-dependent protein kinase can inhibit a calmodulin-stimulated Ca^{2+} pump (ACA2) located in the endoplasmic reticulum of *Arabidopsis*," *Proceedings of the National Academy of Sciences of the United States of America*, vol. 97, no. 11, pp. 6224–6229, 2000.

[153] Z. Wu, F. Liang, B. Hong et al., "An endoplasmic reticulum-bound Ca^{2+}/Mn^{2+} pump, ECA1, supports plant growth and confers tolerance to Mn2+ stress," *Plant Physiologyogy*, vol. 130, no. 1, pp. 128–137, 2002.

[154] J. W. Putney, "Recent breakthroughs in the molecular mechanism of capacitative calcium entry (with thoughts on how we got here)," *Cell Calcium*, vol. 42, no. 2, pp. 103–110, 2007.

[155] V. S. Reddy and A. S. N. Reddy, "Proteomics of calcium-signaling components in plants," *Phytochemistry*, vol. 65, no. 12, pp. 1745–1776, 2004.

[156] H. J. Whalley, A. W. Sargeant, J. F. Steele et al., "Transcriptomic analysis reveals calcium regulation of specific promoter motifs in *Arabidopsis*," *Plant Cell*, vol. 23, pp. 4079–4095, 2011.

[157] C. H. Johnson, M. R. Knight, T. Kondo et al., "Circadian oscillations of cytosolic and chloroplastic free calcium in plants," *Science*, vol. 269, no. 5232, pp. 1863–1865, 1995.

[158] P. L. Tsou, S. Y. Lee, N. S. Allen, H. Winter-Sederoff, and D. Robertson, "An ER-targeted calcium-binding peptide confers salt and drought tolerance mediated by CIPK6 in*Arabidopsis*," *Planta*, vol. 235, pp. 539–552, 2011.

[159] D. Winter, B. Vinegar, H. Nahal, R. Ammar, G. V. Wilson, and N. J. Provart, "An "Electronic Fluorescent Pictograph" browser for exploring and analyzing large-scale biological data sets," *PloS ONE*, vol. 2, no. 1, p. e718, 2007.

[160] M. R. Broadley and P. J. White, "Eats roots and leaves. Can edible horticultural crops address dietary calcium, magnesium and potassium deficiencies?" *Proceedings of the Nutrition Society*, vol. 69, no. 4, pp. 601–612, 2010.

[161] S. Conn and M. Gilliham, "Comparative physiology of elemental distributions in plants," *Annals of Botany*, vol. 105, no. 7, pp. 1081–1102, 2010.

[162] S. J. Conn, M. Gilliham, A. Athman et al., "Cell-specific vacuolar calcium storage mediated by CAX1 regulates apoplastic calcium concentration, gas exchange, and plant productivity in *Arabidopsis*," *Plant Cell*, vol. 23, no. 1, pp. 240–257, 2011.

[163] M. Gilliham, M. Dayod, B. J. Hocking et al., "Calcium delivery and storage in plant leaves: exploring the link with water flow," *Journal of Experimental Botany*, vol. 62, no. 7, pp. 2233–2250, 2011.

[164] M. Kerton, H. J. Newbury, D. Hand, and J. Pritchard, "Accumulation of calcium in the centre of leaves of coriander (*Coriandrum sativum* L.) is due to an uncoupling of water and ion transport," *Journal of Experimental Botany*, vol. 60, no. 1, pp. 227–235, 2009.

[165] V. Demidchik, H. C. Bowen, F. J. M. Maathuis et al., "*Arabidopsis thaliana* root non-selective cation channels mediate calcium uptake and are involved in growth," *The Plant Journal*, vol. 32, no. 5, pp. 799–808, 2002.

[166] W. Y. Song, K. S. Choi, A. Alexis de, E. Martinoia, and Y. Lee, "Brassica juncea plant cadmium resistance 1 protein (BjPCR1) facilitates the radial transport of calcium in the root," *Proceedings of the National Academy of Sciences of the United States of America*, vol. 108, pp. 19808–19813, 2011.

[167] I. Baxter, P. S. Hosmani, A. Rus et al., "Root suberin forms an extracellular barrier that affects water relations and mineral nutrition in *Arabidopsis*," *PLoS Genetics*, vol. 5, no. 5, Article ID e1000492, 2009.

[168] P. J. White, "The pathways of calcium movement to the xylem," *Journal of Experimental Botany*, vol. 52, no. 358, pp. 891–899, 2001.

[169] H. Q. Yang and Y. L. Jie, "Uptake and transport of calcium in plants," *Zhi Wu Sheng Li Yu Fen Zi Sheng Wu Xue Xue Bao*, vol. 31, no. 3, pp. 227–234, 2005.

[170] W. Y. Song, Z. B. Zhang, H. B. Shao et al., "Relationship between calcium decoding elements and plant abiotic-stress resistance," *International Journal of Biological Sciences*, vol. 4, no. 2, pp. 116–125, 2008.

[171] I. Baxter, C. Hermans, B. Lahner et al., "Biodiversity of mineral nutrient and trace element accumulation in*Arabidopsis thaliana*," *PLoS ONE*, vol. 7, Article ID e35121, 2012.

[172] Z. Gao, X. He, B. Zhao et al., "Overexpressing a putative aquaporin gene from wheat, TaNIP, enhances salt tolerance in transgenic *Arabidopsis*," *Plant and Cell Physiology*, vol. 51, no. 5, pp. 767–775, 2010.

[173] R. Aroca, R. Porcel, and J. M. Ruiz-Lozano, "Regulation of root water uptake under abiotic stress conditions," *Journal of Experimental Botany*, vol. 63, pp. 43–57, 2012.

[174] S. Han, R. Tang, L. K. Anderson, T. E. Woerner, and Z. M. Pei, "A cell surface receptor mediates extracellular Ca^{2+} sensing in guard cells," *Nature*, vol. 425, no. 6954, pp. 196–200, 2003.

[175] M. Heinlein, "Plasmodesmata: dynamic regulation and role in macromolecular cell-to-cell signaling," *Current Opinion in Plant Biology*, vol. 5, no. 6, pp. 543–552, 2002.

[176] E. Bayer, C. L. Thomas, and A. J. Maule, "Plasmodesmata in *Arabidopsis thaliana* suspension cells," *Protoplasma*, vol. 223, no. 2–4, pp. 93–102, 2004.

[177] F. Baluška, J. Šamaj, R. Napier, and D. Volkmann, "Maize calreticulin localizes preferentially to plasmodesmata in root apex," *The Plant Journal*, vol. 19, no. 4, pp. 481–488, 1999.

[178] E. B. Tucker and W. F. Boss, "Mastoparan-induced intracellular Ca^{2+} fluxes may regulate cell-to-cell communication in plants," *Plant Physiologyogy*, vol. 111, no. 2, pp. 459–467, 1996.

[179] L. C. Cantrill, R. L. Overall, and P. B. Goodwin, "Cell-to-cell communication via plant endomembranes," *Cell Biology International*, vol. 23, no. 10, pp. 653–661, 1999.

[180] H. J. Martens, A. G. Roberts, K. J. Oparka, and A. Schulz, "Quantification of plasmodesmatal endoplasmic reticulum coupling between sieve elements and companion cells using fluorescence redistribution after photobleaching," *Plant Physiologyogy*, vol. 142, no. 2, pp. 471–480, 2006.

[181] D. Guenoune-Gelbart, M. Elbaum, G. Sagi, A. Levy, and B. L. Epel, "Tobacco mosaic virus (TMV) replicase and movement protein function synergistically in facilitating TMV spread by lateral diffusion in the plasmodesmal desmotubule of Nicotiana benthamiana," *Molecular Plant-Microbe Interactions*, vol. 21, no. 3, pp. 335–345, 2008.

[182] T. L. Holdaway-Clarke, N. A. Walker, P. K. Hepler, and R. L. Overall, "Physiological elevations in cytoplasmic free calcium by cold or ion injection result in transient closure of higher plant plasmodesmata," *Planta*, vol. 210, no. 2, pp. 329–335, 2000.

[183] P. K. Hepler, "Calcium: a central regulator of plant growth and development," *Plant Cell*, vol. 17, no. 8, pp. 2142–2155, 2005.

[184] T. Suzuki, S. Nakajima, A. Morikami, and K. Nakamura, "An *Arabidopsis* protein with a novel calcium-binding repeat sequence interacts with TONSOKU/MGOUN3/BRUSHY1 involved in meristem maintenance," *Plant and Cell Physiology*, vol. 46, no. 9, pp. 1452–1461, 2005.

[185] L. Wordeman, "How kinesin motor proteins drive mitotic spindle function: lessons from molecular assays," *Seminars in Cell and Developmental Biology*, vol. 21, no. 3, pp. 260–268, 2010.

[186] Z. Li and S. Komatsu, "Molecular cloning and characterization of calreticulin, a calcium- binding protein involved in the regeneration of rice cultured suspension cells," *European Journal of Biochemistry*, vol. 267, no. 3, pp. 737–745, 2000.

[187] A. J. Karley and P. J. White, "Moving cationic minerals to edible tissues: potassium, magnesium, calcium," *Current Opinion in Plant Biology*, vol. 12, no. 3, pp. 291–298, 2009.

[188] K. D. Hirschi, "Expression of *Arabidopsis* CAX1 in tobacco: altered calcium homeostasis and increased stress sensitivity," *Plant Cell*, vol. 11, no. 11, pp. 2113–2122, 1999.

[189] T. Punshon, K. Hirschi, J. Yang, A. Lanzirotti, B. Lai, and M. L. Guerinot, "The role of CAX1 and CAX3 in elemental distribution and abundance in*Arabidopsis*seed," *Plant Physiology*, vol. 158, pp. 352–362, 2011.

[190] J. K. Pittman and K. D. Hirschi, "Regulation of CAX1, an *Arabidopsis* Ca^{2+}/H^+ antiporter. Identification of an N-terminal autoinhibitory domain," *Plant Physiologyogy*, vol. 127, no. 3, pp. 1020–1029, 2001.

[191] S. Park, T. S. Kang, C. K. Kim et al., "Genetic manipulation for enhancing calcium content in potato tuber," *Journal of Agricultural and Food Chemistry*, vol. 53, no. 14, pp. 5598–5603, 2005.

[192] C. K. Kim, J. S. Han, H. S. Lee et al., "Expression of an *Arabidopsis* CAX2 variant in potato tubers increases calcium levels with no accumulation of manganese," *Plant Cell Reports*, vol. 25, no. 11, pp. 1226–1232, 2006.

[193] S. Park, N. H. Cheng, J. K. Pittman et al., "Increased calcium levels and prolonged shelf life in tomatoes expressing *Arabidopsis* H^+/Ca^{2+} transporters," *Plant Physiologyogy*, vol. 139, no. 3, pp. 1194–1206, 2005.

[194] H. Mei, N. H. Cheng, J. Zhao et al., "Root development under metal stress in *Arabidopsis thaliana* requires the H+/cation antiporter CAX4," *New Phytologistogist*, vol. 183, no. 1, pp. 95–105, 2009.

[195] N. H. Cheng, J. K. Pittman, B. J. Barkla, T. Shigaki, and K. D. Hirschi, "The *Arabidopsis* cax1 mutant exhibits impaired ion homeostasis, development, and hormonal responses and reveals interplay among vacuolar transporters," *Plant Cell*, vol. 15, no. 2, pp. 347–364, 2003.

[196] N. H. Cheng, J. K. Pittman, J. K. Zhu, and K. D. Hirschi, "The protein kinase SOS_2 activates the *Arabidopsis* H^+/Ca^{2+} antiporter CAX1 to integrate calcium transport and salt tolerance," *Journal of Biological Chemistry*, vol. 279, no. 4, pp. 2922–2926, 2004.

[197] J. Zhao, B. J. Barkla, J. Marshall, J. K. Pittman, and K. D. Hirschi, "The *Arabidopsis* cax3 mutants display altered salt tolerance, pH sensitivity and reduced plasma membrane H^+-ATPase activity," *Planta*, vol. 227, no. 3, pp. 659–669, 2008.

[198] Q. Wu, T. Shigaki, K. A. Williams et al., "Expression of an *Arabidopsis* Ca^{2+}/H^+ antiporter CAX1 variant in petunia enhances cadmium tolerance and accumulation," *Journal of Plant Physiologyogy*, vol. 168, no. 2, pp. 167–173, 2011.

[199] T. Shigaki, H. Mei, J. Marshall, X. Li, M. Manohar, and K. D. Hirschi, "The expression of the open reading frame of *Arabidopsis* CAX1, but not its cDNA, confers metal tolerance in yeast," *Plant Biology*, vol. 12, no. 6, pp. 935–939, 2010.

[200] V. Koren'kov, S. Park, N. H. Cheng et al., "Enhanced Cd^{2+}-selective root-tonoplast-transport in tobaccos expressing *Arabidopsis* cation exchangers," *Planta*, vol. 225, no. 2, pp. 403–411, 2007.

[201] V. Korenkov, B. King, K. Hirschi, and G. J. Wagner, "Root-selective expression of *AtCAX4* and *AtCAX2* results in reduced lamina cadmium in field-grown *Nicotiana tabacum* L.," *Plant Biotechnology Journal*, vol. 7, no. 3, pp. 219–226, 2009.

[202] T. Y. Liu, K. Aung, C. Y. Tseng, T. Y. Chang, Y. S. Chen, and T. J. Chiou, "Vacuolar Ca^{2+}/H^+ transport activity is required for systemic phosphate homeostasis involving shoot-to-root signaling in *Arabidopsis*," *Plant Physiologyogy*, vol. 156, no. 3, pp. 1176–1189, 2011.

[203] P. C. Dekock, D. Vaughan, a. Hall, and C. Ord, "Biochemical studies on blossom end rot [caused mainly by calcium deficiency] of tomatoes," *Plant Physiology*, vol. 48, pp. 312–316, 1980.

[204] H. E. Johnson, D. Broadhurst, R. Goodacre, and A. R. Smith, "Metabolic fingerprinting of salt-stressed tomatoes," *Phytochemistry*, vol. 62, no. 6, pp. 919–928, 2003.

[205] M. D. Taylor and S. J. Locascio, "Blossom-end rot: a calcium deficiency," *Journal of Plant Nutrition*, vol. 27, no. 1, pp. 123–139, 2004.

[206] S. T. de Freitas, M. Padda, Q. Wu, S. Park, and E. J. Mitcham, "Dynamic alternations in cellular and molecular components during blossom-end rot development in tomatoes expressing sCAX1, a constitutively active Ca^{2+}/H^+ antiporter from *Arabidopsis*," *Plant Physiologyogy*, vol. 156, no. 2, pp. 844–855, 2011.

[207] G. Z. Luo, H. W. Wang, J. Huang et al., "A putative plasma membrane cation/proton antiporter from soybean confers salt tolerance in *Arabidopsis*," *Plant Molecular Biology*, vol. 59, no. 5, pp. 809–820, 2005.

[208] K. H. Krause and M. Michalak, "Calreticulin," *Cell*, vol. 88, no. 4, pp. 439–443, 1997.

[209] P. D. Nash, M. Opas, and M. Michalak, "Calreticulin: not just another calcium-binding protein," *Molecular and Cellular Biochemistry*, vol. 135, no. 1, pp. 71–78, 1994.

[210] J. Meldolesi, K. H. Krause, and M. Michalak, "Calreticulin: how many functions in how many cellular compartments?" *Cell Calcium*, vol. 20, no. 1, pp. 83–86, 1996.

[211] M. Michalak, P. Mariani, and M. Opas, "Calreticulin, a multifunctional Ca^{2+} binding chaperone of the endoplasmic reticulum," *Biochemistry and Cell Biology*, vol. 76, no. 5, pp. 779–785, 1998.

[212] M. S. Kwon, C. S. Park, K. R. Choi et al., "Calreticulin couples calcium release and calcium influx in integrin- mediated calcium signaling," *Molecular Biology of the Cell*, vol. 11, no. 4, pp. 1433–1443, 2000.

[213] P. B. Simpson, S. Mehotra, D. Langley, C. A. Sheppard, and J. T. Russell, "Specialized distributions of mitochondria and endoplasmic reticulum proteins define Ca^{2+} wave amplification sites in cultured astrocytes," *Journal of Neuroscience Research*, vol. 52, pp. 672–683, 1998.

[214] M. Opas, M. Szewczenko-Pawlikowski, G. K. Jass, N. Mesaeli, and M. Michalak, "Calreticulin modulates cell adhesiveness via regulation of vinculin expression," *Journal of Cell Biology*, vol. 135, no. 6, pp. 1913–1923, 1996.

[215] L. Perrone, G. Tell, and R. Di Lauro, "Calreticulin enhances the transcriptional activity of thyroid transcription factor-1 by binding to its homeodomain," *Journal of Biological Chemistry*, vol. 274, no. 8, pp. 4640–4645, 1999.

[216] H. Liu, R. C. Bowes, B. Van De Water, C. Sillence, J. F. Nagelkerke, and J. L. Stevens, "Endoplasmic reticulum chaperones GRP78 and calreticulin prevent oxidative stress, Ca^{2+} disturbances, and cell death in renal epithelial cells," *Journal of Biological Chemistry*, vol. 272, no. 35, pp. 21751–21759, 1997.

[217] L. Mery, N. Mesaeli, M. Michalak, M. Opas, D. P. Lew, and K. H. Krause, "Overexpression of calreticulin increases intracellular

Ca^{2+} storage and decreases store-operated Ca^{2+} influx," *Journal of Biological Chemistry*, vol. 271, no. 16, pp. 9332–9339, 1996.

[218] H. L. Roderick, D. H. Llewellyn, A. K. Campbell, and J. M. Kendall, "Role of calreticulin regulating intracellular Ca^{2+} storage and capacitative Ca^{2+} entry in HeLa cells," *Cell Calcium*, vol. 24, no. 4, pp. 253–262, 1998.

[219] C. Fasolato, P. Pizzo, and T. Pozzan, "Delayed activation of the store-operated calcium current induced by calreticulin overexpression in RBL-1 cells," *Molecular Biology of the Cell*, vol. 9, no. 6, pp. 1513–1522, 1998.

[220] J. Denecke, L. E. Carisson, S. Vidal et al., "The tobacco homolog of mammalia calreticulin is present in protein complexes in vivo," *Plant Cell*, vol. 7, pp. 391–406, 1995.

[221] C. Bastianutto, E. Clementi, F. Codazzi et al., "Overexpression of calreticulin increases the Ca^{2+} capacity of rapidly exchanging Ca^{2+} stores and reveals aspects of their lumenal microenvironment and function," *Journal of Cell Biology*, vol. 130, no. 4, pp. 847–855, 1995.

[222] S. Persson, S. E. Wyatt, J. Love, W. F. Thompson, D. Robertson, and W. F. Boss, "The Ca^{2+} status of the endoplasmic reticulum is altered by induction of calreticulin expression in transgenic plants," *Plant Physiologyogy*, vol. 126, no. 3, pp. 1092–1104, 2001.

[223] S. E. Wyatt, P. L. Tsou, and D. Robertson, "Expression of the high capacity calcium-binding domain of calreticulin increases bioavailable calcium stores in plants," *Transgenic Research*, vol. 11, no. 1, pp. 1–10, 2002.

[224] F. Brandizzi, S. Hanton, L. L. Pinto DaSilva et al., "ER quality control can lead to retrograde transport from the ER lumen to the cytosol and the nucleoplasm in plants," *The Plant Journal*, vol. 34, no. 3, pp. 269–281, 2003.

[225] Z. L. Jin, K. H. Joon, A. Y. Kyung et al., "Over-expression of Chinese cabbage calreticulin 1, BrCRT1, enhances shoot and root regeneration, but retards plant growth in transgenic tobacco," *Transgenic Research*, vol. 14, no. 5, pp. 619–626, 2005.

[226] S. Y. Lee, *The involvement of ER calcium in abiotic stress tolerance [Ph.D. thesis]*, 2010.

[227] U. Halfter, M. Ishitani, and J. K. Zhu, "The *Arabidopsis* SOS$_2$ protein kinase physically interacts with and is activated by the calcium-binding protein SOS$_3$," *Proceedings of the National Academy of Sciences of the United States of America*, vol. 97, no. 7, pp. 3735–3740, 2000.

[228] M. Ishitani, J. Liu, U. Halfter, C. S. Kim, W. Shi, and J. K. Zhu, "SOS$_3$ function in plant salt tolerance requires N-myristoylation and calcium binding," *Plant Cell*, vol. 12, no. 9, pp. 1667–1677, 2000.

[229] D. Gong, Y. Guo, K. S. Schumaker, and J. K. Zhu, "The SOS$_3$ family of calcium sensors and SOS$_2$ family of protein kinases in *Arabidopsis*," *Plant Physiologyogy*, vol. 134, no. 3, pp. 919–926, 2004.

[230] M. Nieves-Cordones, F. Caballero, V. Martinez, and F. Rubio, "Disruption of the *Arabidopsis thaliana* inward-rectifier K$^+$ channel AKT1 improves plant responses to water stress," *Plant and Cell Physiology*, vol. 53, pp. 423–432, 2012.

[231] P. H. McCord, *Genetic, genomic, and transgenic approaches to understand internal heat necrosis in potato [Ph.D. thesis]*, 2009.

[232] G. C. Yencho, P. H. McCord, K. G. Haynes, and S. B. R. Sterrett, "Internal heat necrosis of potato—a review," *American Journal of Potato Research*, vol. 85, no. 1, pp. 69–76, 2008.

[233] W. S. Blair, H. P. Bogerd, S. J. Madore, and B. R. Cullen, "Mutational analysis of the transcription activation domain of RelA: identification of a highly synergistic minimal acidic activation module," *Molecular and Cellular Biology*, vol. 14, no. 11, pp. 7226–7234, 1994.

[234] S. T. de Freitas, A. K. Handa, Q. Wu, S. Park, and E. J. Mitcham, "Role of pectin methylesterases in cellular calcium distribution and blossom-end rot development in tomato fruit," *The Plant Journal*, vol. 71, pp. 824–835, 2012.

[235] Q. Wu, T. Shigaki, J. S. Han, C. K. Kim, K. D. Hirschi, and S. Park, "Ectopic expression of a maize calreticulin mitigates calcium deficiency-like disorders in sCAX1-expressing tobacco and tomato," *Plant Molecular Biology*, vol. 80, pp. 609–619, 2012.

[236] R. Reid and J. Hayes, "Mechanisms and control of nutrient uptake in plants," *International Review of Cytology*, vol. 229, pp. 73–114, 2003.

[237] S. J. Conn, P. Berninger, M. R. Broadley, and M. Gilliham, "Exploiting natural variation to uncover candidate genes that control element accumulation in *Arabidopsis thaliana*," *New Phytologistogist*, vol. 193, pp. 859–866, 2012.

[238] B. J. Heyen, M. K. Alsheikh, E. A. Smith, C. F. Torvik, D. F. Seals, and S. K. Randall, "The calcium-binding activity of a vacuole-associated, dehydrin-like protein is regulated by phosphorylation," *Plant Physiologyogy*, vol. 130, no. 2, pp. 675–687, 2002.

[239] K. Yuasa and M. Maeshima, "Purification, properties, and molecular cloning of a novel Ca^{2+}-binding protein in radish vacuoles," *Plant Physiologyogy*, vol. 124, no. 3, pp. 1069–1078, 2000.

[240] B. Dadacz-Narloch, D. Beyhl, C. Larisch et al., "A novel calcium binding site in the slow vacuolar cation channel TPC1 senses luminal calcium levels," *Plant Cell*, vol. 23, pp. 2696–2707, 2011.

[241] J. J. Rios, S. O. Lochlainn, J. Devonshire et al., "Distribution of calcium (Ca) and magnesium (Mg) in the leaves of Brassica rapa under varying exogenous Ca and Mg supply," *Annals of Botany*, vol. 109, pp. 1081–1089, 2012.

[242] E. L. Connolly, "Raising the bar for biofortification: enhanced levels of bioavailable calcium in carrots," *Trends in Biotechnology*, vol. 26, no. 8, pp. 401–403, 2008.

[243] S. Park, M. P. Elless, J. Park et al., "Sensory analysis of calcium-biofortified lettuce," *Plant Biotechnology Journal*, vol. 7, no. 1, pp. 106–117, 2009.

9

Effect of Seed Size and Pretreatment Methods on Germination of *Albizia lebbeck*

Edward Missanjo, Chikumbutso Maya, Dackious Kapira, Hannah Banda, and Gift Kamanga-Thole

Malawi College of Forestry and Wildlife, Private Bag 6, Dedza, Malawi

Correspondence should be addressed to Edward Missanjo; edward.em2@gmail.com

Academic Editors: P. Parolin and S. Satoh

Albizia lebbeck is a multipurpose tree species prioritised for conservation in Malawi. The different plant parts are used in traditional medicine to treat different diseases. However, the seeds are dormant, and the tree species remain undomesticated. A study was conducted to evaluate the effect of seed size and presowing on the germination of *Albizia lebbeck* in a nursery. Seeds were grouped into four categories in regard to their length, small (≤0.5 cm), medium (> 0.5 < 0.8 cm), large (≥0.8 cm), and mixture of small, medium, and large seeds. The seeds were subjected to five main seed pretreatment methods, namely, soaking in sulphuric acid for 2 minutes, nicking, soaking in hot water for 5 minutes, soaking in cold water for 24 hours, and control where seeds were sown without any treatment. The results indicate that combination of nicking and large seeds produced the highest (100%) germination. Hot water treatment was effective in large seeds producing 67.5% germination. The increased germination for mechanically scarified seeds through nicking suggests that seed dormancy in *Albizia lebbeck* is mainly due to its hard seed coat. Therefore, it is recommended to farmers to adopt use of nicking and large seeds, since it is safe and effective.

1. Introduction

Albizia lebbeck is widely spread in the world, and its tree has large leaves and fragrant cluster of green-yellow flowers and long seed pods. Belonging to the family of Leguminosae [1], it is native to tropical Asia and widely cultivated and naturalized in other tropical and subtropical regions including Malawi [2, 3]. *Albizia lebbeck* grows to the height of 18–30 m with a trunk diameter of 50 cm to 1 m at maturity. The leaves are 7 to 15 cm long with one to four pairs of pinnae, and each pinna has 6 to 8 leaflets. The flowers are white with numerous stamens and very fragrant. The fruit pods are 15 to 30 cm long and 2.5 to 5.0 cm broad containing six to twelve seeds [4].

The flowers, bark, fruits, roots, and stems of *Albizia lebbeck* are all used for medicine. A paste of leaves is used to treat skin problems. *Albizia lebbeck* is also known for treating respiratory problems including allergies [5]. Furthermore, other parts of the plants are used to treat eye problem, purify blood, and promote health in teeth. Most importantly, ethanol extract from its pods is effective against some form of

cancer [6]. The leaves are nutritious as they contain proteins, calcium, phosphorous, and amino acids [2, 4]. *Albizia lebbeck* is one of the most promising fodder trees. It has leaves during a large part of the rainy season, and digestibility of the twigs is considerably higher than that of most fodder trees. The concentration of crude protein is about 20% for green leaves, 13% for leaf litter, and 10% for twigs. *In vitro* digestibility is about 45% for mature leaves, 70% for young leaves and 40% for twigs. Leaves, flowers, and pods fall to the ground gradually during the dry season and can be browsed on the ground [3]. It is an excellent fuelwood and charcoal species, and the wood is suitable for construction, furniture, and veneer. The shallow root system makes it a good soil binder and recommendable for soil conservation and erosion control [5, 6].

Despite its importance, the species is becoming scarce in Malawi due to deep seed dormancy. The tree species has been given priority as one of the species for conservation in Malawi to enhance its contribution to health and livelihood of communities. From this point of view, a study was carried out to assess the effects of different seed pretreatment methods and seed size on germination of *Albizia lebbeck*.

2. Materials and Methods

2.1. Study Site. The study was conducted in Malawi located in Southern Africa in the tropical savanna region at Malawi College of Forestry and Wildlife (MCFW) nursery (14°19′S, 34°17′E, and 1591 m above sea level). MCFW receives 1200 mm to 1800 mm rainfall per annum, with annual temperature ranging from 7°C to 25°C. It is situated about 85 km southeast of Lilongwe, the capital.

2.2. Experimental Design and Treatments. A total of 1600 seeds were directly sawn in 10 cm polythene tubes, and one seed was planted per tube. The seeds were subjected to twenty treatments which were completely randomised in four replicates. Each treatment had 20 seeds. The treatment combination consisted of two factors, namely, seed size and pretreatment methods. The first factor of seed size consisted of seeds of length less than 0.5 cm and was denoted as small; 0.51 cm to 0.8 cm seeds were considered medium; seeds of 0.8 cm long and greater were categorized large. A fourth seed size category was a control which consisted of three dimensions (small, large, and medium) of seeds. The second factor of pretreatment methods involved five seed pretreatment methods, namely, cold water soaking for 24 hours, hot water soaking for 5 minutes, and 2 minute immersion in concentrated sulphuric acid (0.3 M H_2SO_4), including mechanical scarification by nicking, and a fifth treatment was a control which consisted of seeds that were left intact.

2.3. Pretreatment Procedure

2.3.1. Cold and Hot Water Treatment. Twenty seeds from each seed size category (small, medium, large, and mixture) were put in similar beaker sizes where cold water at room temperature was poured and the seeds were soaked for 24 hours. Water was then removed and the seeds were planted on the same day. For the hot water treatment, water was heated to approximately 100°C and was then poured into beakers containing twenty seeds from each seed category (small, medium, large, and mixture) and was left to stand for 5 minutes after which the seeds were sown.

2.3.2. Immersion in Concentrated Sulphuric Acid. Seeds of small, medium, large, and mixture seed categories were put into separate beakers. Concentrated sulphuric acid (0.3 M H_2SO_4) was then added to the beakers each containing twenty seeds and were left to soak for 2 minutes. After immersion, the solution was drained off, and seeds were repeatedly rinsed in running tap water until considered safe to handle. Then the seeds were sown.

2.3.3. Nicking. Twenty seeds from each size category were mechanically nicked on one side away from the micropyle using secateurs and then sown immediately. Watering was

done accordingly to keep the beds with adequate moisture. In total, there were twenty treatment combinations and were denoted as follows:

T1: small seeds immersed in 0.3 M sulphuric acid (H_2SO_4) for 2 minutes;

T2: small seeds with nicking;

T3: small seeds soaked in hot water at 100°C for 5 minutes;

T4: small seeds soaked in cold water at room temperature for 24 hours;

T5: small seeds sown without pretreatment;

T6: large seeds immersed in 0.3 M sulphuric acid (H_2SO_4) for 2 minutes;

T7: large seeds with nicking;

T8: large seeds soaked in hot water at 100°C for 5 minutes;

T9: large seeds soaked in cold water at room temperature for 24 hours;

T10: large seeds sown without pretreatment;

T11: medium seeds immersed in 0.3 M sulphuric acid (H_2SO_4) for 2 minutes;

T12: medium seeds with nicking;

T13: medium seeds soaked in hot water at 100°C for 5 minutes;

T14: medium seeds soaked in cold water at room temperature for 24 hours;

T15: medium seeds sown without pretreatment;

T16: mixture of seeds immersed in 0.3 M sulphuric acid (H_2SO_4) for 2 minutes;

T17: mixture of seeds with nicking;

T18: mixture of seeds soaked in hot water at 100°C for 5 minutes;

T19: mixture of seeds soaked in cold water at room temperature for 24 hours;

T20: mixture of seeds sown without pretreatment.

2.3.4. Data Collection and Analysis. Data on germination were recorded on daily basis for a period of eight weeks (56 days) from the day of sowing. Germination was defined as the emergence of radicle from the seed coat. Daily germination percentages were summed up to obtain cumulative germination for each treatment. Data obtained was subjected to analysis of variance (ANOVA) using GenStat for Windows, version 13 [7]. Differences between treatment means were separated using Fischer's least significant difference (LSD) at the 0.05 level. The data was analysed using the following model:

$$Y_{ijk} = \mu + S_i + P_j + (SP)_{ij} + e_{ijk}, \tag{1}$$

where Y_{ijk} is the response variable (germination percentage) of *j*th observation in *i*th treatments, μ is the overall mean, S_i is the fixed effect of seed size ($i = 1, 2, 3, 4$), P_j is the fixed effect of pretreatment methods ($j = 1, 2, 3, 4, 5$),

TABLE 1: Effect of pretreatment methods and size of *Albizia lebbeck* seeds on germination at eight (8) weeks after sowing.

Seed size	Germination percentage (%) for different presowing treatment methods					
	H_2SO_4	Nicking	Hot water	Cold water	Control	Mean
Large	30.0	100	67.5	22.5	22.5	48.5
Medium	54.5	80.0	43.8	28.8	21.2	45.7
Small	49.5	71.2	55.0	10.0	20.0	41.1
Mixture	37.5	70.0	50.0	41.0	33.8	46.5
Mean	42.9[b]	80.3[a]	54.1[b]	25.6[c]	24.4[c]	

Note. Means with different superscripts within a row differ ($P < 0.001$).
SE = 0.243.
LSD = 15.7.
CV = 11.1%.

$(SP)_{ij}$ is the effect of the interaction between seed size and pretreatment methods, and e_{ijk} is the random residual effect, $e_{ijk} \sim N(0, \sigma e^2)$.

3. Results

There were no significant ($P > 0.05$) differences in germination between seed sizes, although larger seeds had a higher germination (48.5%) percentage followed by mixture of seeds (46.5%) and then medium seeds (45.7%). Small seeds had an average of 41.1% germination. However, there were significant ($P < 0.001$) differences in germination among presowing treatments where nicking gave the highest germination (80.3%) followed by immersion in hot water (54.1%) and then immersion in 0.3 M H_2SO_4 acid with 42.9% germination. The combination of nicking and large seeds produced the highest (100%) germination followed by the combination of soaking in hot water and large seeds which produced 67.5% germination (Table 1).

When germination was observed over time, nicking in all the seed size category had the highest rate of germination in the first two weeks, then the rate of germination became constant. However, in hot water treatments germination increased rapidly between the first to sixth week after which it remained constant (Figure 1).

4. Discussion

The results were not significantly different among seed sizes. However, large seeds produced higher germination than other seed size categories. The present findings are in agreement with those reported by [8–10]. Esen et al. [11] reported that large and heavy seeds contain larger amounts of reserves to stimulate germination, seedling survival, and growth.

The results obtained in this study entail the vital role of pretreating *Albizia lebbeck* seeds prior to sowing for enhanced germination and domestication of the species. Germination percentage varied among different pretreatment methods. High seed germination percentage for nicked seeds suggests that this is the best method to be applied before sowing *Albizia lebbeck* seeds. The results reported in this study agreed to those in literature [9, 12–15], in which nicking has been

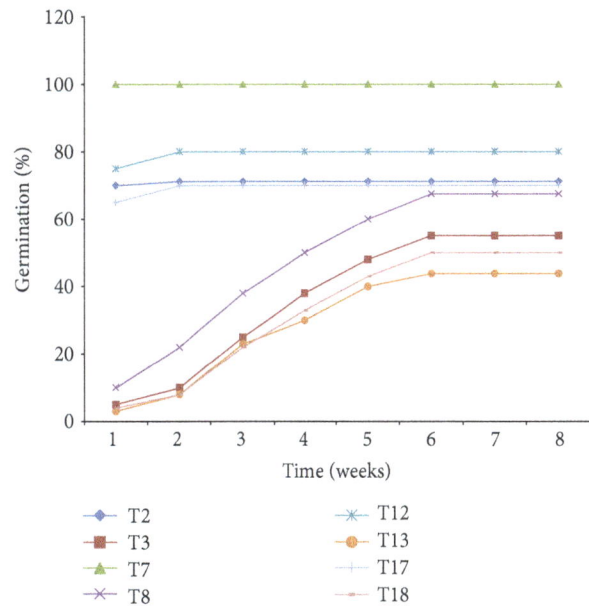

FIGURE 1: Effect of pretreatment methods (nicking and hot water immersion) and size of *Albizia lebbeck* seeds on maximum germination.

shown to enhance germination of different tropical forest tree species. Nicking has been found to be extremely effective for most species, because cracks or cuts made on the seed makes it easier for entry of water and exchange of gases resulting in enzymatic hydrolysis and thus transforming the embryo into a seedling [16–18].

Hot water treatment emerged the second best pre-sowing seed treatment method, producing higher germination percentage of 67.5% for large seeds of *Albizia lebbeck*. Soaking seed in hot water may soften hard seed coats, this makes the seed coats permeable to water and the seeds imbibe and swell as the water cools [9, 19]. Likoswe et al. [15] showed that soaking seeds in hot water leaches out chemical inhibitors resulting in breakage of chemical seed dormancy.

Soaking in sulphuric acid came third with its germination percentage of 41.9%. Whenever seeds are sufficiently soaked in sulphuric acid, it results into over 90% germination percentage [4, 17, 19–21]. Insufficient soaking may not be effective enough as it just makes the seed coat glossy. Furthermore, concentration of the acid and time of exposure are very critical and need to be quantified for each species since seeds exposed for a long time get damaged easily [19, 22]. In this study only one concentration and time of exposure to the sulphuric acid was used.

Less effectiveness of cold water method is due to limited time of soaking. Though the seed coats were softened but the period was not enough to maximize germination percentage and reduce lengthy germination period [12, 23, 24]. Duration more than 24 hours is needed; 48 hours or more per se is required for more effective results [15].

5. Conclusion

Seed pretreatment methods and seed size affected the germination of *Albizia lebbeck*. The larger seeds resulted in higher germination percentage since larger seeds contain more food reserves to support germination. Nicking has shown to be the overall best presowing seed treatment method in *Albizia lebbeck* followed by hot water treatment. Therefore, this study recommends to farmers to adopt use of nicking and large seeds, since it is safe and effective.

Acknowledgments

The authors thank Mr. Anderson Ndema and his colleagues of Malawi College of Forestry and Wildlife (MCFW) for providing them with polythene tubes that were used in the study and also for allowing them to conduct the experiment at MCFW nursery.

References

[1] G. S. Bhat and P. S. Chauhan, "Provenance variation in seed and seedling traits of *Albizia lebbeck* Benth," *Journal of Tree Science*, vol. 21, pp. 52–57, 2002.

[2] M. Faisal, P. P. Singh, and R. Irchhaiya, "Review on *Albizia lebbeck* a potent herbal drug," *International Journal of Pharmaceutics*, vol. 3, no. 5, pp. 63–68, 2012.

[3] D. Jøker, *Seed Leaflet*, Albizia lebbeck *(L.) Benth*, Danida Forest Seed Centre, Humlebaek, Denmark, 2000.

[4] H. P. Msanga, *Seed Germination of Indigenous Trees in Tanzania*, UBC Press, Vancouver, Canada, 2000.

[5] N. Khera and R. P. Singh, "Germination of some multipurpose tree species in five provenances in response to variation in light, temperature, substrate and water stress," *Tropical Ecology*, vol. 46, no. 2, pp. 203–217, 2005.

[6] M. Tigabu and P. C. Odén, "Effect of scarification, gibberellic acid and temperature on seed germination of two multipurpose Albizia species from Ethiopia," *Seed Science and Technology*, vol. 29, no. 1, pp. 11–20, 2001.

[7] R. W. Payne, D. A. Murray, S. A. Harding, D. B. Baird, and D. M. Soutar, *GenStat for Windows Introduction*, VSN International, Hemel Hempstead, UK, 13th edition, 2010.

[8] E. Khurana and J. S. Singh, "Germination and seedling growth of five tree species from tropical dry forest in relation to water stress: impact of seed size," *Journal of Tropical Ecology*, vol. 20, no. 4, pp. 385–396, 2004.

[9] W. F. Mwase and T. Mvula, "Effect of seed size and pre-treatment methods of *Bauhinia thonningii* Schum. on germination and seedling growth," *African Journal of Biotechnology*, vol. 10, no. 26, pp. 5143–5148, 2011.

[10] E. N. Chidumayo, "Growth responses of an African savanna tree, *Bauhinia thonningii* Schumacher, to defoliation, fire and climate," *Trees*, vol. 21, no. 2, pp. 231–238, 2007.

[11] D. Esen, O. Yildiz, M. Sarginci, and K. Isik, "Effects of different pretreatments on germination of *Prunus serotina* seed sources," *Journal of Environmental Biology*, vol. 28, no. 1, pp. 99–104, 2007.

[12] S. Azad, M. R. Manik, S. Hasan, and A. Matin, "Effect of different pre-sowing treatments on seed germination percentage and growth performance of *Acacia auriculiformis*," *Journal of Forestry Research*, vol. 22, no. 2, pp. 183–188, 2011.

[13] B. M. Khan, B. Koirala, and M. K. Hossian, "Effect of different pre-sowing treatments on germination and seedling growth attributes in Ghora Neem (*Melia azedarach*)," *Malaysian Forest*, vol. 64, no. 1, pp. 14–121, 2001.

[14] B. Koirala, M. K. Hossain, and M. S. Hossain, "Effects of different pre-sowing treatments on *Adenanthera pavonia* seeds and initial seedling development in the nursery," *Malaysian Forest*, vol. 63, no. 2, pp. 82–91, 2000.

[15] M. G. Likoswe, J. P. Njoloma, W. F. Mwase, and C. Z. Chilima, "Effect of seed collection times and pretreatment methods on germination of *Terminalia sericea* Burch. ex DC," *African Journal of Biotechnology*, vol. 7, no. 16, pp. 2840–2846, 2008.

[16] B. E. Ayisire, L. A. Akinro, and S. O. Amoo, "Seed germination and in vitro propagation of *Piliostigma thonningii*—an important medicinal plant," *African Journal of Biotechnology*, vol. 8, no. 3, pp. 401–404, 2009.

[17] M. Alamgir and M. K. Hossain, "Effect of pre-sowing treatments on germination and initial seedling development of *Albizia saman* in the nursery," *Journal of Forestry Research*, vol. 16, no. 3, pp. 200–2204, 2005.

[18] M. S. Azad, N. K. Paul, and M. A. Matin, "Do pre-sowing treatments affect seed germination in Albizia richardiana and Lagerstroemia speciosa?" *Frontiers of Agriculture in China*, vol. 4, no. 2, pp. 181–184, 2010.

[19] M. S. Azad, R. K. Biswas, and M. A. Matin, "Seed germination of *Albizia procera* (Roxb.) Benth. in Bangladesh: a basis for seed source variation and pre-sowing treatment effect," *Forestry Studies in China*, vol. 14, no. 2, pp. 124–1130, 2012.

[20] T. Merou, I. Takos, E. Konstantinidou, S. Galatsidas, and G. Varsamis, "Effect of different pretreatment methods on germination of *Albizia julibrissin* seeds," *Seed Science and Technology*, vol. 39, no. 1, pp. 248–252, 2011.

[21] M. S. Azad, M. Zedan-Al-Musa, and M. A. Matin, "Effects of pre-sowing treatments on seed germination of *Melia azedarach*," *Journal of Forestry Research*, vol. 21, no. 2, pp. 193–196, 2010.

[22] L. Schmidt, *Dormancy and Pre-Treatment. Guide To Handling of Tropical and subTropical Forest Seeds*, Danida Forest Seed Centre, Humlebaek, Denmark, 2000.

[23] B. M. Khan, B. Koirala, and M. K. Hossian, "Effect of different pre-sowing treatments on germination and seedling growth attributes in Ghora Neem (*Melia azedarach*L.)," *Malaysian Forest*, vol. 64, no. 1, pp. 14–121, 2001.

[24] F. O. Jimoh and A. T. Oladiji, "Preliminary studies on *Piliostigma thonningii* seeds: proximate analysis, mineral composition and phytochemical screening," *African Journal of Biotechnology*, vol. 4, no. 12, pp. 1439–1442, 2005.

10

Highlights in Seagrasses' Phylogeny, Physiology, and Metabolism: What Makes Them Special?

Jutta Papenbrock

Institute of Botany, Leibniz University Hannover, Herrenhäuser Straße 2, 30419 Hannover, Germany

Correspondence should be addressed to Jutta Papenbrock, jutta.papenbrock@botanik.uni-hannover.de

Academic Editors: M. Kwaaitaal and I. Zarra

The marine seagrasses form an ecological and therefore paraphyletic group of marine hydrophilus angiosperms which evolved three to four times from land plants towards an aquatic and marine existence. Their taxonomy is not yet solved on the species level and below due to their reduced morphology. So far also molecular data did not completely solve the phylogenetic relationships. Thus, this group challenges a new definition for what a species is. Also their physiology is not well understood due to difficult experimental *in situ* and *in vitro* conditions. There remain several open questions concerning how seagrasses adapted secondarily to the marine environment. Here probably exciting adaptation solutions will be detected. Physiological adaptations seem to be more important than morphological ones. Seagrasses contain several compounds in their secondary metabolism in which they differ from terrestrial plants and also not known from other taxonomic groups. Some of these compounds might be of interest for commercial purposes. Therefore their metabolite contents constitute another treasure of the ocean. This paper gives an introduction into some of the most interesting aspects from phylogenetical, physiological, and metabolic points of view.

1. Introduction

Seagrasses are a paraphyletic group of marine hydrophilus angiosperms which evolved three to four times from land plants back to the sea. The following characteristics can be used to define a seagrass species. It lives in an estuarine or in the marine environment, and nowhere else. The pollination takes place underwater with specialized pollen. The seeds which are dispersed by both biotic and abiotic agents are produced underwater. The seagrass species have specialized leaves with a reduced cuticle, an epidermis which lacks stomata and is the main photosynthetic tissue. The rhizome or underground stem is important in anchoring. The roots can live in an anoxic environment and depend on oxygen transport from the leaves and rhizomes but are also important in the nutrient transfer processes [1]. Seagrasses profoundly influence the physical, chemical, and biological environments of coastal waters. Though seagrasses provide invaluable ecosystem services by acting as breeding and nursery ground for a variety of organisms and promote commercial fisheries, many aspects of their physiology are not well investigated. Several studies have indicated that seagrass habitat is declining worldwide [2]. Ten seagrass species are at elevated risk of extinction (14% of all seagrass species) with three species qualifying as endangered. Seagrass loss and degradation of seagrass biodiversity will have serious repercussions for marine biodiversity and the human population that depends upon the resources and ecosystem services that seagrasses provide [3]. This paper aims to highlight some fascinating and sometimes hidden aspects of seagrass physiology and their metabolites and focuses on the distinctiveness of seagrasses also from an evolutionary point of view. Maybe it encourages protecting the invaluable ecosystem of seagrass meadows.

2. Origin of Seagrasses

Seagrasses do not represent the link between marine algae and terrestrial higher plants. This unique ecological group represents the "whales" of the plant kingdom. Similarly to whales seagrasses returned to sea and secondarily colonized marine habitats [4]. The recolonization occurred exclusively from the monocot order Alismatales and appears to be evolutionary unique. The adaption to marine environments is regarded as an analogous adaption and took place at least three times independently during the evolution of seagrasses [5]. Only four families of higher plants, Posidoniaceae, Cymodoceaceae, Hydrocharitaceae, and Zosteraceae, contain exclusively marine species [6] (Figure 1; Table 1). The Posidoniaceae, which are monogeneric, and the Zosteraceae, consisting of the four genera, *Heterozostera, Phyllospadix, Nanozostera,* and *Zostera,* are exclusively marine organisms. Likewise the Cymodoceaceae represents a solely marine family and encompasses the highest variety of genera (*Amphibolis, Cymodocea, Halodule, Syringodium,* and *Thalassodendron*). The Hydrocharitaceae, that mainly comprise genera restricted to freshwater habitats, also include three marine genera (*Enhalus, Halophila,* and *Thalassia*) [5]. Seagrasses can therefore be regarded as an ecological group, occurring worldwide in different climatic zones (Table 1) and sharing various metabolic features with their terrestrial counterparts. However, their metabolism must have undergone several adaptations to survive and colonize shores and oceans worldwide (see Section 5).

Molecular approaches shed light on which seagrass genes have diverged from their terrestrial counterparts via an initial aquatic stage characteristic of the order and to the derived fully-marine stage characteristic of seagrasses. Positively selected genes are associated with general biological pathways such as metabolic pathways, translation, and photosynthesis, probably associated with the Na^+ toxicity of these processes [7].

When did the seagrass genera evolve? There are age estimates for families of monocots published. The Cymodoceaceae have a crown node age of 61 Mya and a stem node age of 67 Mya, the Hydrocharitaceae a crown node age of 75 Mya and a stem node age of 88 Mya, for the Posidoniaceae only a stem node age of 67 Mya can be estimated, and the Zosteraceae appeared only recently with a crown node age of 17 Mya and a stem node age of 47 Mya [8].

3. Morphological Traits and Molecular Markers for Seagrass Differentiation

3.1. Morphological Traits. Although on a global scale seagrasses represent less than 0.1% of the angiosperm taxa, the taxonomical ambiguity in delineating seagrass species is high. The taxonomy of several genera is unsolved. While seagrasses are capable of performing both, sexual and asexual reproduction, vegetative reproduction is common and sexual progenies are always short lived and epimeral in nature (Figure 1; Table 1). This makes species differentiation often difficult, especially for nontaxonomists since the flower as a distinct morphological trait is missing. Some seagrass genera can hardly be distinguished by their morphology at all, for example, examples from the *Halophila* complex (Figure 1; [9]). There are no morphological characters that can distinguish the seagrasses from other aquatic plants. The only character in which most of them differ from the other aquatic plants is the filiform pollen or the strings of spherical pollen. However, it is not known what the special advantage of these may be for life in the marine environment [6].

The *Halophila* section is known as one of the most complex taxonomic challenges [10, 11]. Phenotypic plasticity is a problem not only between populations, but also between species, at least for complex ones. Especially the *Halophila ovalis* complex has little genetic variation but wide morphological plasticity [12]. Difficulties on morphological classification occur during species identification due to overlaps of morphological traits among species of *Halophila* (Figure 1). Current phylogenetic relationships in the order Alismatales based on molecular data compared with morphological were studied by Li and Zhou [13] or more specifically for the Hydrocharitaceae family by Chen et al. [14]. In general, morphological data sets of seagrasses contain poor phylogenetic signals and an incongruence between DNA and morphological results is observed [13, 14].

3.2. Molecular Marker Including DNA Barcoding. Due to the reduced morphological traits and the high phenotypic plasticity molecular methods might help to clarify the taxonomy and also help the nontaxonomists to differentiate among different seagrass taxa. However, so far there is no data set of molecular markers available which resolves all taxonomically accepted seagrass species, and more work has to be done to close this gap of knowledge. The development of a DNA barcoding system assisting also nontaxonomists to identify regional seagrass species was successful. Based on the recommendations of the Consortium for the Barcoding of Life (CBOL), *rbc*L and *mat*K were used. Tree- and character-based approaches demonstrate that the *rbc*L sequence fragment is capable of resolving up to family and genus level. Only *mat*K sequences were reliable in resolving species and partially the ecotype level. Additionally, a plastidic gene spacer was included in the analysis to confirm the identification level. Although the analysis of these three loci solved several nodes, a few complexes remained unsolved, even when constructing a combined tree for all three loci [15]. The addition of a nuclear ITS marker constitutes a good completion of the *rbc*L/*mat*K marker system (Table 2) ([16]; unpublished own data). However, for population studies AFLP or microsatellite analysis has to be used because the resolution of the other marker systems is not sufficient ([16]; unpublished own data).

4. Physiology of Seagrasses

4.1. Conditions for Seagrass Growth. While there are few or no particular structures in seagrass that can be identified as unique in terms of structural adaptation to the marine environment, there is a suite of characters, which together can be taken as representative of seagrasses. These include strap-shaped leaves and anatomical reinforcement to resist wave action, adaptation of leaves to carry out photosynthesis in a seawater environment, osmotic adjustment and other adaptations within the leaf blade and leaf sheath, modifications to rhizomes and roots for different substrates, pollination by hydrophily, reduction in the layers of the pollen wall, and several unique features associated with seed formation and dispersal mechanisms [17]. However, there

FIGURE 1: Schematic illustrations of seagrass members of the four different families. Cymodoceaceae: (a) *Cymodocea serrulata*, (b) *Halodule pinifolia*, (c) *Halodule uninervis*, (d) *Halodule wrightii*. Hydrocharitaceae: (e) *Halophila decipiens*, (f) *Halophila ovalis*, (g) *Halophila ovata*, (h) *Halophila beccarii*. Zosteraceae: (i) *Zostera marina*, (j) *Zostera noltii*. Posidoniaceae: (k) *Posidonia australis* (source of the single schemes: http://ian.umces.edu/symbols/).

TABLE 1: Seagrass species *sensu stricto* according to the definitions by Larkum et al. [1]. The table shows combined data from Ackerman [101] and Short et al. [18].

Family	Genus	Reproductive ecology Mode	Reproductive ecology Decliny	Distribution
Hydrocharitaceae	*Enhalus* (1)	Surface	Monoecious	5*
	Thalassia (2)	Submarine	Dioecious	2, 5
	Halophila (14)	Submarine	Monoecious and dioecious	2, 3, 4, 5, 6
Cymodoceaceae	*Amphibolis* (2)	Submarine	Dioecious	6
	Cymodocea (4)	Submarine	Dioecious	1, 3, 5
	Halodule (8)	Submarine	Dioecious	1, 2, 5
	Syringodium (4)	Submarine	Dioecious	2, 5
	Thalassodendron (2)	Submarine	Dioecious	5, 6
Posidoniaceae	*Posidonia* (1)	Submarine	Bisexual	3, 6
Zosteraceae	*Heterozostera* (1)	Submarine	Monoecious	6
	Phyllospadix (5)	Submarine	Dioecious	4
	Nanozostera (8)	Submarine and surface	Monoecious	1, 3, 4, 5, 6
	Zostera (4)	Submarine and surface	Monoecious	1, 3, 4, 5, 6

*1 Temperate North Atlantic, 2 Tropical Atlantic, 3 Mediterranean, 4 Temperate North Pacific, 5 Tropical Indo-Pacific, 6 Temperate Southern Oceans.

TABLE 2: Overview about molecular studies of seagrasses using different molecular marker systems.

Taxon	Loci used	Source
Alismatales	*rbc*L	Les et al. 1997 [5], Li and Zhou 2009 [13]
Hydrocharitaceae	*rbc*L, *mat*K	Tanaka et al. 1997 [102]
Hydrocharitaceae	18S, *rbc*L, *mat*K, *trn*K 5′ intron, *rpo*B, *rpo*C1, *cob*, *atp*1	Chen et al. 2012 [14]
Halophila	ITS1, 5.8S, ITS2	Uchimura et al. 2008 [11]
Halophila	ITS1, 5.8S, ITS2	Waycott et al. 2002 [103]
Halophila	ITS1, 5.8S, ITS2	Short et al. 2010 [12]
Halodule	*rbc*L, *phy*B, *trn*H-*psb*A	Ito and Tanaka 2011 [104]
Zostera	*rbc*L, *mat*K	Kato et al. 2003 [105]
Zostera	*rbc*L, *trn*K, ITS	Les et al. 2002 [106]
All seagrass genera	*trn*L	Procaccini et al. 1999 [107]
Halodule, Posidonia, Ruppia	*trn*L, ITS	M. Waycott, pers. comm.

are no specific morphological (secondary) adaptations to the marine environment.

Obviously, the physiological adaptations are more important than morphological ones and seagrasses must have evolved very special physiological mechanisms to deal with large fluctuations in salinity. Seagrasses generally have high light requirements, with an average of 10% of surface light. Some species, such as *Halophila*, often grow in deeper water and have been shown to survive at approximately 5% of surface light. Often the distribution of seagrasses is primarily limited by the amount of light that reaches the sediment. Therefore increasing turbidity by resuspension of fine sediment or anthropogenic factors leads to a decrease of seagrass growth and abundance. Seagrasses can be categorized on the basis of their growth forms, which range from small plants with thin leaves (e.g., *Halophila*, *Halodule*) to large plants with thick leaves (e.g., *Thalassia*, *Enhalus*, and *Posidonia*). This gradient in seagrass morphology and turnover rates is also reflected in aspects of distribution, ecophysiology, and ecological interactions. These large variations in morphology

and ecological function of different seagrass species influence, how they interact with higher trophic levels and the type of habitat they provide [18].

Seagrass photosynthesis, particularly in shallow and confined environments, is thus constrained by low CO_2 concentration and low molecular diffusion associated with the boundary layer around the leaves. The greatest physiological and biochemical adaptation is probably the conversion of HCO_3^- in seawater into CO_2 presumably by anhydrase enzymes at the outer tangential walls of epidermal cells and also the presence of a proton pump at the plasmalemma of seagrass leaves. However, this applies also to most freshwater plants. The primary form of dissolved inorganic carbon in the marine environment is bicarbonate (90%), while CO_2 represents a minor fraction (0.5–1%) for seawater at pH 8.1–8.3 (and less than 0.1% at the higher ranges of pH) [19].

The ability to utilize HCO_3^- could be one of the traits evolved in the last common ancestor branch. In contrast, a set of signals of positive selection specific to the *Zostera* lineage could relate to the biochemical mechanism used in carbon fixation. Seagrasses have long been regarded as C_3 plants, but physiological measurements have gathered indications that several seagrass species, including Z. marina, are C_3-C_4 intermediates or have various carbon-concentrating mechanisms to aid the ribulose-1,5-bisphosphate carboxylase/oxygenase (RuBisCO) enzyme in carbon acquisition [20, 21]. Seagrasses are able to activate different mechanisms to cope with conditions of light-limitation and shifted light spectrum through long-lasting metabolic adjustments including downregulation of RuBisCO, enhanced proteolysis and changes in the antenna complex [2]. Seagrasses do not have any stomata, therefore no crassulacean acid metabolism (CAM) is induced. Analyses have also almost excluded the possibility of C_4 metabolism in seagrasses on the evidence from $\delta^{13}C$ experiments [22].

Rising atmospheric CO_2 often triggers the production of plant phenolics [23]. However, it was recently shown that high CO_2 and low pH conditions due to ocean acidification decrease, rather than increase, concentrations of phenolic protective substances in seagrasses and eurysaline marine plants. These responses are different from those exhibited by terrestrial plants. The loss of phenolic substances may explain the higher-than-usual rates of grazing observed near undersea CO_2 vents and suggests that ocean acidification may alter coastal carbon fluxes by affecting rates of decomposition, grazing, and disease. These observations temper recent predictions that seagrasses would necessarily be "winners" in a high CO_2 world [24].

Another aspect to be handled by marine plant organisms is the water and solute transport within the whole organism. By using apoplastic tracers, Barnabas [25] showed that sea water freely enters from the medium and moves amongst the leaf blade and root tissues of Thalassodendron ciliatum and Halodule uninervis. However, water movement is restricted by the suberin of the vascular bundles in the blades and by the hypodermis and endodermis in the roots. Suberin is often deposited as lamellae, either throughout the entire wall or concentrated into bands in the radial walls (i.e., Casparian strips). In leaves of the seagrass Thalassodendron ciliatum suberin is found throughout the cell wall and in the middle lamella between contiguous bundle sheath cells [25]. Furthermore, in contrast to leaf blades, leaf sheaths have a distinct "suberin-like" cuticle that prevents seawater from entering the sheath tissue and acts as a protection of the meristem tissues and developing leaves. Also in the root hypodermis Casparian band-like structures were detected in several seagrass genera (Zostera, Halophila), but not in other submersed species such as Egeria densa, Eichhornia crassipes, and Lemna minor. These bands contained suberin and had an ultrastructure resembling Casparian bands of the endodermis. They blocked apoplastic transport into the interior tissues of the roots. Symplastic transport through the root tissues was not affected by the bands' presence [26]. Therefore, controlled selective transport of ions and water in the presence of osmolytic compounds can be well reconstructed.

Oxygen transport creates another problem for marine (and freshwater) plants due to its low solubility in water. In seagrasses, oxygen is transported to rhizomes and roots of seagrasses during periods of light when photosynthesis releases oxygen into aerenchyma. This kind of tissue is present in all seagrass species to different extents and forms large internal gas spaces [17]. Also air-spaces have been described in below-ground tissues. Around the roots an oxidized zone is formed and oxygen diffuses into the anoxic sediment. By night almost all oxygen transport stops and alcoholic fermentation starts in roots [27]. The oxygenated rhizosphere of seagrasses during photosynthesis might create a special environment for the uptake of limiting nutrients with the help of nutrient/metal-binding and the detoxification of toxic elements by oxygenation and/or binding to chelating compounds. Seagrasses have roots and vascular tissue allowing them to absorb and translocate nutrients from soft sediment. In low nutrient environments this provides seagrass with a competitive advantage over algae as they can access the higher nutrient concentrations available in the sediment compared to the overlying water. On the other hand, in contaminated sediments seagrasses need good strategies to avoid accumulation of toxic compounds in their tissue (see Section 5.6).

Roots and stems of seagrasses trap organic matter and sediment. The decaying organic matter produces a lot of toxic sulfide. So far it was assumed that the amount of oxygen released from the roots is sufficient to detoxify the high amounts of sulfide, and stated therefore another problem successfully solved by seagrasses. However, in most seagrass beds ancient three-stage symbiosis between seagrass (Zostera noltii), lucinid bivalves (Loripes lacteus), and their sulfide-oxidizing gill bacteria reduces sulfide stress for seagrasses. The bivalve-sulfide-oxidizer symbiosis reduces sulfide levels and enhances seagrass production as measured in biomass. In turn, the bivalves and their endosymbionts profit from organic matter accumulation and radial oxygen release from the seagrass roots [28]. Therefore, symbiotic and other types of biotic interactions with organisms also have to be kept in mind when investigating the physiology of seagrasses in the future.

There are still many open questions in seagrass physiology, and therefore more physiological studies are needed to solve the basic problem how they deal with salinity both in seagrasses and in euryhaline aquatics [6]. One reason for the lack of knowledge is the establishment of suitable culturing conditions, a prerequisite for doing reproducible physiological experiments with a sufficient number of repetitions. Culturing of seagrasses is accompanied by a number of difficulties: when grown in climatic chambers using natural sediment, artificial seawater, and high sodium vapor lamps, the temperature needs to be carefully controlled and still the light intensity is rather low inside the water basins. Also artificial seawater does not completely mimic natural seawater conditions producing suboptimal growth or even stress conditions.

More recent approaches also include high throughput methods such as transcriptomics to analyze the responses of seagrasses, at least on the expression level [7]. The resulting EST data are publicly available (Dr. ZOMPO, http://drzompo.uni-muenster.de/) and are very helpful to design comprehensive studies on certain aspects of seagrass physiology.

If these data are combined with metabolomic data we will be able to learn more about the hidden details of seagrass physiology. These approaches will also be speeded up because the complete genome of *Zostera marina* is currently being sequenced and will be published within the next year (http://www.jgi.doe.gov/sequencing/why/Zmarina.html).

4.2. Role of Aquaporines in Seagrasses. Terrestrial plants depend on water supply for their growth and development, and constantly absorb and lose water. The diffusion of water can be driven by concentration gradients of osmotically active solutes or by physical pressure, generating an osmotic or hydrostatic force, respectively. Beyond simple diffusion across a lipid bilayer, the existence of proteinaceous water channels, aquaporins, in plant membranes has been established [29]. While studying mechanisms involved in water transport in marine plants, two aquaporin-encoding genes, *Po*PIP1; 1 and *Po*TIP1; 1, were isolated from *Posidonia oceanica* showing high similarity to plasma membrane- and tonoplast-intrinsic protein-encoding genes, respectively. Hyposalinity induced lower levels of *PIP1* transcripts, while hypersalinity determined more *PIP1* transcripts than normal salinity. *TIP1* transcripts increased in response to both hypo- and hypersalinity after two days of treatment and decreased to control levels after 5 d [30]. The expression was also investigated by *in situ* hybridization [31]. *Po*PIP; 1 transcript was associated with the meristematic region of the apical meristems (shoot and root), whereas the *Po*TIP; 1 was mainly associated with the tissues showing a well-differentiated vacuole compartment. Moreover, *Po*PIP; 1 intensively marked the epidermal and subepidermal cells in the leaves and also in the provascular and vascular tissues. After hypersalinity treatment, the *Po*TIP; 1 tissue expression strongly increased compared to that of *Po*PIP; 1. In contrast to terrestrial plants, where aquaporins are involved in water transport, these *in situ* results suggest a role in the water balance and/or solute transport in the different organs and tissues of *Posidonia oceanica*.

5. Striking Metabolites

5.1. General Introduction. Due to their convergent evolution, seagrasses share a number of analogous acquired metabolic adaptations. But their secondary metabolism varies among the four families that can be considered as true seagrasses. Terrestrial-like species returned to the sea, during the period of the ancient Tethys Sea, surrounded by Africa, Gondwanaland, and Asia, approximately 90 million years ago, thus explaining the "terrestrial-like" chemical profile of the seagrass. Several types of secondary metabolites have been studied in seagrasses, often from a chemotaxonomic viewpoint. Attaway et al. [32] found that the normal alkanes of several genera represented less than 0.01% dry weight but their distribution paralleled current taxonomic schemes of the seagrasses, with *Halodule* and *Syringodium* distinguished from each other and even more clearly from *Thalassia* and *Halophila*. Cluster analysis of high resolution GC-MS analyses of the sterols and fatty acids of a number of species from tropical Australia [33] also confirmed significant

segregation of the genera *Cymodocea* and *Halodule* from the hydrocharitacean genera *Thalassia* and *Enhalus*. However, *Halophila*, a genus from the latter family, but with very different morphology, was separated at a much higher level from all the other seagrasses analyzed. Taxonomic questions at the species level in seagrasses have also been approached chemically by McMillan et al. [34, 35] using secondary products such as the flavonones and their sulfated derivatives from *Amphibolis, Halodule, Halophila, Posidonia,* and *Zostera*. Thus identification of evolutionary unique phytochemicals may elucidate the taxonomic relationships beside existing DNA-based approaches. Chemotaxonomy is not the main criterion to analyze secondary compounds of seagrasses but researchers hope to find new chemical structures of natural compounds which might be used in a different application. Advanced techniques such as LS-MS (MS), GS-MS, and NMR now available to more research groups simplify the analysis of secondary compounds in seagrasses. As recently shown by Wissler et al. [7], phylogenetically interesting genes could be identified by comparing the transcriptional pattern of *Posidonia oceanica* and *Zostera marina*. Transcriptomics combined with metabolomic studies might help to elucidate seagrass-specific metabolomic pathways.

5.2. Phenols in Seagrasses

5.2.1. Occurrence of Phenolic Acids in Seagrasses. The secondary metabolism of seagrasses shares features like the absence of hydrolysable tannins [36]. By taking a rough look at their metabolism, the basic pool of secondary metabolites is similar to their terrestrial relatives from which they have evolved. However, while comparing the variability of single pathways, for example, of the phenolic substances, Vergeer et al. [37] observed that seagrasses can be considered as a rich source for those, including phenolic acids, sulfated phenolic acids, flavones, condensed tannins, and also lignins. The pathways for the production of the huge variability of phenol derivates are interesting from an evolutionary point of view (unpublished own data). The changes in phenol metabolism are the result of continuous evolution: by gene duplication, mutation, subsequent recruitment, and adaptation to specific functions [38]. This reflects the adaption to the marine environment. But some compounds are exclusively found in single seagrass families or species, such as zosteric acid [39]. Some phenolic compounds were isolated from *Zostera marina* and suggested to have an important role in inhibition of microbial growth, amphipod grazing and in the resistance to the so-called waste-disease [40]. The fact that these crude methanolic extracts of *Zostera marina* have been found to inhibit the attachment of marine bacteria, diatoms, barnacles, and polychaetes on artificial surfaces suggests an antifouling potential of *Zostera marina* phenolic compounds. The *p*-(sulfooxy) cinnamic acid (zosteric acid) was isolated for the first time as a natural product from the seagrass *Zostera marina* and was found to prevent attachment of marine bacteria and barnacles to artificial surfaces at nontoxic concentrations [39].

Gallic acid is synthesized and stored in more than 50% of all seagrass species [41]. Also differences in the biosynthesis of phenolic acids can be expected in the different seagrass families. Zosteric acid for example is a hydroxy cinnamic acid derivative, whereas other phenolic acids are caffeic acid derivatives [42]. The occurrence of the sulfated phenolic compounds indicated subgeneric differences in *Zostera* and interspecific differences in *Halophila* [43]. It would be interesting to investigate the biosynthesis of zosteric acid in *Zostera* genera to analyze the biosynthetic enzymes in the pathway. In a patent (http://www.patentgenius.com/patent/6841718.html) a specific sulfotransferase was suggested to be involved. However, no prove was shown to verify the assumption. Harborne and Williams [42] indicated the presence of sulfated flavones in *Halophila, Thalassia,* and *Zostera* species, but they were not recorded in *Syringodium* or *Posidonia*.

At least 23 phenolic compounds were identified in the seagrass *Posidonia oceanica* [44]. Chicoric acid and caftaric acid were identified in detrital and living leaves of the tropical seagrass *Syringodium filiforme* making this abundant renewable raw material of interest for pharmaceutical purposes and food industries [45]. Phytochemical investigations revealed the presence of unidentified sulfated phenolic compounds from nine different species of *Halophila* [43], unidentified sulfated and nonsulfated flavones from the *Halophila ovalis/Halophila minor* complex [35], flavones and flavone glycosides from *Halophila johnsonii* [46], as well as malonylated flavonoid derivatives in the seagrass *Halophila stipulacea* [47]. Production of phenolic compounds depends on the environmental conditions. For example, tannin production can be wound-induced in *Thalassia testudinum* under simulated grazing conditions [48], and the content of phenolic compounds in shoots of *Zostera marina* varied seasonally [49]. The concentration of phenolic compounds was measured in the seagrass *Posidonia oceanica* when interacting with two Bryopsidophyceae, *Caulerpa taxifolia,* and *Caulerpa racemosa*. Several phenolic compounds were identified in *Posidonia oceanica*, with a predominance of caffeic acid in the adult and intermediate leaves. The number of tannin cells, which are assumed to produce the phenolic compounds, increased in the leaves when the degree of interaction with *Caulerpa taxifolia* increased. Therefore, interaction of *Caulerpa taxifolia* with the seagrass *Posidonia oceanica* induces its production of secondary metabolites, probably to limit the invasion of the beds [50].

5.2.2. Polyphenols: Lignin Content and Biosynthesis. One might assume that seagrasses do not contain any lignin because they are supported by the hydraulic forces of the water body. However, in all species analyzed so far lignin was detected in several tissue types and several isoblastic cells. Species variation of the lignin content is dependent on the morphotype and life style of the seagrass. *Posidonia oceanica* contains more lignin than *Zostera marina*, and roots and rhizomes generally contain more lignin than leaves. Obviously, the ability to produce lignin is not lost by the angiosperm ancestors of extant seagrasses upon their colonization of the marine environment. Relative lignin abundances in the different tissues appear to be positively correlated with life

span. Lignification seems to contribute to the longevity of a tissue by protecting it against microbial attack, and deposition of lignin in seagrasses is restricted to tissues that show limited growth [51]. The importance of lignin in making seagrass below-ground organs particularly decay-resistant still needs to be adequately addressed as well as their role as carbon sink. In comparison to lower plants species the seagrass *Posidonia oceanica* shows a broad variability in lignin composition but rather low total lignin content in purified cell walls [52]. The increased heterogeneity of lignin monomer composition might be related to the separation of water transport and support functions, although it has been reported that the presence of syringyl lignin is not necessarily linked to the presence of xylem vessels [53, 54]. Recent data imply that lignification originated as a developmental enabler in the peripheral tissues of protracheophytes and would only later have been coopted for the strengthening of tracheids in eutracheophytes [52]. A pilot study of thermally-assisted hydrolysis and methylation with pyrolysis GC-MS for the analysis of vegetable fibres in forensic science found that the fibre types tended to group into two clusters, with one containing cotton, hemp, and linen; and the other consisting of hessian, sisal, jute, and coir. The fibres of a seagrass sample differed from both groups [55].

Lignins arise from the peroxidase-mediated coupling of *p*-coumaryl, coniferyl, and sinapyl alcohols. In gymnosperms, they are derived from coniferyl alcohol, whereas in angiosperms, lignins are derived from coniferyl and sinapyl alcohols. Until recently, most peroxidases characterized in flowering plants only oxidized coniferyl alcohol. However, recent reports have described the molecular characterization of peroxidases capable of oxidizing sinapyl alcohol (syringyl peroxidases) [56]. Class III peroxidases are members of a large multigene family, only detected in the plant kingdom and absent from green algae sensu stricto (chlorophyte algae or Chlorophyta). Their evolution is thought to be related to the emergence of the land plants. However, class III peroxidases are present in a lower copy number in some basal Streptophytes (Charophyceae), which predate land colonization. Current molecular studies propose that the structural motifs of syringyl peroxidases predate the radiation of tracheophytes, which suggests that syringyl peroxidases existed before the appearance of syringyl lignins [56]. Their high copy number, as well as their conservation could be related to plant complexity and adaptation to increasing stresses. Probably subfunctionalization explains the existence of the different isoforms [57]. In *Arabidopsis thaliana* 73 class III peroxidase genes were clustered in robust similarity groups. Comparison to peroxidases from other angiosperms showed that the diversity observed in *Arabidopsis* preceded the radiation of dicots, whereas some clusters were absent from grasses. Grasses contained some unique peroxidase clusters not seen in dicot plants [58]. The distribution of lignin in the seagrass plant needs to be investigated in more detail. Own microscopical studies using different staining techniques indicate the abundance of different types and numbers of idioblastic lignified cells along the leaf tissue. Their function is so far unknown [59].

5.2.3. Abundance of Flavonoids. The polyphenolic flavonoids are found in either of five chemical structures, like flavones, flavonols, flavanons, flavanols, and anthocyanidins. The presence of sulfated flavones was reported in *Halophila*, *Thalassia*, and *Zostera* species, but they were not recorded in *Syringodium* spp. or *Posidonia oceanica* [42]. Flavonoid sulfates were also detected in *Halophila ovalis* and *Thalassia testudinum* [43, 60]. McMillan et al. [43] extensively studied 43 species of seagrasses and showed that all contained either flavones and/or phenolic acid sulfates. The occurrence of the sulfated phenolic compounds indicated subgeneric differences in *Zostera* and interspecific differences in *Halophila*.

One example demonstrates the antifouling effect of sulfated flavonoides. Significantly fewer thraustochytrid protists (zoosporic fungi) were observed in association with healthy leaf tissue of *Thalassia testudinum* than in association with sterilized samples that were returned to the collection site for 48 h. In support of the hypothesis that seagrass secondary metabolites were responsible for these differences, extracts of healthy *Thalassia testudinum* leaf tissues inhibited the growth of the cooccurring thraustochytrid *Schizochytrium aggregatum* and deterred the attachment of *Schizochytrium aggregatum* motile zoospores to an extract-impregnated substrate. By using *Schizochytrium aggregatum* for bioassay-guided chemical fractionation, luteolin was isolated. These results offered the first complete chemical characterization of a sulfated flavone glycoside from seagrasses and provide evidence that a secondary metabolite chemically defends *Thalassia testudinum* against fouling microorganisms [61]. The four flavones, luteolin, apigenin, luteolin-3-glucoronide, and luteolin-4-O-glucoronide, all of them with antibicrobial potential were identified from the ethanol extract of air-dried *Enhalus acoroides* from South China Sea [62].

Cannac et al. [63] reported flavonoid glycosides and acyl derivatives, which yield after hydrolysis the respective flavonoid aglycones in *Posidonia oceanica* leaves. Later, Cannac et al. [64] found dramatic losses of flavonoids when analyzing freeze-dried and chilled leaves as opposite to fresh and oven-dried leaves of *Posidonia oceanica*. Bitam et al. [47] isolated and identified malonylated flavone glycoside derivatives from *Halophila stipulacea* using HPLC and NMR. Heglmeier and Zidorn [4] compiled and appraised the data of secondary metabolites of *Posidonia oceanica* and they summarized 51 natural products including phenols, phenylmethane, phenylethane, phenylpropane derivatives and their esters, chalkones, and flavonoids. Significantly higher flavonoid amounts were observed in the leaves of intertidal and subtidal *Halophila johnsonii* when compared to the leaves of intertidal *Halophila decipiens* [65]. These functional derivatives of flavonoids are considered to strive against the marine microorganisms exhibiting chemical defense. Takagi et al. [66] identified next to phenols including phenolic acids, lignin and flavonoids, and isoprenoids, also alkaloids in *Phyllospadix iwatensis*. This newly identified flavonoidal alkaloid was called phyllospadin. However, the class of alkaloids is not well represented in the four seagrass families.

5.3. Terpenoids. There is only one detailed report on the occurrence of terpenoids in seagrasses. Despite the important ecological role of *Cymodocea nodosa* in the marine ecosystem, knowledge of its chemical content is limited. Only molecules frequently found in terrestrial plants such as caffeic acid, inositol, sucrose, monoglucoside of quercetin, monoglucoside of isoramnetin, cichoric acid, as well as polyamines like putrescine, spermidine, and spermine, have been reported as constituents of *Cymodocea nodosa*. Furthermore, 24a-ethyl sterols and 24a-methyl sterols along with their 24b-epimers, cymodiene and cymodienol, the first diarylheptanoids isolated from marine organisms, comprise the total number of metabolites isolated from *Cymodocea nodosa* so far. Recently, new terpenoid compounds from the structural class of diarylheptanoids, a new meroterpenoid, and the first briarane diterpene isolated from seagrass, and only the second analog of this class with a tricyclic skeleton. Furthermore this metabolite is the first brominated briarane diterpene [67]. All newly detected compounds were assayed for their antibacterial activity against multidrug resistant (MDR) and methicillin-resistant bacterial strains. The activity reached from weak to strong and therefore opens the field for the formulation of new antibiotics [67] which are urgently needed due the many MDR strains, especially in hospitals. Probably more compounds with antibiotic activity in seagrasses will be found.

5.4. Sugars and Sulfated Polysaccharides in Seagrasses

5.4.1. Remarkable Sugars. The osmoregulation in seagrasses has not been unambiguously elucidated so far. Several sugars might play a role. The sugar chemistry of seagrasses evolved differently from land plants, leading to a broad range of different inositols in addition to the ubiquitous sucrose, glucose, and fructose [68]. Of the nine possible inositols, five are known to occur in plants; *myo-*, *l-chiro-*, *muco-* and *O-methyl-muco*inositol occur in seagrasses. *Myo-*inositol is found in all living cells in amounts usually considerably less than 1% dry weight. It is apparently synthesized by direct cyclization of photosynthetically-produced glucose and it is mainly used in cell wall synthesis. Leaves and rhizomes of some seagrasses, particularly the Zosteraceae, contain relatively large amounts of this compound, up to a maximum of 2.2% dry weight in *Zostera noltii* rhizomes. In most plants the cyclization enzyme is conservative and yields only *myo-*inositol, which then acts as the sole precursor for any other inositols they accumulate. Drew [68, 69] suggested that, since the configuration of the glucose molecule would permit the direct formation during cyclization of all the inositols found in seagrasses, they may be inevitable byproducts of another, less specific, glucose cyclization enzyme.

Taking this as a base, Drew [68] tried to solve the phylogeny of seagrasses, according to the inositol pattern, taking the genus *Halodule* as common ancestor. Only members of Cymodoceaceae accumulate these other inositols, with a preponderance of *l-chiro*-inositol. This compound has been detected in all genera except *Halodule*, with a maximum of 6.8% dry weight in *Cymodocea rotundata* leaves. *Muco-*inositol appears to be slightly less widely distributed in these seagrasses whilst its *O-methyl* ester is restricted to the endemic temperate Australian genus *Amphibolis*. However, those other than *myo*-inositol can probably accumulate to several percent dry weights because they are not subsequently utilized, even after all soluble sugars have been respired away during dark starvation for several days. The possibility that these compounds might be involved in an osmoregulatory role was not supported by studies at high and low salinities, although respiration, and therefore sucrose utilization, was increased at both extremes [68]. Tyerman et al. [70] also implicated sucrose, and possibly amino acids, as minor osmoregulants, in their study of the osmotic environment of *Posidonia australis* and *Zostera capricorni* leaves.

5.4.2. Sulfated Polysaccharides.

Sulfated polysaccharides (SPs) comprise a complex group of macromolecules with a wide range of biological, partly unknown functions. These anionic polymers are widespread in nature, occurring in a large variety of organisms. Although their structures vary among species, their main features are conserved among phyla. Green algal SPs are quite heterogeneous and usually heteropolysaccharides. The red algal SPs (like agar and carrageen) are composed of repeating disaccharide units with different sulfation patterns which vary among species. The SPs from invertebrates such as sea urchins and ascidians (tunicates) are composed of well-defined repetitive units. Chains of 3-linked β-galactoses are highly conserved in some marine taxonomic groups, with a strong tendency toward 4-sulfation in algae and marine angiosperm, and 2-sulfation in invertebrates [71]. SPs of seagrass species are composed of galactose units. Seagrass species contain various amounts dependent on the organ and on the salinity (*Halodule wrightii* 8.5 μg SP and *Halophila decipiens* 7.7 μg SP per mg dry weight), comparably high as some mangrove species, whereas in terrestrial crop plants the values are below 0.001 μg SP per mg dry weight [72, 73].

So far SP biosynthesis and exact physiological role in seagrasses were not clarified. Probably the biosynthesis of sulfated galactans starts with a precursor of lower molecular weight and degree of sulfation suggesting that glycosyltransferases and sulfotransferases may function simultaneously during the biosynthesis of sulfated galactans, at least in *R. maritima* [73]. Until the first seagrass genome will be completely sequenced, the identification of responsible glycosyltransferases and sulfotransferases using the sequenced genome of *Ectocarpus siliculosus* [74] might be a promising approach. Interestingly, green algae, the ancestor of higher plants [75], possess all units of SP also found in all investigated halophytic aquatic plants [72, 73]. This finding suggests that the production of SP is conserved throughout the plant evolution from green algae [73]. It is speculated that the activation and inhibition of glycosyltransferase genes alter the composition of SP among the different phyla [73].

In seaweeds, SPs are found in the extracellular matrix. SP might protect against dehydration occurring at low tide. They are important both in terms of resistance to mechanical stresses and as protection from predators [76]. The function of SP in the plant cell wall in high salt environments is still unclear. It is speculated that SPs increase the Donnan potential [77], supporting ion transport at high salt concentrations. In *Ruppia maritima* SPs were not found when the plant was cultivated in freshwater [73]. Species being able to survive in both saline and freshwater conditions might be well-suited study objects to analyze the function of SP. The current state of knowledge suggests that the presence of SP in plants is an adaptation to high salt environments, which have been conserved during plant evolution from marine green algae.

5.5. Dimethylsulfoniopropionate (DMSP) in Seagrass.

It was shown that next to green, red, and brown algae several angiosperms produce dimethylsulfoniopropionate (DMSP). DMSP is broken down by marine microbes to form two major volatile sulfur products, each with distinct effects on the environment. Its major breakdown product is methanethiol which is assimilated by bacteria into protein sulfur. Its second volatile breakdown product is dimethyl sulfide (DMS). Atmospheric oxidation of DMS, particularly sulfate and methanesulfonic acid, is important in the formation of aerosols in the lower atmosphere. Probably these aerosols act as cloud nucleation sites. Therefore DMS is thought to play a role in the Earth's heat budget by decreasing the amount of solar radiation that reaches the Earth's surface [78].

However, the presence of high concentrations of DMSP in higher plants is limited to a few species such as *Spartina* spp. (>50 μmol DMSP g^{-1} fresh weight in the leaves) [79]. In seagrasses different DMSP concentrations have been found: *Halodule wrightii* 3.3 μmol g^{-1} fresh weight, *Syringodium filiforme* 0.10 μmol g^{-1} fresh weight, *Thalassia testudinum* in epiphytized and nonepiphytized leaves between 0.18 and 4.0 μmol g^{-1} fresh weight, and very low amounts in the rhizome [80]. These results indicate that the degree of epiphytization plays a major role in the contribution of seagrasses to the total DMSP production. The regulation of the biosynthetic pathway of DMSP in seagrasses needs to be elucidated to clarify the overall contribution by seagrasses.

5.6. Peptides and Proteins Involved in Metal Binding.

Heavy metals are taken up by seagrasses and accumulate in different tissues to different extents [81–83]. Several seagrass species such as *Posidonia oceanica* [84], *Cymodocea nodosa* [85], *Cymodocea* spp., *Enhalus acoroides*, *Halodule* spp., *Halophila* spp., *Syringodium isoetifolium*, and *Thalassia hemprichii* [83, 86, 87] have been even used as bioindicator for heavy metal accumulation in contaminated and noncontaminated areas. It was shown that the accumulation of the same metal varies by different species of seagrasses and heavy metals were not homogeneously distributed in all the seagrasses [87]. Also *Thalassia testudinum* was used as bioindicator for trace metal stress and investigated in more detail. For this species, the accumulation of Cd varied in a dose-dependent manner and according to the tissue examined. Cd accumulation of green leaves was higher than other organs (sheaths, roots/rhizomes) after 96 h of treatment. It was found that the higher the Cd concentration in ambient environment the higher the Cd concentration in the tissue.

Besides significant different Cd accumulation in different tissues, results also showed that thiols including cysteine, glutathione and γ-glutamylcysteine, and phytochelatin-(PC-) like peptides were induced after different times of exposure. Thiols were found in the green blades and rhizomes after 24 h of treatment, whereas in the sheaths those thiols accumulated after 114 h of exposure. Moreover, total thiols in green blade tissue showed the highest content, followed by live sheaths and root/rhizome [88].

PCs and metallothioneins (MTs) are Cys-rich metal chelators that represent the two principle groups of metal-binding molecules found across most taxonomic groups [89]. PCs, glutathione-derived metal binding peptides, usually with the structure of $(1'\text{-Glu-Cys})_n\text{-Gly}$ ($n = 2\text{--}11$) are enzymatically synthesized peptides known to be involved in heavy metal detoxification, mainly Cd and As, which has been demonstrated in plants, algae, and some yeast species grown at high heavy metal concentrations [90]. So far, neither the exact composition of PCs in seagrasses nor the biosynthetic enzyme, PC synthase, have been analyzed in seagrasses.

MTs are a group of proteins with low molecular mass and high cysteine content that bind heavy metals and are thought to play a role in their metabolism and detoxification [90]. The criteria that define a protein or peptide as an MT are (i) low molecular weight (<10 kDa), (ii) high metal and sulfur content (>10%), (iii) spectroscopic features typical of M–S bonds, and (iv) absence or scarcity of aromatic amino acids [91]. However, when all criteria are not fulfilled often proteins are called MT-like proteins. There are only a few papers reporting on the metal-binding mechanisms of seagrasses [92, 93]. Three genomic sequences putatively encoding MTs were isolated from *Posidonia oceanica* [92], namely, *Pomt2a*, *Pomt2b* and *Pomt2c* that showed high similarities to putative plants MTs. *Pomt2a* and *Pomt2b* contain a CXXC motif classifying them as type II plant MTs, the remaining *Pomt2c* is considered to be a pseudogene. Moreover, authors indicated that there were at least five MT genes present in *Posidonia oceanica* genome based on Southern blot hybridizations. Results of Giordani et al. [92] showed that based on Northern blot hybridization MT transcript accumulation was increased by Cu and Cd exposure, whereas no apparent effect was observed after Hg treatment. Higher Cu^{2+} concentration (10 μmol) treatment showed higher MT transcript accumulation than low Cu^{2+} concentration (1 μmol). Based on these results Cozza et al. [93] continued to carry out the studies in more detail. Nine MT-like sequences from Cu or Cd treated *Posidonia oceanica* were isolated by RT-PCR. One sequence is similar to *Pomt2b*. Phylogenetic analysis of MT-like protein deduced from isolated MT-encoding genes from *Posidonia oceanica* showed two subgroups. To better understand the functional role of the two MT subgroups one gene representative for each group was used for *in situ* hybridization to discover spatial expression of the plant. Interestingly, the members of these two MT subgroups showed differences in their histological expression, with *Pomt2b* associated with the proliferative tissues whereas *Pomt2f* was associated with the lignified or suberized cell wall [93].

In summary, seagrasses can survive and grow well in environments contaminated with heavy metals. On the one hand, putative MT-like proteins are considered to bind heavy metals. On the other hand, the dominant frequency of putative MT transcripts found in *Zostera marina* under heat stress suggests that other functions of MTs are still unknown. Hence, detailed research on MT function and the number of genes encoding MTs are needed.

6. Economical Use of Seagrasses and Their Products

6.1. Technical Applications. Studies reveal the importance of the seagrass *Zostera marina* for subsistence cultures for many generations, for example, in the North Atlantic [94]. These cultures derive many and varied natural products from this plant species and their recognized historical value contributes to the protection of sites of former gathering activity [94]. Ethical positions regarding the value of resource extraction are common among traditional and indigenous cultures, often leading to the protection of habitats that support valuable resources. Also in Europe the dried flotsam of *Zostera marina* was used as mattrass and padding material, as erosion protection mat and for insulation at the end of the 19th and beginning of the 20th century. Since a few years *Zostera* flotsam is used to press flake boards for insulation purposes (http://tu-dresden.de/Members/soeren.tech/news/Seegras).

Intensive elder literature including ethnobotanical observations and screening of more recent literature revealed many inhibitory activities of seagrass material. Crude extracts showed antibacterial, antifungal, antiviral, antioxidant, antiinflammatory, antidiabetic, anti-cancerogen, and so forth activities. In some studies the researchers aimed to identify and isolate the bioactive compound in the extracts. Fractionation and isolation of compounds by HPLC analysis were partly successful. There are several examples of successful demonstration of bioactive activity of a single compound. One very successful case study is the structure determination of zosteric acid, its subsequent chemical synthesis, and the exploration of its biotechnological applications. This discovery followed the general approach shown in Figure 2.

It was observed that rates of decomposition (usually <1% of dry weight day^{-1}) of *Zostera* flotsam are generally low compared with other vascular macrophyte sources of detritus, but are influenced by many variables. Seagrass detritus undergoes an initial period of leaching, leaving a poor substrate for bacteria because what soluble material remains is deficient in inorganic nutrients, contains inhibitory phenolic compounds, and is protected by cellulose and lignin [49, 95]. One inhibitory compound is zosteric acid (Figure 3) [39].

Recently, zosteric acid was generated by treating trans-4-hydroxycinnamic acid with the sulfur trioxide pyridine complex, a solid, easy to handle compound, in N,N-dimethylformamide (DMF) as solvent. After 2 h at 50°C, trans-4-hydroxycinnamic acid is completely converted to zosteric acid. The latter was isolated as the sodium salt by

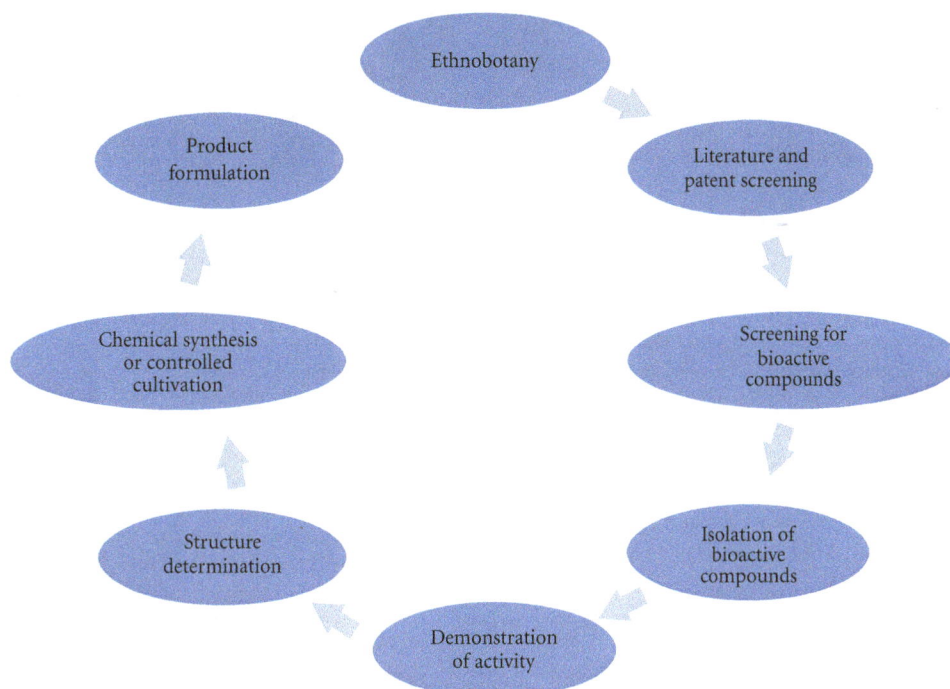

FIGURE 2: General approach from the identification to the application of secondary compounds of economic value.

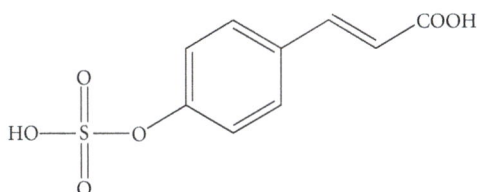

FIGURE 3: Chemical structure of zosteric acid.

adding 30% NaOH to pH 7, followed by extraction with dichloromethane to remove pyridine and DMF [96].

Based on this relatively simple synthesis large amounts of zosteric acid can be produced. Currently, its biofilm inhibiting activity is tested by different research groups. The antifouling effectiveness of zosteric acid has been demonstrated both in static laboratory assays and with zosteric acid directly dispersed in marine water [97]. With respect to the mode of action, it was shown that zosteric acid hinders the biofilm formation by increasing the bacterial motility by 40%. Therefore the phase from the planktonic life form to the sessile by attachment to the surface does not take place [96]. Zosteric acid seems to be an ideal compound because it has low general toxicity but specifically inhibits biofilm formation at early stages. Product development and zosteric acid formulation is underway, for example as an ingredient in antifouling paints. Probably, there will be more compounds isolated from seagrasses with antifouling activity because there are a number of publications reporting inhibition of biofilm formation, such as methanolic extracts of *Cymodocea rotundata* accessions collected in the Gulf of Mannar, India [98].

6.2. Pharmaceutical and Nutraceutical Applications. One of the rare real applications seems to be zosterin, a bioactive pectin from *Zostera asiatica*, which decreases toxicity of antitumor drugs and purges heavy metals from human organisms [99]. These properties led to a patented and marketed drug and food in Russia (Patents RU2128918, RU2129388C1, RU2132696C1, and RU2242217).

It is suggested to use extracts of *Zostera marina* as a material for cosmetics (Patent EP1342468 B1, New cosmetic raw materials from plants of the family Zosteraceae for special cosmetic effects and uses. The use is claimed for the topical cosmetic use of biologically-active ingredients from extracts, seaweed hyaluronates and micropowder obtained from Zosteraceae plants by solvent treatment) and pharmaceutical applications (Patent EP 1338286 A1 Use of extracts from plants of the family Zosteraceae for the prevention and therapy of bacterial and viral infections).

Free L-chiro-inositol was isolated from aqueous extracts of dried detrital *Syringodium filiforme* leaves [100]. The high concentrations found (2.3–2.5% dry weight) offer promise for the exploitation of *Syringodium* flotsam as a new cheap source for nutraceutical or therapeutic applications, considering the demonstrated hypoglycaemic action of L-chiro-inositol.

7. Summary and Outlook

(i) Seagrasses form an ecological group and evolved three to four times towards an aquatic and marine existence.

(ii) Seagrass taxonomy is neither solved on the species level and nor below.

(iii) Their physiology is not well investigated due to difficult *in situ* and *in vitro* growth conditions, and there are still many adaptations to discover.

(iv) Seagrasses contain valuable compounds of economic interest not found in other taxonomic groups.

(v) Seagrass metabolite content is another still buried treasure of the ocean to be lifted.

Acknowledgments

Christina Lucas, Hannover, started the seagrass topic in our laboratory with great enthusiasm and knowledge. Her above-average theoretical and practical abilities greatly speeded up the development of our seagrass projects. Thanks are also to Nguyen Xuan Vy, Hannover, who continues the phylogenetic analysis and seagrass culture with high commitment. Seagrass research in our laboratory is supported by the DAAD (Vigoni project 54644032 and A New Passage to India).

References

[1] A. W. D. Larkum, R. J. Orth, and C. M. Duarte, *Seagrass: Biology, Ecology and Conservation*, Springer, Dordrecht, The Netherlands, 2006.

[2] R. J. Orth, T. J. B. Carruthers, W. C. Dennison et al., "A global crisis for seagrass ecosystems," *BioScience*, vol. 56, no. 12, pp. 987–996, 2006.

[3] F. T. Short, B. Polidoro, S. R. Livingstone et al., "Extinction risk assessment of the world's seagrass species," *Biological Conservation*, vol. 144, no. 7, pp. 1961–1971, 2011.

[4] A. Heglmeier and C. Zidorn, "Secondary metabolites of *Posidonia oceanica* (Posidoniaceae)," *Biochemical Systematics and Ecology*, vol. 38, no. 5, pp. 964–970, 2010.

[5] D. H. Les, M. A. Cleland, and M. Waycott, "Phylogenetic studies in alismatidae, II: evolution of marine angiosperms (seagrasses) and hydrophily," *Systematic Botany*, vol. 22, no. 3, pp. 443–463, 1997.

[6] C. den Hartog and J. Kuo, "Taxonomy and biogeography of seagrasses," in *Seagrass: Biology, Ecology and Conservation*, A. W. D. Larkum, R. J. Orth, and C. M. Duarte, Eds., pp. 1–23, Springer, Dordrecht, The Netherlands, 2006.

[7] L. Wissler, F. M. Codõer, J. Gu et al., "Back to the sea twice: identifying candidate plant genes for molecular evolution to marine life," *BMC Evolutionary Biology*, vol. 11, no. 1, article 8, 2011.

[8] T. Janssen and K. Bremer, "The age of major monocot groups inferred from 800+ *rbcL* sequences," *Botanical Journal of the Linnean Society*, vol. 146, no. 4, pp. 385–398, 2004.

[9] M. Waycott, G. Procaccini, D. H. Les, and T. Reusch, "A genetic perspective in seagrass evolution, ecology and conservation," *Springer Academic*, vol. 2, pp. 25–50, 2006.

[10] J. Kuo and C. den Hartog, "Seagrass taxonomy and identification key," in *Global Seagrass Research Methods*, F. T. Short, C. A. Short, and R. G. Coles, Eds., vol. 33, pp. 31–58, Elsevier Science, Amsterdam, The Netherlands, 2001.

[11] M. Uchimura, E. Jean Faye, S. Shimada, T. Inoue, and Y. Nakamura, "A reassessment of *Halophila* species (Hydrocharitaceae) diversity with special reference to Japanese representatives," *Botanica Marina*, vol. 51, no. 4, pp. 258–268, 2008.

[12] F. T. Short, G. E. Moore, and K. A. Peyton, "*Halophila ovalis* in the Tropical Atlantic Ocean," *Aquatic Botany*, vol. 93, no. 3, pp. 141–146, 2010.

[13] X. Li and Z. Zhou, "Phylogenetic studies of the core Alismatales inferred from morphology and *rbcL* sequences," *Progress in Natural Science*, vol. 19, no. 8, pp. 931–945, 2009.

[14] L. Y. Chen, J. M. Chen, R. Wahiti Gituru, and Q. F. Wang, "Generic phylogeny, historical biogeography and character evolution of the cosmopolitan aquatic plant family Hydrocharitaceae," *BMC Evolutionary Biology*, vol. 12, article 30, 2012.

[15] C. Lucas, T. Thangaradjou, and J. Papenbrock, "Development of a DNA barcoding system for seagrasses: successful but not simple," *Public Library of Science ONE*, vol. 7, no. 1, Article ID e35107, 2012.

[16] J. L. Olsen, W. T. Stam, J. A. Coyer et al., "North Atlantic phylogeography and large-scale population differentiation of the seagrass *Zostera marina* L.," *Molecular Ecology*, vol. 13, no. 7, pp. 1923–1941, 2004.

[17] J. Kuo and C. den Hartog, "Seagrass morphology, anatomy and ultrastructure," in *Seagrass: Biology, Ecology and Conservation*, A. W. D. Larkum, R. J. Orth, and C. M. Duarte, Eds., pp. 51–87, Springer, Dordrecht, The Netherlands, 2006.

[18] F. Short, T. Carruthers, W. Dennison, and M. Waycott, "Global seagrass distribution and diversity: a bioregional model," *Journal of Experimental Marine Biology and Ecology*, vol. 350, no. 1-2, pp. 3–20, 2007.

[19] N. Marba, M. Holmer, E. Gacia, and C. Barron, "Seagrass beds and coastal biogeochemistry," in *Seagrass: Biology, Ecology and Conservation*, A. W. D. Larkum, R. J. Orth, and C. M. Duarte, Eds., pp. 135–157, Springer, Dordrecht, The Netherlands, 2006.

[20] S. Beer, A. Shomer-ilan, and Y. Waisel, "Carbon metabolism in seagrasses: II. Patterns of photos ynthetic CO_2 incorporation," *Journal of Experimental Botany*, vol. 31, no. 4, pp. 1019–1026, 1980.

[21] H. Frost-Christensen and K. Sand-Jensen, "The quantum efficiency of photosynthesis in macroalgae and submerged angiosperms," *Oecologia*, vol. 91, no. 3, pp. 377–384, 1992.

[22] J. A. Raven, A. M. Johnston, J. E. Kübler et al., "Seaweeds in cold seas: evolution and carbon acquisition," *Annals of Botany*, vol. 90, no. 4, pp. 525–536, 2002.

[23] P. Stiling and T. Cornelissen, "How does elevated carbon dioxide (CO_2) affect plant-herbivore interactions? A field experiment and meta-analysis of CO_2-mediated changes on plant chemistry and herbivore performance," *Global Change Biology*, vol. 13, no. 9, pp. 1823–1842, 2007.

[24] T. Arnold, C. Mealey, H. Leahey et al., "Ocean acidification and the loss of phenolic substances in marine plants," *Public Library of Science ONE*, vol. 7, no. 4, Article ID e35107, 2012.

[25] A. D. Barnabas, "Apoplastic tracer studies in the leaves of a seagrass. II. Pathway into leaf veins," *Aquatic Botany*, vol. 35, no. 3-4, pp. 375–386, 1989.

[26] A. D. Barnabas, "Casparian band-like structures in the root hypodermis of some aquatic angiosperms," *Aquatic Botany*, vol. 55, no. 3, pp. 217–225, 1996.

[27] O. Pedersen, J. Borum, C. M. Duarte, and M. D. Fortes, "Oxygen dynamics in the rhizosphere of *Cymodocea rotundata*," *Marine Ecology Progress Series*, vol. 169, pp. 283–288, 1998.

[28] T. van der Heide, L. L. Govers, J. Fouw et al., "A three-stage symbiosis forms the foundation of seagrass ecosystems," *Science*, vol. 336, no. 6087, pp. 1432–1434, 2012.

[29] I. Johansson, M. Karlsson, U. Johanson, C. Larsson, and P. Kjellbom, "The role of aquaporins in cellular and whole plant water balance," *Biochimica et Biophysica Acta*, vol. 1465, no. 1-2, pp. 324–342, 2000.

[30] P. Maestrini, T. Giordani, A. Lunardi, A. Cavallini, and L. Natali, "Isolation and expression of two aquaporin-encoding genes from the marine phanerogam *Posidonia oceanica*," *Plant and Cell Physiology*, vol. 45, no. 12, pp. 1838–1847, 2004.

[31] R. Cozza and T. Pangaro, "Tissue expression pattern of two aquaporin-encoding genes in different organs of the seagrass *Posidonia oceanica*," *Aquatic Botany*, vol. 91, no. 2, pp. 117–121, 2009.

[32] D. H. Attaway, P. L. Parker, and J. A. Mears, "Normal alkanes of five coastal spermatophytes," *Publications of the Institute of Marine Science, University of Texas*, vol. 15, pp. 13–19, 1970.

[33] F. T. Gillan, R. W. Hogg, and E. A. Drew, "The sterol and fatty acid compositions of seven tropical seagrasses from North Queensland, Australia," *Phytochemistry*, vol. 23, no. 12, pp. 2817–2821, 1984.

[34] C. McMillan, S. C. Williams, L. Escobar, and O. Zapata, "Isoenzymes, secondary compounds and experimental cultures of Australian seagrasses in *Halophila, Halodule, Zostera, Amphibolis* and *Posidonia*," *Australian Journal of Botany*, vol. 29, pp. 247–260, 1981.

[35] C. McMillan, "Sulfated flavonoids and leaf morphology of the Halophila ovalis-H. minor complex (hydrocharitaceae) in the Pacific Islands and Australia," *Aquatic Botany*, vol. 16, no. 4, pp. 337–347, 1983.

[36] T. M. Arnold and N. M. Targett, "Marine tannins: the importance of a mechanistic framework for predicting ecological roles," *Journal of Chemical Ecology*, vol. 28, no. 10, pp. 1919–1934, 2002.

[37] L. H. Vergeer, T. L. Aarts, and J. D. De Groot, "The "wasting disease" and the effect of abiotic factors (light intensity, temperature, salinity) and infection with *Labyrinthula zosterae* on the phenolic content of *Zostera marina* shoots," *Aquatic Botany*, vol. 52, no. 1-2, pp. 35–44, 1995.

[38] M. Boudet, "Evolution and current status of research in phenolic compounds," *Phytochemistry*, vol. 68, no. 22–24, pp. 2722–2735, 2007.

[39] J. S. Todd, R. C. Zimmerman, P. Crews, and R. S. Alberte, "The antifouling activity of natural and synthetic phenolic acid sulphate esters," *Phytochemistry*, vol. 34, no. 2, pp. 401–404, 1993.

[40] R. C. Quackenbush, D. Bunn, and W. Lingren, "HPLC determination of phenolic acids in the water-soluble extract of *Zostera marina* L. (eelgrass)," *Aquatic Botany*, vol. 24, no. 1, pp. 83–89, 1986.

[41] O. Zapata and C. McMillan, "Phenolic acids in seagrasses," *Aquatic Botany*, vol. 7, pp. 307–317, 1979.

[42] J. B. Harborne and C. A. Williams, "Occurrence of sulphated flavones and caffeic acid esters in members of the fluviales," *Biochemical Systematics and Ecology*, vol. 4, no. 1, pp. 37–41, 1976.

[43] C. McMillan, O. Zapata, and L. Escobar, "Sulphated phenolic compounds in seagrasses," *Aquatic Botany*, vol. 8, pp. 267–278, 1980.

[44] S. Agostini, J. M. Desjobert, and G. Pergent, "Distribution of phenolic compounds in the seagrass *Posidonia oceanica*," *Phytochemistry*, vol. 48, no. 4, pp. 611–617, 1998.

[45] G. Nuissier, B. Rezzonico, and M. Grignon-Dubois, "Chicoric acid from *Syringodium filiforme*," *Food Chemistry*, vol. 120, no. 3, pp. 783–788, 2010.

[46] Y. Meng, A. J. Krzysiak, M. J. Durako, J. I. Kunzelman, and J. L. C. Wright, "Flavones and flavone glycosides from *Halophila johnsonii*," *Phytochemistry*, vol. 69, no. 14, pp. 2603–2608, 2008.

[47] F. Bitam, M. L. Ciavatta, M. Carbone, E. Manzo, E. Mollo, and M. Gavagnin, "Chemical analysis of flavonoid constituents of the seagrass Halophila stipulacea: first finding of malonylated derivatives in marine phanerogams," *Biochemical Systematics and Ecology*, vol. 38, no. 4, pp. 686–690, 2010.

[48] T. M. Arnold, C. E. Tanner, M. Rothen, and J. Bullington, "Wound-induced accumulations of condensed tannins in turtlegrass, *Thalassia testudinum*," *Aquatic Botany*, vol. 89, no. 1, pp. 27–33, 2008.

[49] P. G. Harrison, "Detrital processing in seagrass systems: a review of factors affecting decay rates, remineralization and detritivory," *Aquatic Botany*, vol. 35, no. 3-4, pp. 263–288, 1989.

[50] O. Dumay, J. Costa, J. M. Desjobert, and G. Pergent, "Variations in the concentration of phenolic compounds in the seagrass *Posidonia oceanica* under conditions of competition," *Phytochemistry*, vol. 65, no. 24, pp. 3211–3220, 2004.

[51] V. A. Klap, M. A. Hemminga, and J. J. Boon, "Retention of lignin in seagrasses: angiosperms that returned to the sea," *Marine Ecology Progress Series*, vol. 194, pp. 1–11, 2000.

[52] J. M. Espiñeira, E. Uzal, L. V. Gómez Ros et al., "Distribution of lignin monomers and the evolution of lignification among lower plants," *Plant Biology*, vol. 13, no. 1, pp. 59–68, 2011.

[53] Z. Jin, S. Shao, K. S. Katsumata, and K. Iiyama, "Lignin characteristics of peculiar vascular plants," *Journal of Wood Science*, vol. 53, no. 6, pp. 520–523, 2007.

[54] P. T. Martone, J. M. Estevez, F. Lu et al., "Discovery of lignin in seaweed reveals convergent evolution of cell-wall architecture," *Current Biology*, vol. 19, no. 2, pp. 169–175, 2009.

[55] R. Kristensen, S. Coulson, and A. Gordon, "THM PyGC-MS of wood fragment and vegetable fibre forensic samples," *Journal of Analytical and Applied Pyrolysis*, vol. 86, no. 1, pp. 90–98, 2009.

[56] A. R. Barceló, L. V. G. Ros, and A. E. Carrasco, "Looking for syringyl peroxidases," *Trends in Plant Science*, vol. 12, no. 11, pp. 486–491, 2007.

[57] C. Mathé, A. Barre, C. Jourda, and C. Dunand, "Evolution and expression of class III peroxidases," *Archives of Biochemistry and Biophysics*, vol. 500, no. 1, pp. 58–65, 2010.

[58] L. Duroux and K. G. Welinder, "The peroxidase gene family in plants: a phylogenetic overview," *Journal of Molecular Evolution*, vol. 57, no. 4, pp. 397–407, 2003.

[59] C. Lucas, *Classification and characterization of seagrass species using DNA barcoding and physiological parameters [M.S. thesis]*, Leibniz University Hannover, Hannover, Germany, 2011.

[60] D. C. Rowley, M. S. T. Hansen, D. Rhodes et al., "Thalassiolins A–C: new marine-derived inhibitors of HIV cDNA integrase," *Bioorganic and Medicinal Chemistry*, vol. 10, no. 11, pp. 3619–3625, 2002.

[61] P. R. Jensen, K. M. Jenkins, D. Porter, and W. Fenical, "Evidence that a new antibiotic flavone glycoside chemically defends the sea grass *Thalassia testudinum* against *Zoosporic fungi*," *Applied and Environmental Microbiology*, vol. 64, no. 4, pp. 1490–1496, 1998.

[62] S. H. Qi, S. Zhang, P. Y. Qian, and B. G. Wang, "Antifeedant, antibacterial, and antilarval compounds from the South China Sea seagrass *Enhalus acoroides*," *Botanica Marina*, vol. 51, no. 5, pp. 441–447, 2008.

[63] M. Cannac, L. Ferrat, C. Pergent-Martini, G. Pergent, and V. Pasqualini, "Effects of fish farming on flavonoids in *Posidonia oceanica*," *Science of the Total Environment*, vol. 370, no. 1, pp. 91–98, 2006.

[64] M. Cannac, L. Ferrat, T. Barboni, G. Pergent, and V. Pasqualini, "The influence of tissue handling on the flavonoid content of the aquatic plant *Posidonia oceanica*," *Journal of Chemical Ecology*, vol. 33, no. 5, pp. 1083–1088, 2007.

[65] N. M. Gavin and M. J. Durako, "Localization and antioxidant capacity of flavonoids from intertidal and subtidal *Halophila johnsonii* and *Halophila decipiens*," *Aquatic Botany*, vol. 95, no. 3, pp. 242–247, 2011.

[66] M. Takagi, S. Funahashi, K. Ohta T, and T. Nakabayashi, "Phyllosadine, a new flavonoidal alkaloid from the sea-grass *Phyllospadix iwatensis*," *Agricultural and Biological Chemistry*, vol. 44, no. 12, pp. 3019–3020, 1980.

[67] I. Kontiza, M. Stavri, M. Zloh, C. Vagias, S. Gibbons, and V. Roussis, "New metabolites with antibacterial activity from the marine angiosperm *Cymodocea nodosa*," *Tetrahedron*, vol. 64, no. 8, pp. 1696–1702, 2008.

[68] E. A. Drew, "Sugars, cytolitols and seagrass phylogeny," *Aquatic Botany*, vol. 15, no. 4, pp. 387–408, 1983.

[69] E. A. Drew, "Factors affecting photosynthesis and its seasonal variation in the seagrasses *Cymodocea nodosa* (Ucria) Aschers, and *Posidonia oceanica* (L.) Delile in the Mediterranean," *Journal of Experimental Marine Biology and Ecology*, vol. 31, no. 2, pp. 173–194, 1978.

[70] S. D. Tyerman, A. I. Hatcher, R. J. West, and A. W. D. Larkum, "*Posidonia australis* growing in altered salinities: leaf growth, regulation of turgor and the development of osmotic gradients," *Australian Journal of Plant Physiology*, vol. 11, no. 1-2, pp. 35–47, 1984.

[71] V. H. Pomin, "Structural and functional insights into sulfated galactans: a systematic review," *Glycoconjugate Journal*, vol. 27, no. 1, pp. 1–12, 2010.

[72] R. S. Aquino, A. M. Landeira-Fernandez, A. P. Valente, L. R. Andrade, and P. A. S. Mourão, "Occurrence of sulfated galactans in marine angiosperms: evolutionary implications," *Glycobiology*, vol. 15, no. 1, pp. 11–20, 2005.

[73] R. S. Aquino, C. Grativol, and P. A. S. Mourão, "Rising from the sea: correlations between sulfated polysaccharides and salinity in plants," *PLoS ONE*, vol. 6, no. 4, Article ID e18862, 2011.

[74] G. Michel, T. Tonon, D. Scornet, J. M. Cock, and B. Kloareg, "The cell wall polysaccharide metabolism of the brown alga *Ectocarpus siliculosus*. Insights into the evolution of extracellular matrix polysaccharides in Eukaryotes," *New Phytologist*, vol. 188, no. 1, pp. 82–97, 2010.

[75] L. A. Lewis and R. M. McCourt, "Green algae and the origin of land plants," *American Journal of Botany*, vol. 91, no. 10, pp. 1535–1556, 2004.

[76] J. M. Cock, L. Sterck, P. Rouzé et al., "The *Ectocarpus* genome and the independent evolution of multicellularity in brown algae," *Nature*, vol. 465, pp. 617–621, 2010.

[77] F. G. Donnan, "The theory of membrane equilibria," *Chemical Reviews*, vol. 1, no. 1, pp. 73–90, 1924.

[78] G. E. Shaw, "Bio-controlled thermostasis involving the sulfur cycle," *Climatic Change*, vol. 5, no. 3, pp. 297–303, 1983.

[79] J. W. H. Dacey and N. V. Blough, "Hydroxide decomposition of dimethysulfoniopropionate to form dimethylsulfide," *Geophysical Research Letters*, vol. 14, no. 12, pp. 1246–1249, 1987.

[80] J. W. H. Dacey, G. M. King, and P. S. Lobel, "Herbivory by reef fishes and the production of dimethylsulfide and acrylic acid," *Marine Ecology Progress Series*, vol. 112, no. 1-2, pp. 67–74, 1994.

[81] P. H. Nienhuis, "Background levels of heavy metals in nine tropical seagrass species in Indonesia," *Marine Pollution Bulletin*, vol. 17, no. 11, pp. 508–511, 1986.

[82] V. A. Catsiki and P. Panayotidis, "Copper, chromium and nickel in tissues of the Mediterranean seagrasses *Posidonia oceanica* and *Cymodocea nodosa* (Potamogetonaceae) from Greek coastal areas," *Chemosphere*, vol. 26, no. 5, pp. 963–978, 1993.

[83] C. Govindasamy, M. Arulpriya, P. Ruban, J. L. Francisca, and A. Ilayaraja, "Concentration of heavy metals in seagrasses tissue of the Palk Strait, Bay of Bengal," *Environmental Sciences*, vol. 2, no. 1, pp. 145–153, 2011.

[84] C. Lafabrie, C. Pergent-Martini, and G. Pergent, "Metal contamination of *Posidonia oceanica* meadows along the Corsican coastline (Mediterranean)," *Environmental Pollution*, vol. 151, no. 1, pp. 262–268, 2008.

[85] L. Marín-Guirao, A. M. Atucha, J. L. Barba, E. M. López, and A. J. García Fernández, "Effects of mining wastes on a seagrass ecosystem: metal accumulation and bioavailability, seagrass dynamics and associated community structure," *Marine Environmental Research*, vol. 60, no. 3, pp. 317–337, 2005.

[86] L. Li and X. Huang, "Three tropical seagrasses as potential bio-indicators to trace metals in Xincun Bay, Hainan Island, South China," *Chinese Journal of Oceanology and Limnology*, vol. 30, no. 2, pp. 212–224, 2012.

[87] T. Thangaradjou, S. Raja, P. Subhashini, E. P. Nobi, and E. Dilipan, "Heavy metal enrichment in the seagrasses of Lakshadweep group of islands—a multivariate statistical analysis," *Environmental Monitoring and Assessment*. In press.

[88] T. Alvarez-Legorreta, D. Mendoza-Cozatl, R. Moreno-Sanchez, and G. Gold-Bouchot, "Thiol peptides induction in the seagrass *Thalassia testudinum* (Banks ex König) in response to cadmium exposure," *Aquatic Toxicology*, vol. 86, no. 1, pp. 12–19, 2008.

[89] A. K. Grennan, "Metallothioneins, a diverse protein family," *Plant Physiology*, vol. 155, no. 4, pp. 1750–1751, 2011.

[90] C. Cobbett and P. Goldsbrough, "Phytochelatins and metallothioneins: roles in heavy metal detoxification and homeostasis," *Annual Review of Plant Biology*, vol. 53, pp. 159–182, 2002.

[91] C. A. Blindauer and O. I. Leszczyszyn, "Metallothioneins: unparalleled diversity in structures and functions for metal ion homeostasis and more," *Natural Product Reports*, vol. 27, no. 5, pp. 720–741, 2010.

[92] T. Giordani, L. Natali, B. E. Maserti, S. Taddei, and A. Cavallini, "Characterization and expression of DNA sequences encoding putative type-II metallothioneins in the seagrass *Posidonia oceanica*," *Plant Physiology*, vol. 123, no. 4, pp. 1571–1582, 2000.

[93] R. Cozza, T. Pangaro, P. Maestrini, T. Giordani, L. Natali, and A. Cavallini, "Isolation of putative type 2 metallothionein encoding sequences and spatial expression pattern in the seagrass *Posidonia oceanica*," *Aquatic Botany*, vol. 85, no. 4, pp. 317–323, 2006.

[94] S. Wyllie-Echeverria, P. Arzel, and P. A. Cox, "Seagrass conservation: lessons from ethnobotany," *Pacific Conservation Biology*, vol. 5, no. 4, pp. 333–335, 2000.

[95] P. G. Harrison, "Control of microbial growth and of amphipod grazing by water-soluble compounds from leaves of *Zostera marina*," *Marine Biology*, vol. 67, no. 2, pp. 225–230, 1982.

[96] F. Villa, D. Albanese, B. Giussani, P. S. Stewart, D. Daffonchio, and F. Cappitelli, "Hindering biofilm formation with zosteric acid," *Biofouling*, vol. 26, no. 6, pp. 739–752, 2010.

[97] C. A. Barrios, Q. Xu, T. Cutright, and B. M. Z. Newby, "Incorporating zosteric acid into silicone coatings to achieve its slow release while reducing fresh water bacterial attachment," *Colloids and Surfaces B*, vol. 41, no. 2-3, pp. 83–93, 2005.

[98] S. H. Bhosale, V. L. Nagle, and T. G. Jagtap, "Antifouling potential of some marine organisms from India against species of *Bacillus* and *Pseudomonas*," *Marine Biotechnology*, vol. 4, no. 2, pp. 111–118, 2002.

[99] Y. N. Loenko, A. A. Artyukov, E. P. Kozlovskaya, V. A. Miroshnichenko, and G. B. Elyakov, *Zosterin*, Dal'nauka, Vladivostok, Russia, 1997.

[100] G. Nuissier, F. Diaba, and M. Grignon-Dubois, "Bioactive agents from beach waste: *Syringodium* flotsam evaluation as a new source of l-chiro-inositol," *Innovative Food Science and Emerging Technologies*, vol. 9, no. 3, pp. 396–400, 2008.

[101] J. F. Ackerman, "Sexual reproduction of seagrasses: pollination in the marine context," in *Seagrass: Biology, Ecology and Conservation*, A. W. D. Larkum, R. J. Orth, and C. M. Duarte, Eds., pp. 89–109, Springer, Dordrecht, The Netherlands, 2006.

[102] N. Tanaka, H. Setoguchi, and J. Murata, "Phylogeny of the family hydrocharitaceae inferred from *rbcL* and *matK* gene sequence data," *Journal of Plant Research*, vol. 110, no. 1099, pp. 329–337, 1997.

[103] M. Waycott, D. W. Freshwater, R. A. York, A. Calladine, and W. J. Kenworthy, "Evolutionary trends in the seagrass genus *Halophila* (Thouars): insights from molecular phylogeny," *Bulletin of Marine Science*, vol. 71, no. 3, pp. 1299–1308, 2002.

[104] Y. Ito and N. Tanaka, "Hybridisation in a tropical seagrass genus, *Halodule* (Cymodoceaceae), inferred from plastid and nuclear DNA phylogenies," *Telopea*, vol. 13, no. 1-2, pp. 219–231, 2011.

[105] Y. Kato, K. Aioi, Y. Omori, N. Takahata, and Y. Satta, "Phylogenetic analyses of *Zostera* species based on *rbcL* and *matK* nucleotide sequences: implications for the origin and diversification of seagrasses in Japanese waters," *Genes and Genetic Systems*, vol. 78, no. 5, pp. 329–342, 2003.

[106] D. H. Les, M. L. Moody, S. W. L. Jacobs, and R. J. Bayer, "Systematics of Seagrasses (Zosteraceae) in Australia and New Zealand," *Systematic Botany*, vol. 27, no. 3, pp. 468–484, 2002.

[107] G. Procaccini, L. Mazzella, R. S. Alberte, and D. H. Les, "Chloroplast tRNA(Leu) (UAA) intron sequences provide phylogenetic resolution of seagrass relationships," *Aquatic Botany*, vol. 62, no. 4, pp. 269–283, 1999.

11

Disturbance of Opportunistic Small-Celled Phytoplankton in Lake Kinneret

Yury Kamenir and Zvy Dubinsky

The Mina and Everard Goodman Faculty of Life Sciences, Bar-Ilan University, Ramat-Gan 52900, Israel

Correspondence should be addressed to Yury Kamenir, kamenir@mail.biu.ac.il

Academic Editors: R. Dolferus, K. Pawlowski, and A. Singh

In spite of the chaotic dynamics of specific populations, similarity of annual species-abundance distributions was proven for phytoplankton assemblage during a "stable" period (1985–1994) of Lake Kinneret (Israel). This similarity declined during the "extreme" years (1995–1999) that followed, characterized by explicit changes in the phytoplankton annual-succession pattern. The rank-abundance distributions of species exhibit a pronounced difference between the taxonomically rich central region, producing the reliable assemblage backbone and highly variable tails of a few species. Therefore, the distribution pattern comparison enhances the importance of ubiquitous small disturbances valuable for diagnostics. Some phyla (in this case, Cyanophyta) were especially vulnerable to structural changes. A simple disturbance index was constructed, based on opportunistic small-celled species. The fine-structure disturbances, which can provide early-warning information, are discussed.

1. Introduction

Ever-growing anthropogenic pressure causes pronounced changes in natural ecosystems on both local and global scales. Some of these changes are undesirable and even dangerous. Such a situation demands scientific tools for quantitative estimations, ecological forecast, and diagnostics of aquatic assemblage structural changes.

Size-spectrum (Figure 1) analysis [1] is one of the tools capable of encompassing the whole assemblage and supporting various quantitative studies. Large-scale size-spectrum studies have demonstrated the inherent typical patterns in the biomass size spectrum and its normalized variant, describing aquatic communities of very different natures and spatial scales [1, 2]. While "ataxonomic" biomass spectra ignore the taxonomic affiliation of organisms, another type of size structure, specifically, size-frequency distributions of species and other taxonomic units, have also been studied for a long time and exhibit common patterns [3]. These patterns also seem to be capable of surviving strong environmental stress and some forms of anthropogenic regulations [4]. The interspecific frequency distributions of animal body sizes have attracted much attention of ecologists and evolutionary biologists. Broad-scale comparisons of invariant size-frequency distributions of species have already been established for several terrestrial and aquatic assemblages [3, 5]. Common patterns, following one of the oldest and most universal laws, were also found for species-abundance distributions, including the rank-abundance distribution and the respective histogram based on the species-abundance classes [6, 7].

Comparisons of several types of species distributions enhance better appreciation of the vulnerability of the aquatic assemblage. We have already applied several types of taxonomic and ataxonomic distributions to analyze the phytoplankton of subtropical Lake Kinneret, Israel [8, 9]. Our comparative studies have shown (Figure 1) that a few small-celled species dominate the cell-abundance spectra, and several large-cell species make up the bulk of the assemblage biomass, whereas many rare species with low biomass and population abundances increase assemblage diversity [8, 9]. Three zones, in which 2 opposite tails include a small number of species, while the central size region encompasses the community main taxonomical store, are discernible

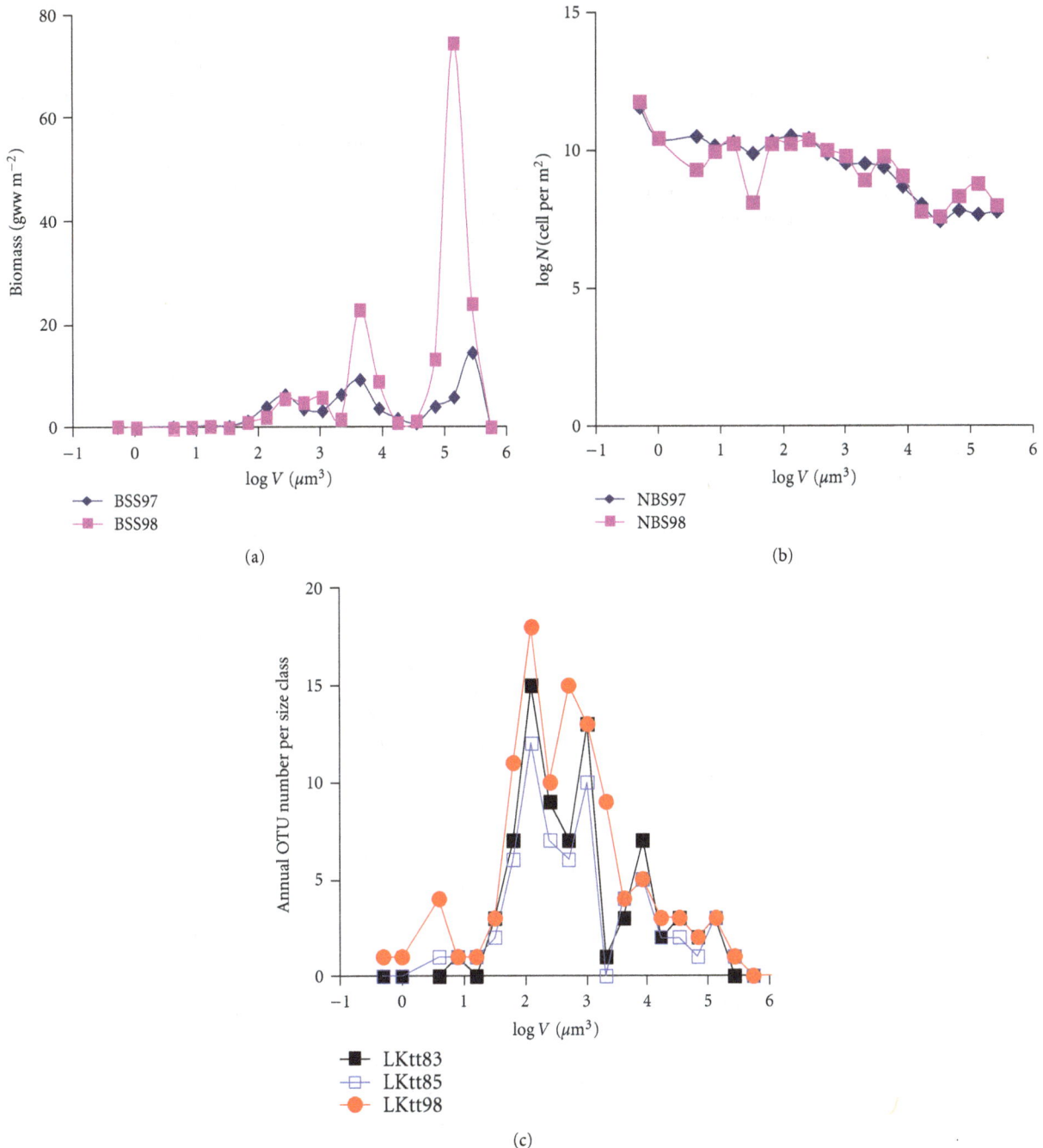

FIGURE 1: Comparisons of annual size spectra. (a, b) The "Sheldon" and normalized biomass size spectra, respectively, for extremely disturbed years, 1997 and 1998 (after [8]). The biomass size spectrum describes the assemblage biomass distribution to size classes of cell volume logarithm, and the normalized biomass size spectrum can be interpreted as the cell abundance (N cell) distribution approximation. (c) The traditional taxonomic size-spectrum (TTSS) change for representative years of the stable (1983, 1985) and extremely disturbed (1998) periods (after [9]). The curves are marked by the year they represent (yy). All phytoplankton species with cell size $\geq 2\,\mu$m and colonial species of smaller cell size are included, but not unicellular picoplankton. Size classes were created by doubling cell volume (V) and are presented on a logarithmic scale ($\log_{10} V$).

(Figure 1(c)). While the central region shows rather stable cell-abundance curve (Figure 1(b)), both peripheral zones exhibit pronounced variability of the curve. The right tail of the cell-size scale, describing very large cells ($\log V > 4$), contained the main part of the biomass (Figure 1(a)). In contrast, the left-hand tail seemed to be nonsignificant if the biomass was the main criterion; however, both the cell abundance and species number demonstrated high vulnerability at that size range (Figures 1(b) and 1(c)).

The notable similarity of the assemblage taxonomic structure patterns (Figure 1(c)) agrees well with the stability of the assemblage ataxonomic size structure (Figure 1(b)), while it is in contrast to the well-known empirical data and mathematical models considering a very high variability and chaotic dynamics of the assemblage populations [10–12]. Whereas the response of the average lake communities to environmental impacts is often quite predictable, the algal community dynamics on the species level usually seems to be erratic [12]. Therefore, the similarity of the whole-assemblage taxonomic or ataxonomic spectrum general pattern looks especially valuable for purposes of modeling and prediction. At the same time, comparisons of small details in the pattern fine-structure can highlight diagnostically important traits.

A fundamental problem of environmental science is to characterize the anthropogenic impacts on aquatic communities in a quantitative way, providing detection of disturbances at an early stage [13, 14]. While biomass is used as a traditional criterion, some species often demonstrate very high and very variable abundances [15]. Such cell-abundance variability seems suitable for providing early-warning information. The cell-abundance distributions can serve to obtain insights into disturbance analysis and perturbation diagnostics [7]. The departure from the lognormal distribution [6] was suggested as providing indices of pollution [13]. The assemblage-perturbation index, based on opportunistic species [14], was produced from a small number of species.

The well-documented stable pattern of Lake Kinneret annual succession was maintained during many years (1969–1992), accompanied with the 1993–94 anthropogenic impacts and subsequent disturbed years [15–17]. The long-term record of the Kinneret can be used to validate hypotheses considering the phytoplankton composition reaction on increased anthropogenic stress. Some parts of phytoplankton (phyla) positioned at specific regions of taxonomic distributions can be especially sensitive under disturbances and, therefore, valuable for diagnostics.

In this new work, we attempted to select and compare the species distribution patterns of the most interesting parts of a lake phytoplankton assemblage, as seen from long-term monitoring.

The general aim of the study was a search for diagnostically valuable quantitative changes in specific taxonomic structure patterns based on species abundance of Lake Kinneret (Israel) phytoplankton.

The working hypothesis was that some distinctions between consistent backbone and diagnostically valuable parts of phytoplankton taxonomic structure can be ascertained. A simple quantitative index can be developed to estimate community disturbance.

2. Methods

2.1. Site Description.
Lake Kinneret, Israel, situated ca. 210 m below sea level at 32°45′N, 35°30′E [16], is a warm, monomictic lake with a surface area of 170 km², maximum depth of 44 m, and mean depth of 26 m. The lake, which is used for recreation and fishery, is also the main source of drinking water in Israel, and its water quality is of prime

national importance. Therefore, a routine program to monitor numerous biotic and abiotic parameters has been carried out since 1969. From the beginning of this monitoring program, Kinneret phytoplankton have exhibited distinct stability of species succession and biomass-dynamics annual patterns [17]. However, numerous human-induced regulations and climatic changes have led to drastic modifications in phytoplankton succession. Two major perturbations that affected the Kinneret ecosystem were the collapse of the Kinneret fisheries (1993) as a result of overfishing, and the reflooding of the dried peat soils of the Hula Valley in the lake's catchment (1994), leading to changes in amounts and contents (nutrients, heavy metals, and organic matter) of water flowing into the lake [15]. After 1994, these modifications were expressed in the pronounced interannual variability in phytoplankton biomass dynamics. The most notable changes included the absence of the prevailing spring *Peridinium gatunense* blooms in some years, their record-high intensification during some other years, intensification of winter *Aulacoseira granulata* blooms, replacement of the summer species assemblage of mostly nanoplanktonic palatable cells with less palatable forms, proliferation in summer-fall of N_2-fixing, toxic cyanobacteria, and so forth [15].

For this study, we chose to focus on a long (15 years), continuous time interval consisting of two parts, referred to as the "stable" period (1985–1994) and the "extreme" period that followed (1995–1999). The extreme period and one of the previous years (1985) had been already subjected to size-spectrum analyses. The data, methods, and revealed typical patterns of biomass and taxonomic size-spectra of the whole phytoplankton assemblage were described in [8, 9].

2.2. Phytoplankton Data Acquisition and Processing.
As part of the routine monitoring program, phytoplankton samples were collected biweekly using a 5-L bathometer at a fixed pelagic station at the deepest part of the lake, from 9–11 discrete depths throughout the water column. Microscopic counting of Lugol-preserved samples was produced using inverted microscope [15]. All phytoplankton species with individual cells greater than 2 μm in diameter were identified and counted according to species, and for the more abundant species with variable cell size—also according to size categories. Since our individual taxon is not strictly a species but in some cases also a size category within a species, we refer to each as an operational taxonomic unit (OTU) [18]. From the smaller cell range, only the relatively common colony-forming cyanobacteria were included. Sample processing was described in detail in [9, 15].

2.3. Species-Abundance Distributions.
Species-abundance distributions of all OTUs encountered within a community were developed. We applied the annual maximal abundance estimates (N_j, cell per mL) in order to amplify sensitivity while looking for opportunistic and other disturbance-sensitive species. The dependence of maximal annual species abundance (N_j) on its typical cell volume (V_j) have been described via scatter plots and linear regressions. Each annual rank-abundance distribution was created of all OTU

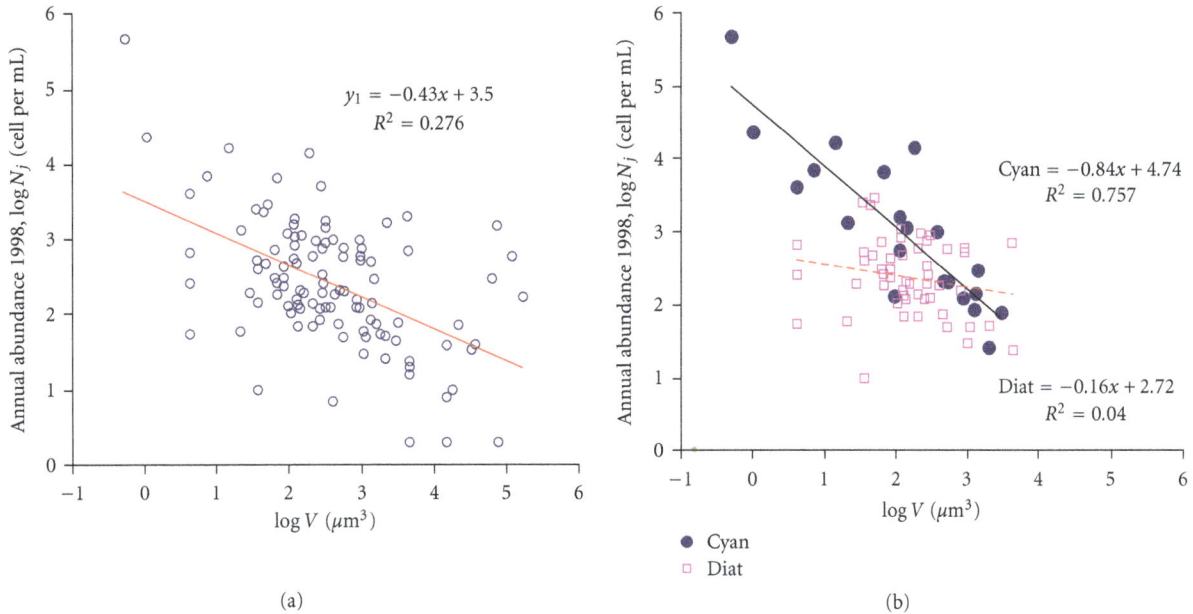

FIGURE 2: Scatter plots of maximal cell abundance of each species (N_j) versus its cell volume (V_j), for a disturbed year (1998). (a) All phytoplankton. (b) Smallest algae—Cyanophyta (Cyan) and diatoms (Diat). Both axes are log-transformed.

registered during one year, sorted in the descending order of the species abundances (N_j). Each annual species abundance distribution was also plotted as a histogram of species on the y-axis versus cell abundance using logarithmic x-axis [7]. The size class corresponds to an octave, that is, duplication of the abundance [6].

A modified assemblage disturbance index, based on opportunistic species [14], was produced. The disturbance index was calculated as the average value of 5 log-transformed abundance dominants (ranks 1–5). The difference between the index values of the 1st year (1985) and each following year (i.e., first 5 pairs of log-transformed abundances) was compared by paired t-test. The difference between the disturbance indices of 2 groups of 5 years was tested by two-sample two-sided t-test. The SPSS program, version 15.0 (SPSS Inc., Chicago, IL 60606, USA), was used for all statistical procedures.

3. Results

Scatter plots of maximal annual cell abundance of each species (N_j) versus cell volume (V_j) have shown a very high variability—almost 6 orders of N_j magnitude for the whole phytoplankton assemblage, including 4 orders for 1 size fraction (Figure 2(a)). Only very few species had $\log N_j > 4$, all of them being small species, $\log V_j < 2$. The smallest algae were presented by two phyla—Cyanophyta and Bacillariophyta. These 2 phyla demonstrated very different N_j versus V_j trends: an almost horizontal trend for Bacillariophyta and a very steep negative trend for Cyanophyta (Figure 2(b)). Almost horizontal trends were also shown to several small phyla in the central cell-size region (Chlorophyta, Cryptophyta, Prasynophyta, and Haptophyta); the largest algae (Dinophyta) demonstrated even an opposite trend.

Two 5-year (stable and extreme) periods (1990–1994 and 1995–1999, Figures 3(a) and 3(b), resp.) represent annual histograms of the species number distribution between abundance octaves (i.e., cell-abundance duplications), where each OTU was presented by its annual maximal abundance value (N_j). While the general pattern was almost unchanged, each annual histogram presented slightly different shape (Figure 3(a)). The extreme period (1995–1999, Figure 3(b)) produced more different and symmetrical bells; the last year (1999) exhibited a higher and thinner bell. The most clear distinction between the two periods (Figures 3(a) and 3(b)) was the length of the histogram right tail.

The increase of the highest abundances was also especially conspicuous on the rank-abundance distributions of the stable and extreme periods presented for each annual dataset (Figures 3(b) and 3(c), resp.). The rank-abundance curves produced almost the same shape during the stable years, while the extreme years exhibited much more variable shapes. A notable peculiarity was a sharp division of each curve to 3 parts of considerably different steepness. The middle region exhibited a very gentle, almost straight line suitable for a linear regression. At the same time, the right tail was convex and long, while the very steep left-hand tail was formed by the first 5–7 leading species. The middle part was exceptionally long for years 1998 and 1999, while unusually short for 1990.

After that, the histograms of the 2 periods (a) and (b) were compared using the 5-year average estimates (Figure 3(e)). The general pattern of the 2 histograms was almost the same, however, for the extreme period, the bell was broader due to the left slope movement, and the right-hand tail was longer. The disturbance index was calculated for each annual dataset. The index values for 15 consecutive years were compared with the horizontal

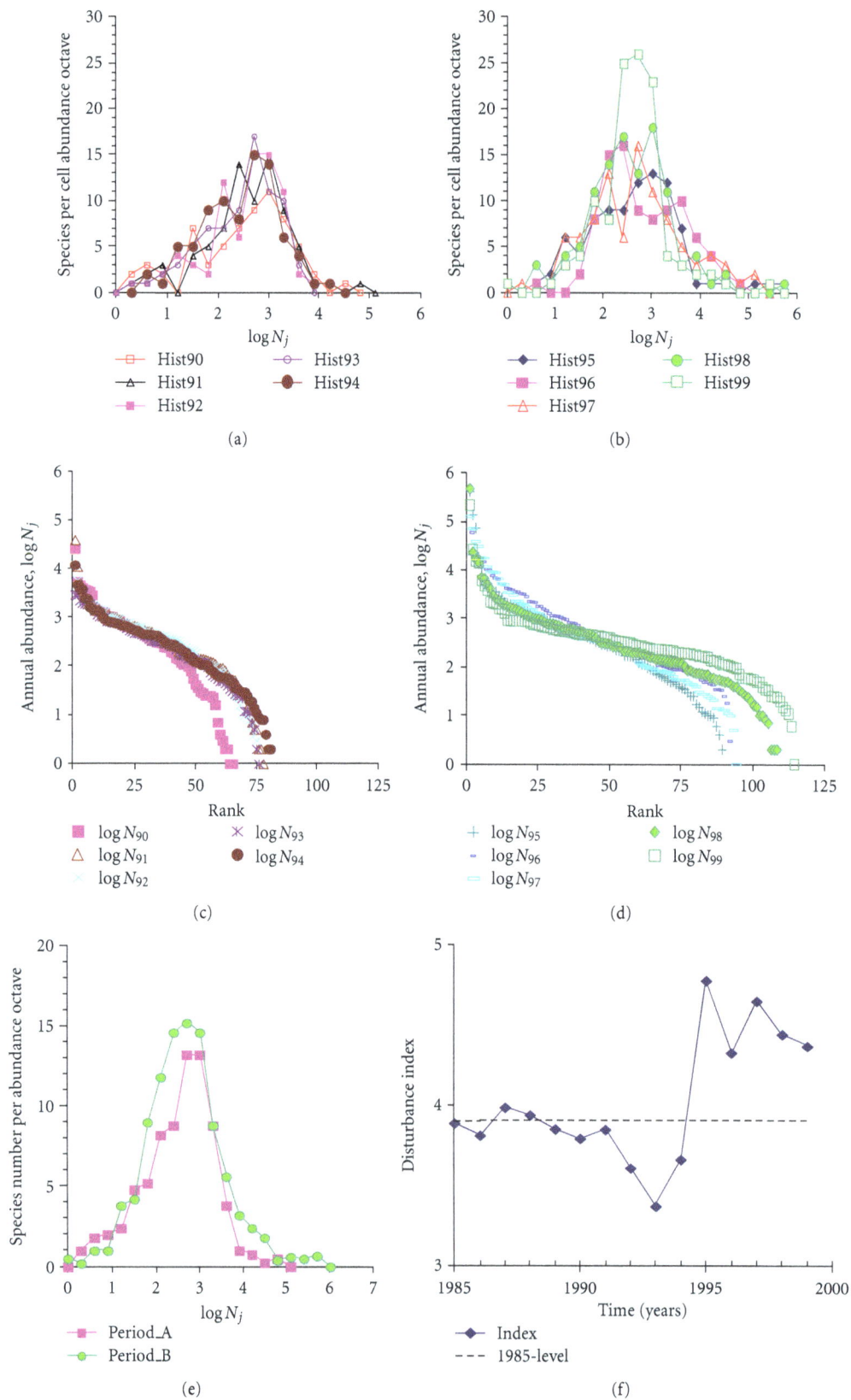

FIGURE 3: Species-abundance distributions of Lake Kinneret phytoplankton. Histograms of species abundance, representing species-number distribution according to octaves (i.e., cell abundance duplications), for 2 periods (a, b). (c, d) Rank-abundance distributions of species for the stable and the following disturbed periods. (e) The 5-year (see plates (a, b)) average histograms of species abundance. (f) Phytoplankton disturbance index, representing the average value of 5 leading abundances (see the Methods), in comparison with the index level for 1985.

line corresponding to the year 1985 (Figure 3(f)). A drastic change of the index values characterized the extreme period of 1995–1999, and even the preceding years (1992–1994). The difference between 2 groups of 5 years (1989–1993 and 1995–1999) was statistically significant (two-sample t-test, $P < 0.001$, $n = 10$). The difference between index values (i.e., 5 log-transformed abundance dominants) of year 1985 and each following year was significant (paired t-test, $n = 5$) for the years 1993 and 1998 ($P < 0.05$) and 1992, 1995–1997 and 1999 ($P < 0.01$).

4. Discussion

The abundance-versus-volume scatter plot shows that the majority of species occupy the center of the phytoplankton cell-volume range (Figure 2(a)). Only several large species (mainly Dinophyta) occupy the right tail ($\log V > 4$), where the main part of total biomass is concentrated (Figure 1(a)). The opposite side of the size range also contains very few species, including the abundance dominants. One of the phyla—Cyanophyta—looks like conquerors, trying to capture the left-side extremity (from $\log V = 3$ to $\log V <$ 0) by very few species (Figure 2(b)), producing a very low biomass of a species and size class (Figure 1(a)), but record high species abundances (Figure 2). While the biomass serves as a traditional criterion for the assemblage disturbance analysis (e.g., bloom of Dinophyta)—especially conspicuous at the right tail ($\log V > 4$)—the opposite tail, describing small cells ($\log V < 1$), can also be used to diagnose the assemblage disturbance.

Statistically significant differences between the annual average estimates of the stable and extreme periods were found for several environmental variables: the lake water level, its phosphorous load, chloride concentration, and surface layer temperature [15]. The nutrient stoichiometry and the rising water temperature seem to be plausible factors favoring some Cyanophyta species.

The increase of the abundance dominants was especially conspicuous while comparing the stable and extreme periods (1990–1994 and 1995–1999, respectively, Figures 3(c), and 3(d)). The species distribution histograms (Figure 3(e)) had almost the same general pattern for 2 periods; hence, the fine structure change seems to be meaningful. The histogram-B left slope moved leftwards due to the addition of a number of new OTUs with relatively small abundances ($\log N_j < 3$). This change can be seen on the rank-abundance distributions (Figure 3(d)) since their middle region length and slope changed due to the addition of several species with low abundances ($\log N_j \sim 2$-3) supplemented by very few abundance champions ($\log N_j > 4$). Such changes can be interpreted as the morphological diversity growth obtained by very few opportunistic populations supported by several populations with much lower abundances. The new species are mainly Cyanophyta which flourished after 1994 (Figure 3(d)). The length of the histogram right-hand tail can serve as a quantitative disturbance indicator.

Species-abundance distributions were considered by numerous specialists as promising for significant insights into basic and applied ecological science [7]. Considering

the pattern changes, we can calculate several characteristics of each distribution (slopes of each region, the central part length, and its linear regression determination coefficient). Then, such indices (e.g., the distribution top-end height) can be easily estimated, therefore, their changes can be used for diagnostics and require a more detailed analysis. The rank-abundance curve left-hand tail was especially steep and short (Figure 3(d)), consisting of only 5–7 species. As a rule, these were small diatoms and Cyanophyta, mainly *Microcystis* spp., widely known as common harmful algae. A drastic change of the disturbance index values characterized the extreme period of 1995–1999 (Figure 3(f)). The early warning information can be seen for several preceding years (1992–1994). The causes of the disturbance-index changes for several years before the apparent changes of the biomass succession deserve additional studies.

5. Conclusions

The whole-assemblage species abundance distributions demonstrated a high level of similarity; the distribution pattern changes become evident only during periods of extreme changes of the phytoplankton assemblage annual succession. At the same time, the species-list variation is notable even during the stable period. In such a system, the phylum appears to be an intermediate level of optimal sensitivity, suitable for the aims of assemblage structural-similarity estimation. Some phyla (Cyanophyta, in our case) demonstrate especially high sensitivity at Lake Kinneret.

The species-abundance distributions can be helpful in measuring phytoplankton-structure disturbances. Distinction between the rank-abundance distribution backbone and its two extremities helps construct simple quantitative indices suitable for diagnosis of assemblage disturbances.

Acknowledgments

The authors thank Eva Feldman and Tatiana Fishbein, who conducted the phytoplankton counts as part of the Kinneret monitoring program funded by the Israel Water Commission. They are grateful to Kirill Khailov, Danara Krupatkina, Oleg Makarov, Giuseppe Morabito, and Yosef Yakobi for valuable discussions and advice.

References

[1] R. W. Sheldon, A. Prakash, and W. H. Sutcliffe, "The size distribution of particles in the ocean," *Limnology and Oceanography*, vol. 17, pp. 327–340, 1972.

[2] S. R. Kerr and L. M. Dickie, *The Biomass Spectrum: A Predator-Prey Theory of Aquatic Production*, Columbia University, New York, NY, USA, 2001.

[3] J. H. Brown, *Macroecology*, Chicago University, Chicago, Ill, USA, 1995.

[4] T. D. Havlicek and S. R. Carpenter, "Pelagic species size distributions in lakes: are they discontinuous?" *Limnology and Oceanography*, vol. 46, no. 5, pp. 1021–1033, 2001.

[5] L. L. Chislenko, *Structure of Fauna and Flora as Related to the Sizes of Organisms*, Moscow University, Moscow, Ruissa, 1981.

[6] F. W. Preston, "The commonness, and rarity, of species," *Ecology*, vol. 29, pp. 254–283, 1948.

[7] B. J. McGill, R. S. Etienne, J. S. Gray et al., "Species abundance distributions: moving beyond single prediction theories to integration within an ecological framework," *Ecology Letters*, vol. 10, no. 10, pp. 995–1015, 2007.

[8] Y. Kamenir, "Stability of Lake Kinneret phytoplankton structure as evidenced by several types of size spectra," *Fundamental and Applied Limnology*, vol. 168, no. 4, pp. 345–354, 2007.

[9] Y. Kamenir, Z. Dubinsky, and T. Zohary, "Consistent annual patterns of water mass occupancy are revealed by taxonomic units of Lake Kinneret phytoplankton," *Israel Journal of Plant Sciences*, vol. 56, no. 1-2, pp. 91–101, 2008.

[10] R. Heerkloss and G. Klinkenberg, "A long-term series of a planktonic foodweb: a case of chaotic dynamics," *Internationale Vereinigung für Theoretische und Angewandte Limnologie*, vol. 26, pp. 1952–1956.

[11] J. Huisman and F. J. Weissing, "Fundamental unpredictability in multispecies competition," *American Naturalist*, vol. 157, no. 5, pp. 488–494, 2001.

[12] M. Scheffer, S. Rinaldi, J. Huisman, and F. J. Weissing, "Why plankton communities have no equilibrium: solutions to the paradox," *Hydrobiologia*, vol. 491, pp. 9–18, 2003.

[13] J. S. Gray, "Detecting pollution induced changes in communities using the log-normal distribution of individuals among species," *Marine Pollution Bulletin*, vol. 12, no. 5, pp. 173–176, 1981.

[14] K. I. Ugland, A. Bjørgesæter, T. Bakke, B. Fredheim, and J. S. Gray, "Assessment of environmental stress with a biological index based on opportunistic species," *Journal of Experimental Marine Biology and Ecology*, vol. 366, pp. 169–174, 2008.

[15] T. Zohary, "Changes to the phytoplankton assemblage of Lake Kinneret after decades of a predictable, repetitive pattern," *Freshwater Biology*, vol. 49, no. 10, pp. 1355–1371, 2004.

[16] C. Serruya, Ed., *The Kinneret. Monographia Biologica*, Junk, The Hague, The Netherlands, 1978.

[17] T. Berman, L. Stone, Y. Z. Yacobi et al., "Primary production and phytoplankton in Lake Kinneret: a long-term record (1972–1993)," *Limnology and Oceanography*, vol. 40, no. 6, pp. 1064–1076, 1995.

[18] P. H. A. Sneath and R. R. Sokal, *Numerical Taxonomy*, Freeman, San Francisco, Calif, USA, 1973.

12

A Comparative Study of Two Endemic *Isoëtes* Species from South Italy

Paola Ernandes and Silvano Marchiori

Laboratory of Systematic Botany and Plant Ecology, Di.S.Te.B.A., University of the Salento, 73100 Lecce, Italy

Correspondence should be addressed to Paola Ernandes, paola.ernandes@unile.it

Academic Editors: M. Adrian, S. Rachmilevitch, S. Satoh, and N. Uehlein

Two *Isoëtes* taxa (Isoëtaceae, Pteridophyta) have recently been described in Puglia and Sicily (south Italy). Though morphologically similar (Ernandes, 2011), they differ in diagnostic characters and habitat preferences. In this paper, we highlight the differences and similarities between *Isoëtes iapygia* Ernandes, Beccarisi *et* Zuccarello (Ernandes et al., 2010) and *Isoëtes todaroana* Troia *et* Raimondo (Troia and Raimondo, 2009). Individuals are described in terms of selected diagnostic characters. Morphometric differences and other distinguishing characteristics suggest that *I. iapygia* and *I. todaroana* should be considered as separate species.

1. Introduction

The Isoëtaceae are a small, cosmopolitan family of aquatic, heterosporous pteridophytes which are related to the extant Selaginellaceae and Lycopodiaceae families. The family is monotypic and comprises 350 species [1, 2], which are classified mainly on the basis of megaspore surface morphology [3–5]. In recent years, a number of investigations have sought to determine the phylogenetic relationships among *Isoëtes* species based on morphological, ecological [6–12], karyological [13, 14] and biomolecular [1, 10, 15, 16].

Macrospore surface morphology provides a ready means for species sorting and is of great importance in species identification [17–22]. Surface features differ markedly among species and therefore have long been used to distinguish taxa. Also known as "sculpturing," spore ornamentation is the most frequently used taxonomic criterion in flora [23]. Together with other features and habitat preferences, macrospore characteristics are also useful for determining natural species alliances [4]. Microspores, on the other hand, have been largely neglected in taxonomic schemes until recently [24].

In the last few years, interest in the "quillwort" genus has led to investigation of other morphological features including leaf anatomy, ligules, corm cross-sections and lobes, velum cover, and scales [24]. Plant morphology is considered to be relevant to plant systematic, ecology, genetics, and physiology [25].

Italy is home to 7 species of *Isoëtes*, 3 of which are protected at national level and listed as endangered due to habitat loss, extension of agricultural land, and invasion by exotic species [26–28].

Two new taxa have been recently described in southern Italy. They appear morphologically similar [29] but differ in terms of diagnostic characters and habitat preferences.

In this paper we highlight the differences and similarities between these two taxa: *Isoëtes iapygia* Ernandes, Beccarisi *et* Zuccarello [30] and *Isoëtes todaroana* Troia *et* Raimondo [31].

2. Materials and Methods

The study area is located in two regions in the south of Italy (Puglia and Sicily), considered the most important hotspots in the Mediterranean Basin (Medàil and Quèzel, [32]). Especially in terms of climate and geological characteristics, the area where *I. iapygia* is located is classified as "Hills of Murge and Salento" [33, 34]. The climate and pedoclimate are Mediterranean-subcontinental to continental. The mean annual air temperature is 14–20°C and mean annual precipitation is 420–700 mm. The rainiest months are October and November while the driest months are from June to August. The soil moisture and temperature regimes are xeric, and to a lesser extent dry xeric or thermic. The geological substrate is characterized by Mesozoic limestone, marl, and residual deposits. The mean altitude is 191 metres a.s.l. and the mean slope is 3% [34]. The main soil types are: shallow and eroded soils, soils with carbonates, clays, and sandy soils [34]. The area where *I. todaroana* is located is classified as "Hills of Sicily on Tertiary clayey flysch, limestone, sandstone and gypsum, and coastal plains" [33, 34]. The climate and pedoclimate are Mediterranean-subtropical. The mean annual air temperature is 16–20°C. The mean annual precipitation is 450–670 mm, and the rainiest months are November and January while the driest months are from May to September. The soil moisture and temperature regimes are xeric and dry xeric or thermic. The geological substrate is characterized by Tertiary clayey flysch, sandstone, and gypsum. The mean altitude is 247 meters a.s.l. while the mean slope is 12% [34]. The main soils are characterised by accumulation of carbonates and more soluble salts, clay, and alluvial deposits [34].

Individuals were described on the basis of the following 20 diagnostic characters: number of corm lobes, corm width, shape of corm section, root shape, number of leaves, leaf length and shape, presence and shape of phyllopodia or scales, air chambers, stomata, shape and length of ligule, velum, spore diameter, spore morphology, spore ornamentation, perine microornamentation, laesura arms, and equatorial ring. In addition, information was collected on habitus, substrate characteristics (pH of soil, type of soil, and geological substrate), type of habitat, community species, and geographical range (sites of presence, chorology). Both fresh and dried individuals of *I. iapygia* and *I. todaroana* were analysed, and all data are shown in Table 1.

Macrosporangia and microsporangia were selected for each individual, randomly selecting spores in accordance with Ernandes et al. [30]. All morphological and anatomical characteristics were observed by SEM, after being treated in accordance with protocols [30]. Sporal characteristics were defined in accordance with Ferrarini et al. [35], Musselmann [24], Prada [6], and Hickey [4].

3. Results and Discussion

The most obvious visible differences between *I. iapygia* and *I. todaroana* are in spore morphology and ornamentation. The SEM analysis showed that *I. iapygia* has larger tuberculate macrospores than *I. todaroana*. These are also round in polar view, with numerous tubercles attached to each other and a rudimentary, undulate equatorial ridge (Figures 1(a) and 1(b)). The laesura arms are flattened and do not form a prominent girdle. In contrast, *I. todaroana* has tuberculate macrospores with aculeate tubercles, triangular in polar view, with a well developed equatorial ridge, raised laesura arms, and a prominent girdle, giving the plant its characteristic shape (Figures 1(c) and 1(d)).

Of great interest is the comparison at higher magnification (reported here for the first time), which shows the microornamentation of the macrospores. The perine of *I. iapygia* has densely fimbriate structures (Figures 2(a) and 2(b)) that differ from those described in the literature [6, 10, 24, 36, 37]; the perine of *I. todaroana* has elongated filaments, welded together, similar to what has been reported for *Isoëtes histrix* Bory (Figures 2(c) and 2(d)) [8, 37, 38].

The two taxa also differ in the size and ornamentation of microspores, ligule shape, and length, leaf shape and number, and corm structure (Table 1). The corm of *I. iapygia* is larger than that of *I. todaroana* and the shape of its cross-section is significantly different (Figure 3(a)): in the *I. iapygia* cross-section the secondary cortex is hexagonal, the lateral meristem becoming trilobed at the base. In *I. todaroana* the shape of the secondary cortex is that of an irregular polygon and the lateral meristem seems to be more rounded than lobed (Figure 3(b)).

Another obvious difference is in the scales and phyllopodia. These two structures, which are sometimes confused in the literature, represent two independently derived mechanisms for resistance to desiccation (Taylor and Hyckey, [39]). Phyllopodia are the sclerified remnants of the bases of fully developed leaves. The presence of phyllopodia was in part the basis for assigning some *Isoëtes* species to the *Terrestres* section [40]. Scales are complete leaf primordia which become arrested early in their development and are in general unsclerified [40]. By this definition, *I. iapygia* has a small number of minute, brown, translucent scales [30] at the base of the naked corm (Figure 4(a)) while *I. todaroana* has black, hardened scales that are similar to phyllopodia, with two rounded lateral lobes and one short central spine-like lobe [31] (Figure 4(b)).

The geographical range of *I. iapygia* is well defined, limited to 10 sites in the southern part of Puglia [29] that do not overlap with the range of other congeners. The information on *I. todaroana* is insufficient to define its true geographical range because it has so far only been found in the *locus classicus*, in an area of about 200 × 100 m [31].

The habitat of both taxa is Mediterranean temporary ponds (3170*), considered an international priority [41]. The type locality of *I. todaroana* is described as temporary wetland with a mosaic of communities characterized by aquatic macrophytes such as *Bolboschoenus maritimus* (L.) Palla, *Eleocharis palustris* (L.) Roem. and Schult., *Scirpus cernuus* Vahl, *Mentha pulegium* L., *Damasonium alisma* subsp. *bourgaei* (Coss.), *Oenanthe* sp., *Lythrum* sp., *Tamarix* sp., and *Romulea* sp. [31, 42]. *I. iapygia* is found in very small areas, inside rock pools or on a thin layer of mosses (such as *Pleurochaete squarrosa* (Brid.) Lindb. and *Cheilotela chloropus* (Brid.) Broth.), in which the community is dominated by microphytes characteristic of the *Isoëto-Nanojuncetea* phytosociological class, including *Ranunculus sardous* Crantz, *Polypogon maritimus* Willd., *Romulea bulbocodium* (L.) Sebast. *et* Mauri, and *Romulea columnae* Sebast. *et* Mauri.

The different specificity of the habitat of the two varieties is also seen in the type of soil and geological substrate: *I. todaroana* grows on a distinctive geological substrate of calcareous sandstone with a thin layer of clay on top and an alkaline pH (Table 1) (Troìa and Raimondo [43]).

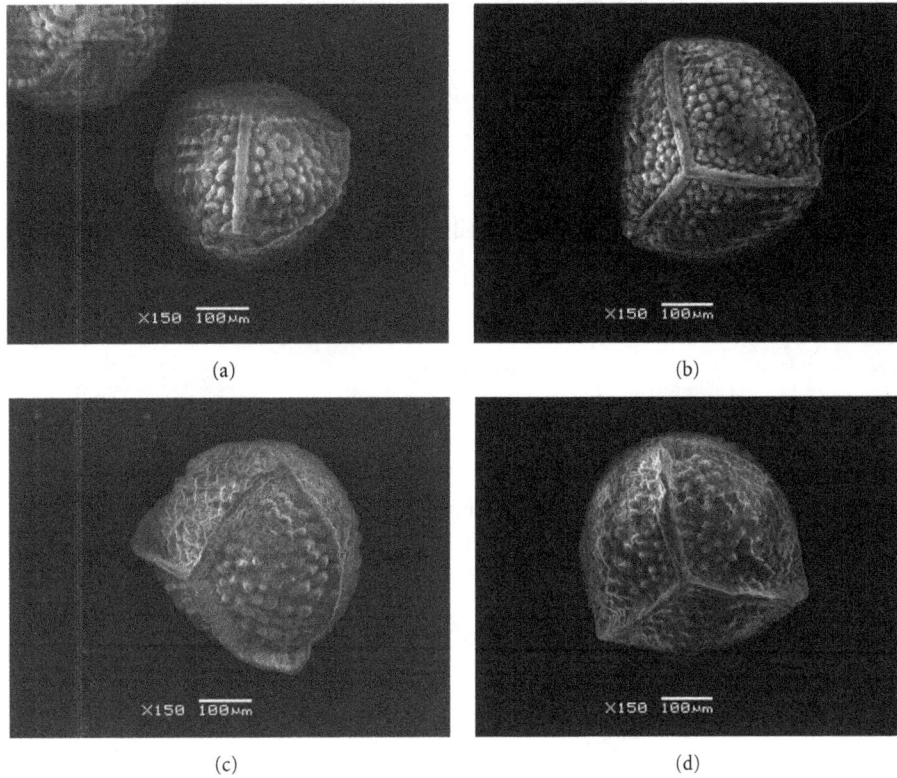

FIGURE 1: Macrospore of *I. iapygia* viewed by SEM ((a): lateral view, (b): proximal view) and *I. todaroana* ((c): lateral view, (d): proximal view). Scale bar: 100 μm.

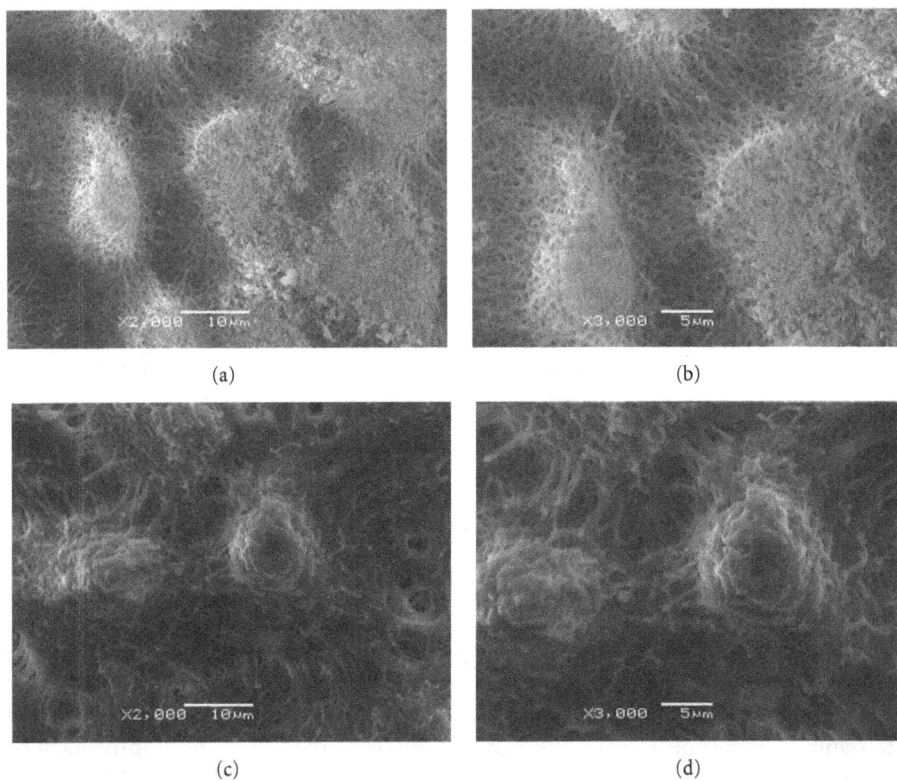

FIGURE 2: Macrospore microornamentation types viewed by SEM: *I. iapygia* ((a), scale bar: 10 μm; (b), scale bar: 5 μm) and *I. todaroana* ((c), scale bar: 10 μm; d, scale bar: 5 μm).

TABLE 1: Characters used to compare and describe individuals.

Morphometric characteristics	Isoëtes iapygia	Isoëtes todaroana
Corm lobes	Trilobed	Rounded
Corm width (cm)	0.8	0.5
Shape of corm section	Hexagonal	Irregular
Root shape	Dichotomous	Dichotomous
Leaf shape	Filiform, arched	Lanceolate, erect
Leaf number	20	25
Velum	Complete	Complete
Stomata	Present	Present
Number of air chambers	2	2
Scales	Present	Absent
Phyllopodia	Absent	Present
Macrospore macroornamentation	Tuberculate	Tuberculate
Microspore macroornamentation	Coniculate-echinate	Aculeate
Macrospore microornamentation	Densely fimbriate	Filamentous
Macrospore diameter (μm)	450	440
Macrospore profile	Rounded	Triangular
Microspore lenght (μm)	30	25
Ligule shape	Ovate	Lanceolate
Ligule lenght (mm)	1.2	1.0
Laesura arms	Flattened	Raised
Equatorial ring	Little pronounced	Pronounced
Ecological characters		
Habitus	Terrestrial	Amphibious
pH of soil	7.3	8.7
Type of soil	clay	clay
Geological substrate	Limestone	Sandstone
Habitat		
Type (Natura 2000 Code)	3170*	3170*
Characteristic species of community	Pleurochaete squarrosa (Brid.) Lindb., Cheilotela chloropus (Brid.) Broth., Ranunculus sardous Crantz, Polypogon maritimus Willd., Romulea bulbocodium (L.) Sebast. et Mauri, Romulea columnae Sebast. et Mauri.	Bolboschoenus maritimus (L.) Palla, Eleocharis palustris (L.) Roem. and Schult., Scirpus cernuus Vahl, Mentha pulegium L., Oenanthe sp., Lythrum sp., Tamarix sp., Romulea sp.,
Geographical range		
Sites of presence	10	1
Chorology	Southern Puglia	Insufficient data

In contrast, *I. iapygia* grows on a thin layer of soil or mosses on a limestone substrate with a neutral or subbasic pH (Table 1).

The two taxa share a number of traits including dichotomous roots, complete velum, presence of stomata, and two air chambers. However it has been widely documented that at least for *I. iapygia*, the reduction in leaf area and the presence of two air chambers represent an adaptation to xeric environments [30]. Indeed, *I. iapygia* is a terrestrial species, never submerged by water. In all sites where it is recorded the characteristic layer of mosses is episodically soaked, retaining sufficient moisture for the development of the plant, though for several months of the year the habitat is characterised by aridity and high temperatures [30].

In contrast, the type locality of *I. todaroana* is a temporary wetland that dries out only in summer. It is a remnant

(a)   (b)

FIGURE 3: Cross section of corm ((a): *I. iapygia*, (b): *I. todaroana*). Scale bar: 0.25 cm.

(a)   (b)

FIGURE 4: Enlargement of scales and phyllopodia of (a): *I. iapygia* and (b): *I. todaroana*.

of a wider wetland, most of which has been reclaimed and converted to farmland that now surrounds the "type locality." Thus it has an amphibious habitus, with both emergent and submerged growth in temporary ponds [31].

4. Conclusions

The morphometric differences and other distinguishing characteristics suggest that *I. iapygia* and *I. todaroana* should be considered as separate species. The presence of scales and phyllopodia, megaspore ornamentation, and habitus suggest a connection between these two species and the Mediterranean "terrestrial" section of the genus, including *I. histrix, I. subinermis, I olympica, I. setacea, I. duriei,* and the recently identified *I. libanotica* [40, 44–49].

On the other hand, the fibrillose surface of the megaspores suggests a relationship with the amphibious *I. velata* group, although the terrestrial *I. histrix* also has this characteristic [8, 37].

The formulation of hypotheses is impeded by the documented morphological convergence and phenotypic plasticity of the genus [50]. *I. iapygia* has probably evolved gradually, as the result of spatial isolation and low genetic exchange with congeners, due to the low dispersal ability of macro- and microspores. This is reflected in its characteristic morphological traits, including densely fimbriate macrospores, arched and filiform leaves, translucent scales, two air chambers, and the extreme specificity of its habitat.

It would be interesting to compare the relationships between these two species and other Mediterranean taxa using the cladistic approach, which may help to identify a common ancestor or clarify their evolution, and to evaluate other original hypotheses.

Targeted studies in this direction are underway and it is likely that new species will be found in the future.

References

[1] R. J. Hickey, "Genus Isoëtes in the new world: an overview," *American Journal of Botany*, vol. 84, article 162, 1997.

[2] R. J. Hickey, C. Macluf, and W. C. Taylor, "A re-evaluation of Isoetes savatieri franchet in Argentina and Chile," *American Fern Journal*, vol. 93, no. 3, pp. 126–136, 2003.

[3] H. P. Fuchs-Eckert, "Isoëtes palmeri HP Fuchs, eine neue Isoëtes-Art des Pàramo," *Mededeelingen van het Botanisch Museum en Herbarium van de Rijks Universiteit te Utrecht*, vol. 10, pp. 165–174, 1981.

[4] R. J. Hickey, "Isoëtes megaspore surface morphology: nomenclature, variation, and systematic importance," *American Fern Journal*, vol. 76, pp. 1–16, 1986.

[5] N. E. Pfeiffer, "Monograph of the Isoetaceae," *Annals of the Missouri Botanical Garden*, vol. 9, pp. 79–232, 1922.

[6] C. Prada, "Estudio de la anatomia foliar de las especies espanolas del genero Isoetes L.," *Lagascalia*, vol. 9, pp. 107–113, 1979.

[7] C. Prada, "Estudios palinológicos y cromosomicos en las especies espanolas del genero Isoetes (Isoetaceae)," *Palinologia*, vol. 1, pp. 211–225, 1980.

[8] C. Prada, "El genero Isoetes L. en la Peninsula Iberica," *Acta Botanica Malacitana*, vol. 8, pp. 73–100, 1983.

[9] C. H. Rolleri and C. Prada, "Endodermis foliares en el genero Isoetes L. (Isoetaceae)," *Acta Botanica Malacitana*, vol. 29, pp. 191–201, 2004.

[10] C. Prada and C. H. Rolleri, "A new species of Isoetes (Isoetaceae) from Turkey, with a study of microphyll intercellular pectic protuberances and their potential taxonomic value," *Botanical Journal of the Linnean Society*, vol. 147, no. 2, pp. 213–228, 2005.

[11] M. I. Romero and C. Real, "A morphometric study of three closely related taxa in the European Isoetes velata complex," *Botanical Journal of the Linnean Society*, vol. 148, no. 4, pp. 459–464, 2005.

[12] H. Liu, Q. F. Wang, and W. C. Taylor, "Morphological and anatomical variation in sporophylls of Isoetes sinensis palmer (Isoetaceae), an endangered quillwort in China," *American Fern Journal*, vol. 96, no. 3, pp. 67–74, 2006.

[13] R. J. Hickey, "Chromosome numbers of Neotropical Isoetes," *American Fern Journal*, vol. 74, pp. 9–13, 1984.

[14] A. Troia, "The genus Isoetes L. (Lycophyta, Isoetaceae): synthesis of kariological data," *Webbia*, vol. 56, pp. 201–218, 2001.

[15] S. B. Hoot and W. C. Taylor, "The utility of nuclear ITS, a LEAFY homolog intron, and chloroplast atpB-rbcL spacer region data in phylogenetic analyses and species delimitation in Isoetes," *American Fern Journal*, vol. 91, no. 3, pp. 166–177, 2001.

[16] S. B. Hoot, W. C. Taylor, and N. S. Napier, "Phylogeny and biogeography of Isoetes (Isoëtaceae) based on nuclear and chloroplast DNA sequence data," *Systematic Botany*, vol. 31, no. 3, pp. 449–460, 2006.

[17] W. C. Taylor, R. H. Mohlenbrock, and J. A. Murphy, "The spores and taxonomy of Isoetes butleri and I. melanopoda," *American Fern Journal*, vol. 65, pp. 33–38, 1975.

[18] J. R. Croft, "A taxonomic revision of Isoëtes L. (Isoëtaceae) in Papuasia," *Blumea*, vol. 26, pp. 177–190, 1980.

[19] R. J. Hickey, "A new Isoëtes from Jamaica," *American Fern Journal*, vol. 71, pp. 69–74, 1981.

[20] B. M. Boom, "Synopsis of Isoetes in the southeastern United States," *Castanea*, vol. 47, pp. 38–59, 1982.

[21] L. Kott and D. M. Britton, "Spore morphology and taxonomy of Isoetes in northeastern North America," *Canadian Journal of Botany*, vol. 61, no. 12, pp. 3140–3163, 1983.

[22] R. G. Stolze and R. J. Hickey, "Isoetaceae," in *The Flora of Guatemala. Part. III. Fieldiana*, R. G. Stolze, Ed., vol. 2 of *Botany*, pp. 62–67, 1983.

[23] C. A. Brown, "What is the role of spores in Fern taxonomy?" *American Fern Journal*, vol. 50, pp. 6–14, 1960.

[24] L. J. Musselman, "Ornamentation of Isoetes (Isoetaceae, Lycophyta) microspores," *Botanical Review*, vol. 68, no. 4, pp. 474–487, 2002.

[25] D. R. Kaplan, "The science of plant morphology: definition, history, and role in modern biology," *American Journal of Botany*, vol. 88, no. 10, pp. 1711–1741, 2001.

[26] F. Conti, G. Abbate, A. Alessandrini, and C. Blasi, Eds., *An Annotated Checklist of the Italian Vascular Flora*, Palombi Editori, Roma, Italy, 2005.

[27] S. Frattini, E. Somaschini, R. Gentili et al., "Isoetes echinospora Durieu. Schede per la Lista Rossa della Flora vascolare e crittogamica Italiana," *Informatore Botanico Italiano*, vol. 42, no. 2, pp. 598–601, 2010.

[28] E. Barni, C. Minuzzo, C. Siniscalco et al., "Schede per la Lista Rossa della Flora vascolare e crittogamica Italiana," *Informatore Botanico Italiano*, vol. 42, no. 2, pp. 602–604, 2010.

[29] P. Ernandes, "2011. Il genere Isoëtes L. (Pteridophyta, Lycophyta): note tassonomiche, ecologia e distribuzione in Puglia," *Annali del Museo Civico di Rovereto*, vol. 26, pp. 347–358, 2010.

[30] P. Ernandes, L. Beccarisi, and V. Zuccarello, "A new species of Isoëtes (Isoëtaceae, Pteridophyta) for the Mediterranean," *Plant Biosystems*, vol. 144, no. 4, pp. 805–813, 2010.

[31] A. Troia and F. M. Raimondo, "Isoëtes todaroana (isoëtaceae, lycopodiophyta), a new species from Sicily (Italy)," *American Fern Journal*, vol. 99, no. 4, pp. 238–243, 2009.

[32] F. Médail and P. Quézel, "Hot-spots analysis for conservation of plant biodiversity in the Mediterranean Basin," *Annals of the Missouri Botanical Garden*, vol. 84, pp. 112–127, 1997.

[33] E. A. C. Costantini, F. Urbano, and G. L'Abate, Soils Regions of Italy, 2004, http://www.soilmaps.it/.

[34] G. L'Abate, R. Barbetti, E. A. C. Costantini, S. Magini, and M. Fantappiè, *Italian Soil Information System*, 2011, http://www.soilmaps.it/.

[35] E. Ferrarini, F. Ciampolini, R. E. G. Pichi Sermolli, and D. Marchetti, "Iconographia Palynologica Pteridophytorum Italiae. Isoëtes L. (Isoëtaceae, Lycophyta)," *Webbia*, vol. 40, pp. 1–202, 1986.

[36] H. K. Choi, J. Jung, and C. Kim, "Two new species of Isoetes (Isoetaceae) from Jeju Island, South Korea," *Journal of Plant Biology*, vol. 51, no. 5, pp. 354–358, 2008.

[37] S. Bagella, M. C. Caria, A. Molins, and J. A. Rosselló, "Different spore structures in sympatric Isoetes histrix populations and their relationship with gross morphology, chromosome number, and ribosomal nuclear ITS sequences," *Flora*, vol. 206, no. 5, pp. 451–457, 2011.

[38] A. Troia, A. Orlando, and R. Schicchi, "Approccio micromorfologico alla sistematica del genere Isoëtes (Isoëtaceae, Lycopodiophyta): analisi della superficie delle megaspore," *Bollettino dei Musei e degli Istituti Biologici dell'Universita di Genova*, pp. 73–100, 2011.

[39] W. C. Taylor and R. J. Hyckey, "Habitat, evolution and speciation in Isoetes," *Annals of the Missouri Botanical Garden*, vol. 79, pp. 613–622, 1992.

[40] J. Bolin, R. D. Bray, M. Keskin, and L. J. Musselman, "The genus Isoëtes L., (Isoëtaceae, Lycophyta) in South-Western Asia," *Turkish Journal of Botany*, vol. 32, pp. 447–457, 2008.

[41] European Commission DG Environment, Interpretation Manual of European Union Habitats, EUR27, 2007.

[42] A. Troia, C. Bazan, and R. Schicchi, "Nuove aree di rilevante interesse naturalistico nella Sicilia centro-occidentale: proposte di tutela," *Naturalista Siciliano*, vol. 35, no. 2, pp. 257–293, 2011.

[43] A. Troìa and F. M. Raimondo, "Isoëtes todaroana (Isoëtaceae, Lycopodiophyta), a new species from Sicily (Italy)," *American Fern Journal*, vol. 99, no. 4, pp. 238–243, 2010.

[44] J. Bolin, R. D. Bray, and L. J. Musselman, "A new species of diploid Quillwort (Isoëtes Isoëtaceae, Lycophyta) from Lebanon," *Novon*, vol. 21, pp. 295–298, 2011.

[45] A. C. Jermy and J. R. Akeroyd, "Isoetes L.," in *Flora Europaea 1*, T. G. Tutin, V. H. Heywood, N. A. Burges, D. H. Valentine, S. M. Walters, and D. A. Webb, Eds., p. 581, Cambridge University Press, 2nd edition, 1993.

[46] O. de C. Bolòs, *La Vegetacio de les Illes Balears*, Istitut d'Estudis Catalans, Barcelona, Spain, 1996.

[47] H. Coste, *Flore Descriptive et Illustreè de la France, de la Corse et des Contrèes Limitrophes*, 1937, Edited by: Blanchard.

[48] W. Greuter, H. M. Burdet, and G. Long, Eds., *Med-Checklist 1.Geneve: Conservatoire et Jardin Botaniques*, 1984.

[49] L. N. Derrick, A. C. Jermy, and A. M. Paul, "Checklist of European Pteridophytes," *Sommerfeltia*, vol. 6, pp. 1–94, 1987.

[50] R. J. Hickey, W. C. Taylor, and N. T. Luebke, "The species concept in Pteridophyta with special reference to Isoëtes," *American Fern Journal*, vol. 79, pp. 78–89, 1989.

13

Germination of Primed Seed under NaCl Stress in Wheat

Michael P. Fuller,[1] Jalal H. Hamza,[1,2] Hail Z. Rihan,[1] and Mohammad Al-Issawi[1]

[1] School of Biomedical and Biological Sciences, Faculty of Science and Technology, Plymouth University, Plymouth PL4 8AA, UK
[2] Department of Agronomy, College of Agriculture, University of Baghdad, Baghdad, Iraq

Correspondence should be addressed to Michael P. Fuller, mfuller@plymouth.ac.uk

Academic Editors: F. A. Culianez-Macia and A. W. Woodward

Soil salinity affects a large and increasing amount of arable land worldwide, and genetic and agronomic solutions to increasing salt tolerance are urgently needed. Experiments were conducted to improve wheat seed performance under salinity stress conditions after priming. An experiment was conducted using a completely randomized design of four replications for germination indices in wheat (*Triticum aestivum* L. cv. Caxton). Normal and primed seed with PEG_{6000} at -1 MPa and five concentrations of NaCl (0, 50, 100, 150, and 200 mM) were tested. Results indicate that priming seed significantly ($P < 0.05$) increased germination percentage at first count and final count, coefficient of velocity of germination, germination rate index, and mean germination time, while increasing of NaCl concentration significantly reduced it. Priming seed improved germination attributes at all NaCl concentration levels. The priming appeared to be able to overcome the effect of salt stress at 50 to 100 mM and reduce the effect of NaCl at higher concentrations up to 200 mM. The primed seed gave both faster germination and led to higher germination when under salt stress. We conclude that using priming techniques can effectively enhance the germination seed under saline condition.

1. Introduction

Salinity is a major limiting factor in crop productivity all over the world. Salinity affects plant growth at all developmental stages; however, sensitivity varies from one growth stage to another. Germination is a critical stage of the plant cycle and improved tolerance of high salinity could improve the stability of plant production [1]. Water is osmotically held in salt solutions. Therefore, the salt concentration completely inhibits germination at higher levels or induces a state of dormancy at low levels, it also reduces imbibition of water because of lowered osmotic potentials of the medium and causes changes in metabolic activity [2].

For the maintenance of high yield of crops under salt stressed conditions, various research tools are being tried to counteract the effects of salinity. Seed priming treatments are simply applied practices that can reduce the effects of salinity with small inputs of capital and energy. Many such seed priming or invigoration treatments are being used to improve the rate and speed of germination under stressed conditions or with substandard seed lots [3]. Pre-sowing

treatments such as priming techniques with different salts, water, and osmoprotectants assist the germination and establishment process in the field [4]. Priming of seeds with $CaCl_2$, followed by priming with KCl and NaCl, were found to be effective in alleviating the adverse effects of salt stress on wheat plants through their effects on altering the levels of different plant phytohormones [5]. Khan et al. [6] observed that priming of seeds using NaCl improved seedling vigour and establishment under salt stress conditions. Seed priming improves seed performance by encouraging rapid, uniform, and vigorous germination which helps seedlings to grow in stressed conditions [4, 7, 8]. Similarly, Jafar et al. demonstrated that seed priming could be used successfully to assist the germination of wheat in the field [9].

The analysis of physiological changes in plants associated with seed-priming may be useful for advancing the understanding of plant salt tolerance and may suggest strategies by which plants acquire or alter their salt tolerance. The primary objective of the present study was to determine the effect of priming on seedling growth and to determine if the mechanism of activation of seeds may be used as a technique to increase the viability of the seed to perform better under excess salinity.

2. Methods

The experiment was conducted at the School of Biomedical and Biological Sciences, Plymouth University, UK, in 2011, to determine if a seed priming technique can enhance the seed tolerance to salinity stress. An experiment was conducted with wheat (*Triticum aestivum* L. cv. Caxton) using a completely randomized design of four replications. Normal and primed seed with PEG_{6000} at -1 MPa and five concentrations of NaCl (0, 50, 100, 150, and 200 mM) were tested.

The primed solution (-1 MPa) was prepared by dissolving 21.9 g PEG/100 mL of PEG_{6000} in distilled water at 20°C and confirmed using an OSMOMET device. Seeds (250 g) for each treatment were immersed in 300 mL of the priming solution for 6 h in plastic containers covered with lids to prevent evaporation loss. After priming, the seed samples were removed and rinsed several times in tap water and then dried back to the original moisture level of 12% by subjecting the seed to 20 ± 1°C and 50% RH for 24 h. A digital Protimeter was used for measuring and monitoring the moisture content during the drying period. Fifty seeds from each treatment were placed in 140 mm diameter petri dishes on two layers of Whatman no.1 filter paper moistened with 40 mL of the appropriate salt test solutions (0, 50, 100, 150, and 200 mM of NaCl). The petri dishes were placed in an incubator at 20 ± 1°C under dark conditions to germinate. Seeds were considered to be germinated when they exhibited a radicle extension longer than 2 mm. The germination count was recorded daily up to 8 days. From the germination counts several germination attributes were studied to investigate the effect of the priming on salt tolerance including germination percentage (%) as first count after 4 days (FG), germination percentage (%) as final count after 8 days (LG), coefficient of velocity of germination (CVG), and germination rate index (GRI), mean germination time (MGT) as follow:

$$\text{CVG} \left(\% \text{ day}^{-1} \right) = \frac{\sum Ni}{\sum (NiTi)} \times 100, \qquad (1)$$

(see [10]),

$$\text{GRI} \left(\% \text{ day}^{-1} \right) = \sum \left(\frac{Ni}{i} \right), \qquad (2)$$

(see [11]),

$$\text{MGT} \left(d \right) = \frac{\sum (NiTi)}{\sum Ni}, \qquad (3)$$

(see [11]),
where N is the number of seeds germinated on day i, and Ti is the number of days from sowing.

The CVG gives an indication of the rapidity of germination. It increases when the number of germinated seeds increases and the time required for germination decreases. Theoretically, the highest CVG possible is 100. This would

occur if all seeds germinated on the first day [10]. The GRI reflects the percentage of germination on each day of the germination period. Higher GRI values indicate higher and faster germination [11]. The lower MGT, the faster a population of seeds has germinated [11].

2.1. Statistical Analysis. All data were subjected to analysis of variance (ANOVA) using SPSS software (version 17), and comparisons of means were made using the least significant difference test (LSD) at $P < 0.05$ level of confidence. Calculated coefficients of simple correlation between attributes were also studied [12].

3. Results

3.1. ANOVA. Analyses of variance showed significant effects of priming and NaCl concentrations on attributes of germination (Table 1). The effects of the NaCl concentrations accounted for a high proportion of the variance in all analyses. There was no significant interaction effect in any of the attributes of germination.

3.2. The Effect of Priming and NaCl Concentration. Priming seed with PEG_{6000} at -1 MPa significantly increased the attributes of germination compared with normal seed (Table 2) and increasing NaCl concentration significantly reduced the attributes of germination compared with control treatment (0 NaCl).

Priming seed improved germination at all of NaCl concentration levels compared with normal seed (Figure 1) and showed that the pattern of response to NaCl concentrations were the same for normal and primed seed. The primed seed gave both faster germination (FG, CVG, GRI, and MGT) and led to higher germination (LG) when under salt stress (Figure 1).

Increasing salt stress reduced the speed of germination and the germination rate. For normal seed this was evident at 50 mM NaCl and got progressively worse at each increment of salt applied (Figure 1). Whilst speed of germination was affected by each increment of salt (FG, CVG, GRI, and MGT) and salt significantly reduced the final germination (LG) above 100 mM NaCl. For primed seed, speed of germination was affected by each increment of salt above 50 mM NaCl (FG, CVG, GRI, and MGT), and salt significantly reduced the final germination (LG) above 100 mM NaCl.

3.3. Correlation between Attributes. Positive highly significant correlations ($P < 0.001$) were found between FG, LG, CVG, and GRI, and negative highly significant correlations ($P < 0.001$) were found between each of FG, LG, CVG, and GRI, and MGT (Table 3).

4. Discussion

The application of increasing salt stress had substantial negative effects on all the attributes investigated with an overall negative effect on germination. The values of FG, LG, CVG, and GRI were decreased, while the time required

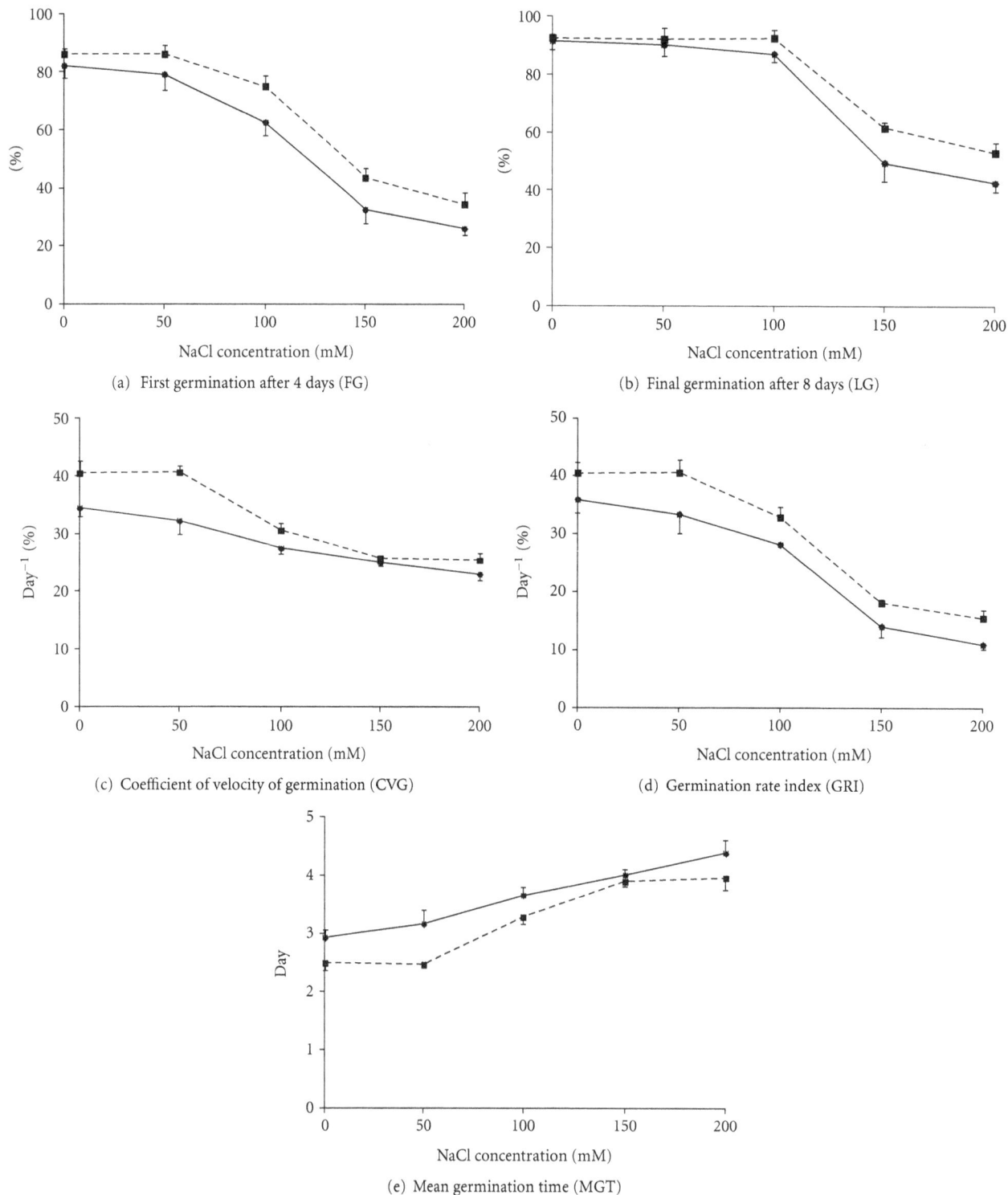

(a) First germination after 4 days (FG)

(b) Final germination after 8 days (LG)

(c) Coefficient of velocity of germination (CVG)

(d) Germination rate index (GRI)

(e) Mean germination time (MGT)

FIGURE 1: Interaction priming × NaCl concentration for attributes of germination, (a) FG (b) LG, (c) CVG, (d) GRI, and (e) MGT. Primed seed with PEG_{6000} at -1 MPa (dotted lines) and normal seed (solid lines). Vertical bars indicate SE.

to obtain a faster germination (MGT) was increased. This study showed significant negative relationships between the concentration of sodium chloride and between most of the germination attributes and a positive relationship with MGT. This agrees with other observations in wheat, where salinity has been shown to negatively affect the rate of starch remobilization by causing a decrease in α-amylase activity [13].

The results also showed that priming of wheat seeds with PEG_{6000} at -1 MPa can have significant improvements

TABLE 1: Analysis of variance for effect of priming and NaCl concentration on attributes of germination.

	df	Mean square				
		FG	LG	CVG	GRI	MGT
Priming	1	739.60**	384.40**	171.304**	250.50**	1.63642**
NaCl concentration	4	5050.85**	3770.35**	302.104**	1089.44**	3.45680**
Priming × NaCl concentration	4	22.35^{ns}	48.65^{ns}	18.698^{ns}	3.00^{ns}	0.08384^{ns}
Error	30	58.53	47.80	7.489	13.85	0.09563
CV (%)		12.6	9.2	9.0	13.8	9.0

**Significant at $P < 0.01$, ^{ns}Not significant.

TABLE 2: Means comparison of the effect of priming and NaCl concentration on attributes of germination.

Treatment	Attributes of germination				
	FG	LG	CVG	GRI	MGT
Primed seed (PEG$_{6000}$ at −1 MPa)	65.0^a	78.3^a	32.5^a	29.5^a	3.2^a
Normal seeds	56.4^b	72.1^b	28.4^b	24.5^b	3.6^b
NaCl concentration (mM)					
0	84.0^a	92.0^a	37.4^a	38.2^a	2.7^a
50	82.5^a	91.0^a	36.4^a	37.0^a	2.8^a
100	68.8^b	89.8^a	29.0^b	30.5^b	3.5^b
150	38.0^c	55.5^b	25.4^c	16.1^c	4.0^c
200	30.2^c	47.8^c	25.2^c	13.3^c	4.2^c

Means within a column with common characters are not significant at $P < 0.05$.

TABLE 3: Correlation coefficients between attributes of germination.

	MGT	GRI	CVG	LG
FG	−0.926***	0.975***	0.885***	0.938***
LG	−0.807***	0.949***	0.753***	
CVG	−0.979***	0.914***		
GRI	−0.933***			

***Significant at $P < 0.001$.

on certain attributes of germination. Primed seeds were superior in FG, LG, CVG, and GRI, and it also decreased MGT such that the required time was faster with primed seed compared with normal seeds. Priming seed appeared to be able to completely overcome the effect of salt stress at 50 to 100 mM and reduce the effect of salt at higher concentrations up to 200 mM. Seed priming can activate the metabolic processes within the seed to get early emergence of the radicle even under saline conditions. Our findings agree with Cantliffe [4] who found that priming seed in an osmotic solution may improve germination through metabolic activation involving the synthesis of nucleic acids, proteins, and enzymes, and increasing respiratory activity and energy reserve utilization. These findings are important for field situations where soil salinity is suspected as the use of priming could make the difference between successful field germination and establishment or substantial crop failure.

5. Conclusion

We conclude that the use of priming techniques can enhance the germination of wheat seed under saline conditions and under conditions of mild salt stress priming can entirely overcome the effect salt. We recommend conducting similar studies on other wheat varieties to investigate whether this is a general phenomenon. Additional work also needs to evaluate germination and early seedling growth under field conditions.

References

[1] V. Jajarmi, "Effect of water stress on germination indices in seven wheat cultivar," *Proceedings of World Academy of Science, Engineering and Technology*, vol. 49, pp. 105–106, 2009.

[2] S. Rafiq, T. Iqbal, A. Hameed, R. Zulfiqar Ali, and N. Rafiq, "Morphobiochemical analysis of salinity stress response of wheat," *Pakistan Journal of Botany*, vol. 38, no. 5, pp. 1759–1767, 2006.

[3] S. S. Lee and J. H. Kim, "Total sugars, α-amylase activity, and germination after priming of normal and aged rice seeds," *Korean Journal of Crop Science*, vol. 45, no. 2, pp. 108–111, 2000.

[4] D. J. Cantliffe, "Seed enhancements," *Acta Horticulturae*, vol. 607, pp. 53–62, 2003.

[5] M. Iqbal, M. Ashraf, A. Jamil, and S. ur-Rehman, "Does seed priming induce changes in the levels of some endogenous plant hormones in hexaploid wheat plants under salt stress?" *Journal of Integrative Plant Biology*, vol. 48, no. 2, pp. 181–189, 2006.

[6] H. A. Khan, C. M. Ayub, M. A. Pervez, R. M. Bilal, M. A. Shahid, and K. Ziaf, "Effect of seed priming with NaCl on salinity tolerance of hot pepper (*Capsicum annuum* L.) at seedling stage," *Soil & Environment*, vol. 28, no. 1, pp. 81–87, 2009.

[7] M. Ashraf and M. R. Foolad, "Pre-sowing seed treatment-A shotgun approach to improve germination, plant growth, and crop yield under saline and non-saline conditions," *Advances in Agronomy*, vol. 88, pp. 223–271, 2005.

[8] F. Carbineau and D. Come, "Priming: a technique for improving seed quality," *Seed Testing International*, no. 132, 2006.

[9] M. Z. Jafar, M. Farooq, M. A. Cheema et al., "Improving the performance of wheat by seed priming under saline conditions," *Journal of Agronomy & Crop Science*, vol. 198, pp. 38–45, 2012.

[10] M. A. Kader and S. C. Jutzi, "Effects of thermal and salt treatments during imbibition on germination and seedling

growth of sorghum at 42/19°C," *Journal of Agronomy and Crop Science*, vol. 190, no. 1, pp. 35–38, 2004.

[11] M. A. Kader, "A comparison of seed germination calculation formulae and the associated interpretation of resulting data," *Journal and Proceeding of the Royal Society of New South Wales*, vol. 138, pp. 65–75, 2005.

[12] R. G. Steel, J. H. Torrie, and D. A. Dickey, *Principles and Procedures of Statistics, a Biometrical Approach*, McGraw Hill Book, Singapore, 3rd edition, 1997.

[13] M. Almansouri, J. M. Kinet, and S. Lutts, "Effect of salt and osmotic stresses on germination in durum wheat (*Triticum durum* Desf.)," *Plant and Soil*, vol. 231, no. 2, pp. 243–254, 2001.

14

To What Extent Do Protected Areas Determine the Conservation of Native Flora? A Case Study in the Sudanian Zone of Burkina Faso

Lassina Traoré, Amadé Ouédraogo, and Adjima Thiombiano

Laboratory of Plant Biology and Ecology, University of Ouagadougou, 03 BP 7021, Ouagadougou 03, Burkina Faso

Correspondence should be addressed to Lassina Traoré, ltraorej@gmail.com

Academic Editors: A. Culham and S. Jansen

Natural vegetation contributes significantly to the daily needs of local people especially in the developing countries. This exerts a high pressure on freely accessible natural savannas and jeopardizes the conservation of protected areas. In Burkina Faso, conservation measures, such as the creation of protected forests, have been taken to safeguard the remaining indigenous vegetation. However, little is known about the effectiveness of these protected areas in conserving biodiversity. This study compared the diversity and structural characteristics of the vegetation communities in protected and unprotected areas in the Sudanian zone of Burkina Faso. A total of 208 species representing 41 families and 145 genera were found. Significant differences were found between the species richness in the north Sudanian sector for tree savannas and in the south Sudanian sector for the shrub savannas, tree savannas, savanna woodlands, and the woodlands of land use types. All tree size-class distributions in each vegetation type formed a reverse J-shaped curve, indicating vegetation dominated by juvenile individuals. Similarity in tree species composition between management regimes was found to be low, which reflects differences in habitat conditions, disturbance, and topography. Urgent measures are needed to ensure effective and efficient management and conservation of biodiversity in the protected areas of Burkina Faso.

1. Introduction

In West Africa, the degradation of savanna woodlands due to agricultural expansion, overgrazing, bush fire and wood cutting is a serious environmental concern [1, 2]. The current mosaic of these savanna-woodland ecosystems is the result of combined climate pejoration and human pressure impacts. The overexploitation of the parklands is reported as a major cause of the degradation of vegetation cover and loss of biodiversity [3]. In general, this phenomenon is observed in many areas of West Africa [4–6] and particularly in Burkina Faso where the dynamics of land occupation for agriculture is extensive. In this way the natural vegetation declines and the human pressure on the protected areas increases. Western Burkina Faso is the principal area of the cotton production in the country [7]. This agricultural basin is regarded as a space with high economic potentialities and consequently attracts more and more people from other parts of the country. The increasing demand of agriculture lands is to the detriment of the ancient traditional fallow practices which favor the conservation and restoration of parklands [8].

In the 1930s, a large part of the Sudanian zone of West Africa was delimited and protected by the colonial administration to create wildlife sanctuaries and prevent the expansion of shifting cultivation [9]. Following political independence, these forests and woodland reserves were conserved through the status of state forests for wood production, ecotourism, and biodiversity conservation [9]. In Burkina Faso, natural land reserves represent 26% of the country's total area [9, 10].

A consensus emerges with regard to the need for preserving the protected areas. However, a better conservation management depends initially on the understanding of specific factors and thorough causes which are responsible for the degradation process. The aim of this study was to assess the impact of protected areas on the woody plants

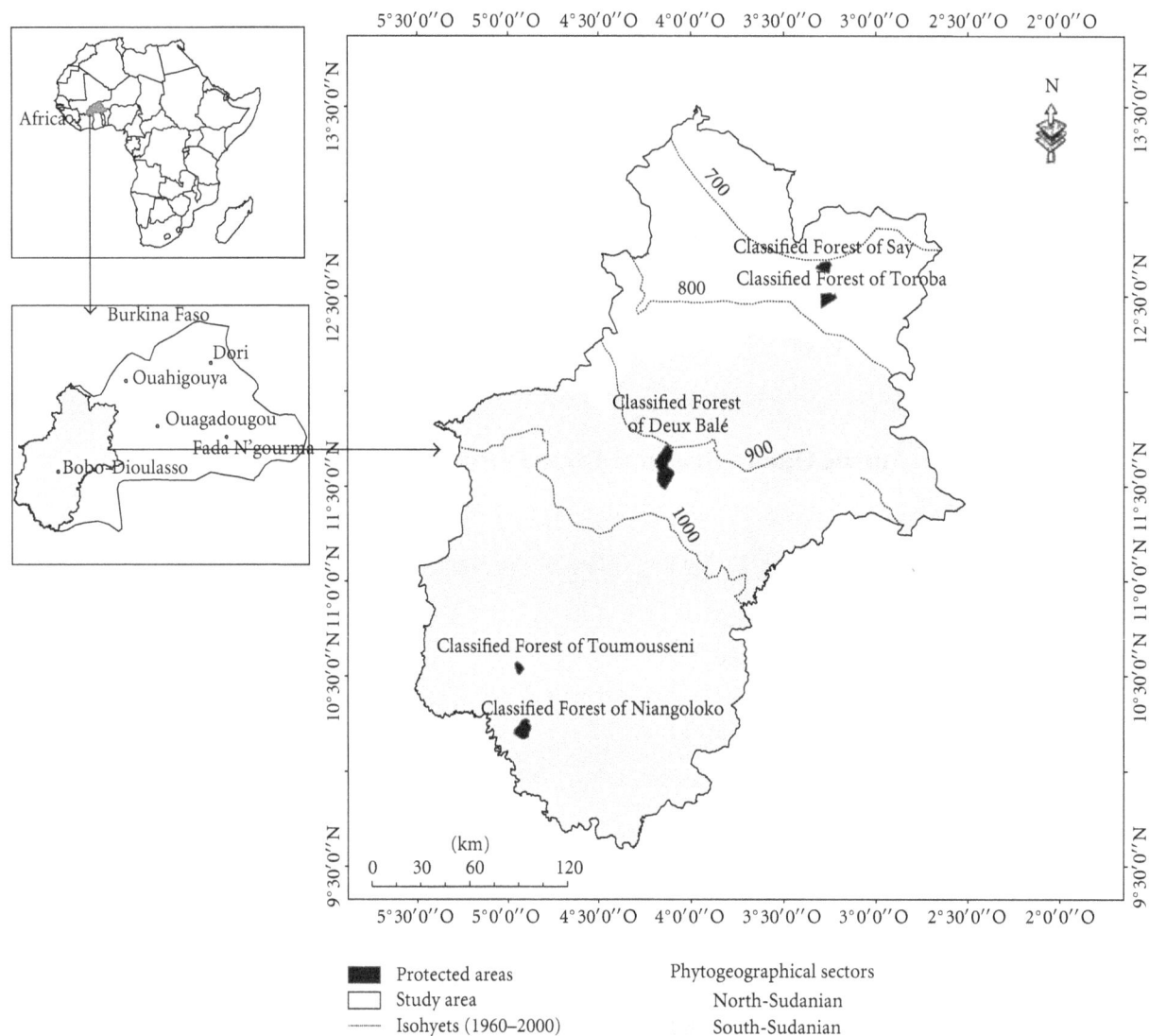

FIGURE 1: The location of the study sites within the sudanian zone, western Burkina Faso.

communities in the different vegetation types between two land use regimes in Western Burkina Faso. We hypothesize that the diversity and structural characteristics of woody vegetation would be enhanced within protected areas. The results of this study will provide information that could help to conciliate land use regimes with sustainable management and conservation of woody species communities.

2. Materials and Methods

2.1. Study Area. This paper was carried out in Western Burkina Faso across two phytogeographical sectors, the north-Sudanian (NS) and the south-Sudanian (SS) sectors [11]; this area is situated between 9°30′–13°30′N and 5°30′–2°30′W (Figure 1). The rainy season lasts from May to October and the annual rainfall range is 700–800 m and 900–1000 m in the north and south sudanian sectors, respectively. The mean annual temperature ranges from 15 to 40.8°C and 17 to 37.9°C, in the NS and the SS, respectively. The

dominant soils are ferralsols and lixisols ranged from poorly to completely leached ones [1, 12]. Specific common woody species are *Acacia seyal* Del., *Combretum micranthum* G. Don, and *Guiera senegalensis* J. F. Gmel. in the north-Sudanian zone and *Acacia sieberiana* DC., *Burkea africana* Hook. F, *Daniellia oliveri* (R.) Hutch. and Dalz. C, and *Isoberlinia doka* Craib. and Stapf. for the south-Sudanian sector.

In this study, protected areas (PAs) are defined as clearly delimited areas of natural vegetation that have been officially classified—with appropriate legal status—by public authorities with the aim of ensuring protection of natural resources as well as ecosystem functions and services. Data were collected in five PAs across the study area. All the five PAs have a status of classified forests: classified forests of Say and "Toroba" (Dédougou) were sampled in the north-Sudanian sector, and classified forests of "deux Balé" (Boromo), of "Toumousséni" (Banfora) and "Niangoloko" in the south Sudanian sector. Say and Toroba state forests

To What Extent Do Protected Areas Determine the Conservation of Native Flora? A Case Study in
the Sudanian Zone of Burkina Faso

117

were classified in 1937 and 1938, respectively. They cover 5400 ha and 2700 ha, respectively. Balé, Niangoloko, and Toumousséni were classified in 1937, 1936, and 1954 and cover 57000 ha, 124500 ha, and 2500 ha, respectively. Toroba, Balé, Toumousséni, and Niangoloko are managed by the State Forest Department Services. Say is under a private management by a concessionaire [13]. Illegal logging, grazing, and harvesting of nonwoody forest products are frequent in all these protected areas. Unprotected areas (UAs) in this study are fallows, freely accessible natural savannas, and unsuitable lands for agriculture but used for pasture (e.g., rocky lands). Unprotected areas are strongly affected by various anthropogenic activities, for example, extensive livestock grazing, bush fires, and various harvestings of timber and nontimber forest products including wood, leaves, bark, flowers, and fruits.

2.2. Data Collection. The considered plant communities are shrub savannas, tree savannas, woodland savannas, woodland and gallery forests as defined by Aubreville [14]. The inventory design followed a random sampling scheme and was applied systematically in and outside selected protected areas across the north and south Sudanian zones. Sampling units were $1000 \, m^2$ ($20 \, m \times 50 \, m$) size plots at the rate of 10–15 plots per vegetation type and per phytogeographical sector. Data collection consisted in recording and identifying all woody species and measuring their diameter at breast height (DBH). A measuring tape was used to measure girth at breast height (GBH, at height 1.3 m), which was then converted to the following diameter at breast height (DBH) using the formula: DBH = GHB/π. A total of 256 plots were reached and distributed as follows: NS-PAs: 51, NS-UAs: 52, SS-PAs: 82, and SS-UAs: 71.

2.3. Data Analysis. Mean population density was calculated for each vegetation community in protected areas (PAs) and in unprotected areas (UAs). t-tests were used to compare mean population density and basal areas between the two types of management regimes. Results were considered significant at $P < 0.05$.

2.3.1. Species Richness and Diversity. For each vegetation community, we calculate the following ecological parameters which are widely used indices to measure biological diversity [15]:

(i) Shannon's diversity index (H'):

$$H' = -\sum_{i=1}^{s} p_i \ln p_i, \tag{1}$$

where s is the total number of species in the community (richness) and p_i is the relative abundance of the ith species in a plot.

(ii) Simpsons' Index (D):

$$D = \sum_{i=1}^{s} \frac{Ni(Ni-1)}{N(N-1)}, \tag{2}$$

where Ni is the number of individual of the ith species and N is the total number of species in vegetation types. $D = 0$ indicates high diversity, and 1 indicates low diversity. In order to obtain "more intuitive" values, $1 - D$ was calculated, the maximum of diversity being represented by value 1, and the minimum of diversity by value 0.

(iii) Hill's Diversity Index:

$$\text{Hill} = \frac{(1/D)}{e^{H'}}, \tag{3}$$

where $1/D$ represents Simpson's index and $e^{H'}$ is Shannon's exponential index. Hill's index of diversity makes it possible to obtain a more precise appreciation of observed diversity. In order to facilitate interpretation, it is then possible to use the index 1-Hill; maximum diversity will be represented by value 1 and minimal diversity by value 0. Hill's index seems most relevant as so far it integrates the two other indices (Shannon and Simpson) and allows comparisons between vegetation communities.

To evaluate β-diversity (similarity between vegetation communities), Jaccard's similarity index and Horns' modification of Morisita's index were computed. Jaccard's coefficient of similarity was calculated based on presence/absence data of the species while Horn's modification of Morisita's index takes the species abundance into account. Both indices vary between 0 and 1: a value close to 1 indicates greater similarity between sites and hence low β-diversity [15].

2.3.2. Population Structure and Size-Class Distribution. For each plant community the following structural parameters were calculated.

(i) the tree density of the stands (N), that is, the average number of trees per plot expressed in trees/ha

$$N = \frac{n}{s}, \tag{4}$$

n is the overall number of trees in the plot, and s is the area of plot ($s = 0.10$ ha).

(ii) the basal area of stands (G, in m^2/ha), that is, the sum of the cross-sectional area at 1.3 m above the ground level of all trees on a plot, expressed in m^2/ha as follows:

$$G = \pi \, x \frac{D_i^2}{4}, \tag{5}$$

D_i is the diameter of tree at 1.3 m above the ground level.

Mean population density was calculated for each species in protected areas (PAs) and in unprotected areas (UAs). t-tests were used to compare mean population density and basal areas between the two types of management regimes. Results were considered significant at $P < 0.05$.

To establish the size-class distributions (SCDs), diameters of all trees (dbh $\geq$5 cm) were used to construct histograms with size classes of 5 cm interval. The observed shape was adjusted to the 3-parameter-Weibull theoretical distribution because of its flexibility [16]:

$$f(x) = \frac{c}{b} \left[\frac{(x-a)}{b} \right]^{c-1} e^{-[x-a/b]^c}, \tag{6}$$

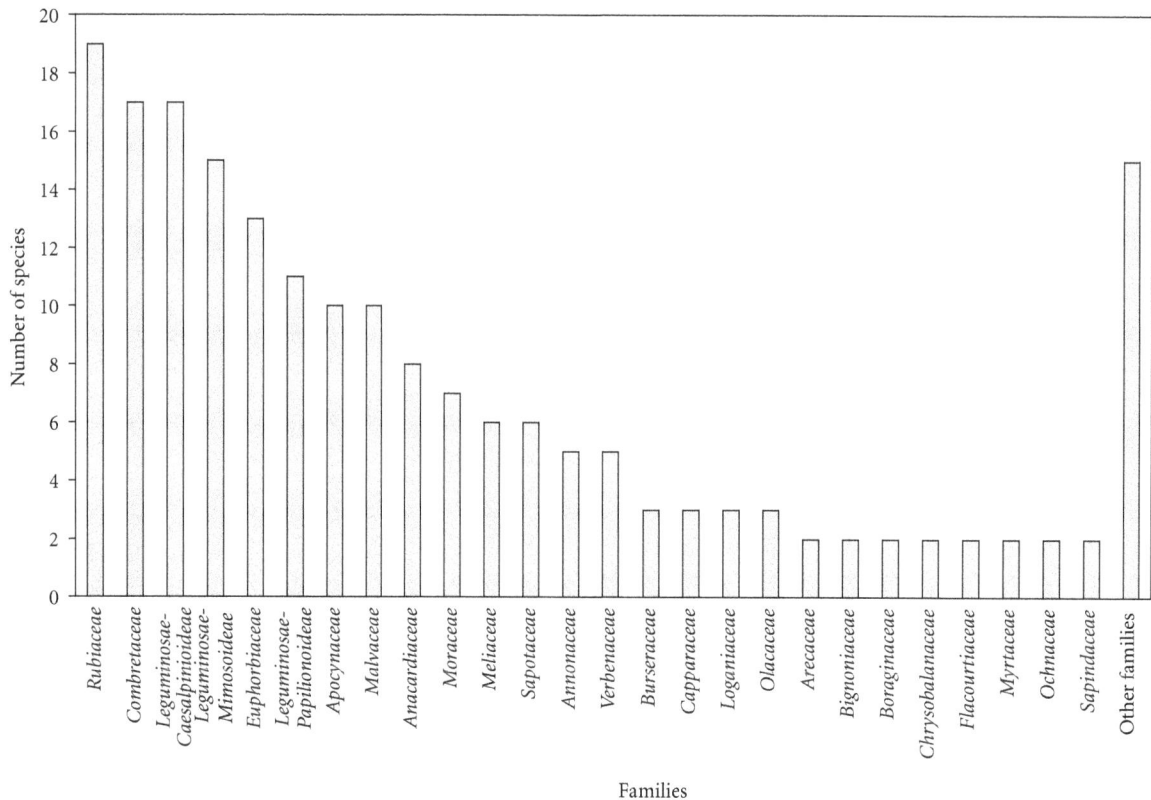

FIGURE 2: Plant's families.

where x = tree diameter, a = 5 cm, the threshold trees diameter structure, b = scale parameter linked to the central value of diameters, and c = shape parameter of the structure. For each community, tree diameter was used to estimate the parameters b and c based on the maximum likelihood method [17]. The Weibull distribution reveals a wide variety of shapes expressed by the shape parameter c [18]. When $c <$ 1 the distribution has a reversed J shape; whilst when $c = 1$ it is a negative exponential distribution. For values of $c > 1$ the distribution is unimodal. If $1 < c < 3.6$ the distribution has a positive skew; when $c = 3.6$ it is approximately normal and when $c > 3.6$ the distribution has a negative skew. This adjustment covers diameter distributions that are strongly positively or negatively skewed as well as normally distributed [19, 20].

In each case the log-linear analysis [21] was performed in SAS [22] to test the adequacy of the observed structure to the Weibull distribution. The following model, described by Caswell [21], was used:

$$\text{Log Frequency} = F + F_{\text{-Class}} + F_{\text{-adjustment}} + \varepsilon, \quad (7)$$

where F = mean frequency of the classes, $F_{\text{-Class}}$ = nonrandom gap linked to the differences in frequency between classes, $F_{\text{-adjustment}}$ = nonrandomly gap linked to differences between observed and theoretical frequencies, and ε = error of the model.

The hypothesis of adequacy between both distributions is accepted if the probability value of the test is higher than 0.05.

3. Results

3.1. Species Richness and Woody Plant Community Structures.
A total of 208 species, distributed within 145 genera and 41 families, were recorded in both phytogeographical zones. The *Rubiaceae, Combretaceae, Leguminosae-Caesalpinioideae,* and *Leguminosae-Mimosoideae* families showed the highest number of species in both management regimes (Figure 2).

In the north sudanian area, the t-test revealed that the mean specific richness differed significantly only in tree savannas between PAs and UAs ($P = 0.006$), being highest in the PAs (57 species) than in the UAs (19 species). *Acacia dudgeoni, Acacia seyal, Boswellia dalzielii, Burkea africana, Detarium microcarpum, Combretum molle, Loeseneriella africana, Pseudocedrela kotschyi,* and *Xeroderris stuhlmannii* were the exclusive species in the tree savannas of the PAs.

As far as the structural parameters of stands are considered, no significant difference was observed between land use types in all vegetation communities. The mean stem density and the mean basal area did not significantly increase from UAs to the PAs for all plant communities in both phytogeographical sectors (Table 3). The observed SCDs for all plant communities showed a reverse "J" shape; with a Weibull shape parameter smaller than or close to 1; this is an indication of a steeply descending monotonic distribution function (Figure 3(a)).

In the south sudanian zone, the mean values of specific richness showed significant differences between the two land

To What Extent Do Protected Areas Determine the Conservation of Native Flora? A Case Study in the Sudanian Zone of Burkina Faso

119

TABLE 1: Woody species diversity and composition between vegetation communities in the two management regimes in western Burkina Faso.

Sectors	Vegetation communities	Land use types	F	G	SR	RSM	P value	H'	P value	D	1 − D	P value	Hill	1 − "Hill"	P value
NS	Gallery forests	PAs	10	15	16	4.50 ± 1.24	0.701	1.08 ± 0.32	0.764	0.41 ± 0.15	0.58 ± 0.14	0.542	0.44 ± 0.04	0.56 ± 0.04	0.880
		UAs	11	16	19	4.81 ± 2.52		1.14 ± 0.49		0.46 ± 0.26	0.53 ± 0.24		0.44 ± 0.04	0.56 ± 0.04	
	Shrub savannas	PAs	16	28	43	9.6 ± 5.70	0.812	1.31 ± 0.84	0.992	0.49 ± 0.33	0.50 ± 0.30	0.836	0.37 ± 0.05	0.63 ± 0.05	0.073
		UAs	20	41	55	9.06 ± 5.47		1.30 ± 0.91		0.46 ± 0.35	0.53 ± 0.33		0.41 ± 0.05	0.59 ± 0.05	
	Tree savannas	PAs	23	43	57	12.33 ± 2.44	0.006*	1.74 ± 0.70	0.387	0.23 ± 0.25	0.76 ± 0.23	0.065	0.52 ± 0.27	0.48 ± 0.24	0.311
		UAs	9	17	19	7.83 ± 2.78		1.17 ± 0.71		0.49 ± 0.29	0.50 ± 0.24		0.40 ± 0.10	0.60 ± 0.08	
	Woodland savannas	PAs	21	40	53	9.62 ± 6.89	0.493	1.50 ± 0.89	0.560	0.37 ± 0.31	0.62 ± 0.29	0.602	0.42 ± 0.04	0.58 ± 0.04	0.497
		UAs	17	30	37	7.90 ± 4.60		1.29 ± 0.77		0.43 ± 0.28	0.56 ± 0.25		0.40 ± 0.07	0.60 ± 0.07	
SS	Gallery forests	PAs	25	39	42	35.16 ± 8.61	0.941	2.05 ± 0.16	0.665	0.14 ± 0.03	0.85 ± 0.03	0.327	0.45 ± 0.06	0.55 ± 0.05	0.405
		UAs	20	40	46	35.50 ± 11.55		1.97 ± 0.55		0.19 ± 0.17	0.80 ± 0.15		0.50 ± 0.18	0.50 ± 0.16	
	Shrub savannas	PAs	24	45	57	17.18 ± 4.81	0.038*	2.16 ± 0.53	0.231	0.18 ± 0.15	0.81 ± 0.13	0.429	0.37 ± 0.06	0.63 ± 0.05	0.931
		UAs	24	43	54	13.00 ± 3.97		1.87 ± 0.56		0.24 ± 0.17	0.75 ± 0.16		0.37 ± 0.04	0.63 ± 0.03	
	Tree savannas	PAs	27	52	63	17.27 ± 6.82	0.026*	2.31 ± 0.51	0.061	0.14 ± 0.08	0.85 ± 0.07	0.179	0.42 ± 0.07	0.58 ± 0.07	0.973
		UAs	20	38	49	11.69 ± 4.55		1.95 ± 0.39		0.19 ± 0.08	0.80 ± 0.07		0.42 ± 0.09	0.58 ± 0.08	
	Woodland savannas	PAs	24	59	74	16.92 ± 4.35	0.036*	2.28 ± 0.31	0.010*	0.16 ± 0.10	0.83 ± 0.09	0.015*	0.36 ± 0.08	0.64 ± 0.07	0.673
		UAs	22	54	66	12.15 ± 6.69		1.81 ± 0.54		0.29 ± 0.15	0.70 ± 0.14		0.35 ± 0.10	0.65 ± 0.09	
	Woodlands	PAs	32	68	83	17.72 ± 4.49	0.043*	2.16 ± 0.45	0.209	0.19 ± 0.12	0.80 ± 0.10	0.444	0.34 ± 0.06	0.66 ± 0.08	0.686
		UAs	24	56	64	13.30 ± 4.87		1.90 ± 0.47		0.24 ± 0.14	0.75 ± 0.13		0.35 ± 0.06	0.65 ± 0.07	

F: family, G: genera, SR: species richness, RSM: mean species richness, H': Shannon's index, D: Simpson's index, P: P value in RSM, H', D, $1 − D$, Hill, and $1 −$ Hill were given with standard deviation.
* Significant value.

(a)

(b)

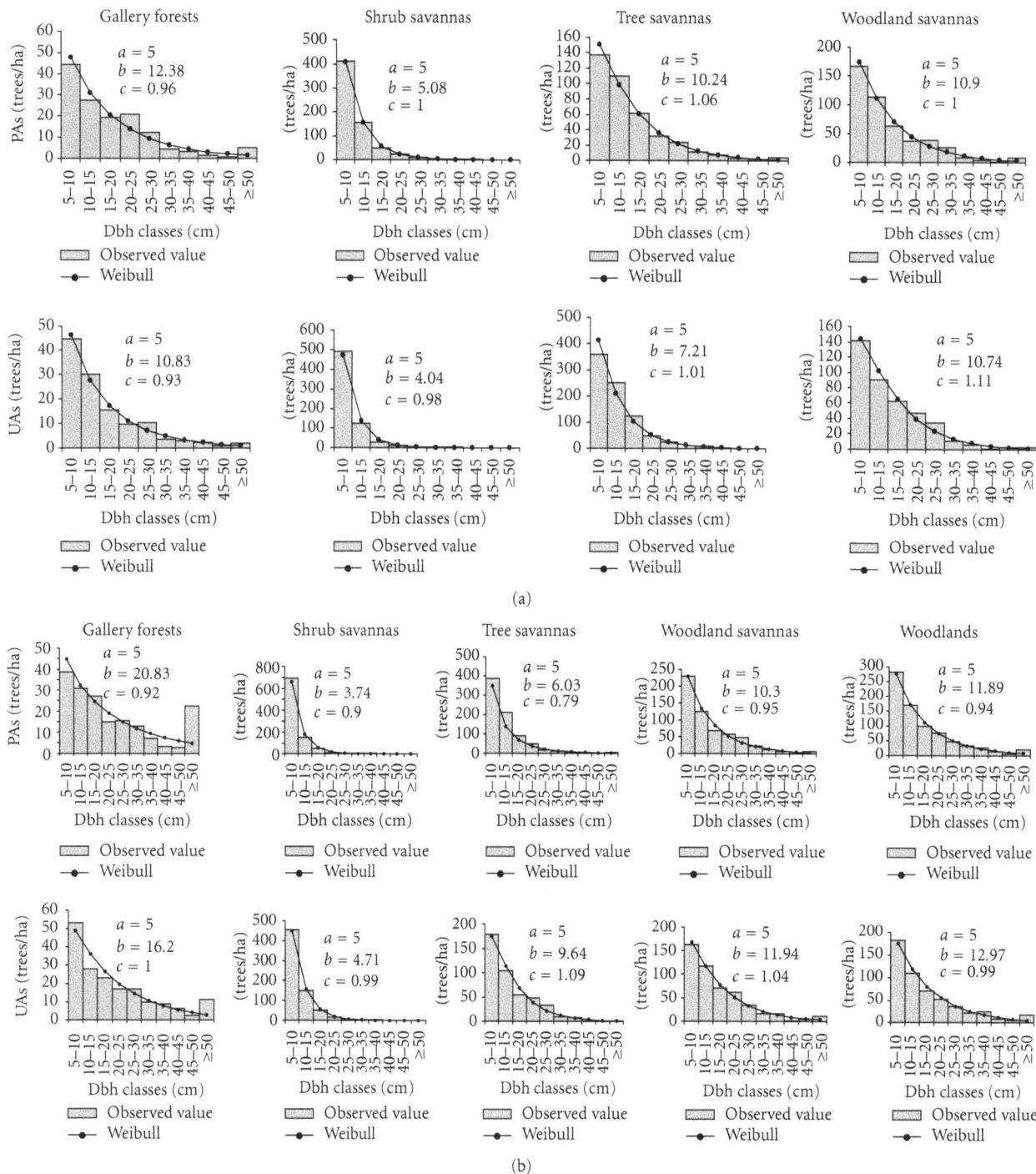

FIGURE 3: The size-class distribution (SCDs) of vegetation communities in protected areas (PAs) and unprotected areas (UAs): (a) for the north and (b) for the south Sudanian sectors of Burkina Faso. a, b and c are the three parameters of the Weibull theoretical distribution.

use types in the shrub savannas ($P = 0.038$), tree savannas ($P = 0.026$), savanna woodlands ($P = 0.036$), and woodlands ($P = 0.043$). For each vegetation community, species richness was highest in the PAs than in the UAs (Table 1). *Ficus sur, Flacourtia flavescens, Monotes kerstingii, Syzygium guineense,* and *Xeroderris stuhlmannii* in the shrub savannas; *Malacantha alnifolia, Mimusops kummel, Monotes kerstingii,*

Ekebergia senegalensis, Sericanthe chevalieri, and *Syzygium guineense* in the tree savannas; *Ekebergia senegalensis, Erythrophleum africanum, Hexalobus monopetalus, Manilkara multinervis, Lannea barteri,* and *Swartzia madagascariensis* in the savanna woodlands, and *Capparis tomentosa, Carissa edulis, Cola cordifolia, Erythrophleum suaveolens, Lonchocarpus cyanescens, Syzygium guineense,* and *Mimusops kummel*

To What Extent Do Protected Areas Determine the Conservation of Native Flora? A Case Study in
the Sudanian Zone of Burkina Faso

121

TABLE 2: Similarity in the species composition between vegetation communities in two management regimes in western Burkina Faso.

Indices	North sudanian								South sudanian									
	Gf		Shs		Ts		Ws		Gf		Shs		Ts		Ws		W	
	PAs	UAs	PAs	UAs	PAs	UAs	PAs	UAs	PAs	UAs	PAs	UAs	PAs	UAs	PAs	UAs	PAs	UAs
Jaccard	0.402		0.515		0.333		0.475		0.275		0.541		0.435		0.463		0.475	
Morisita-Horn	0.585		0.589		0.288		0.615		0.375		0.565		0.491		0.605		0.786	

PAs: protected areas, UAs: unprotected areas, and Gf: gallery forests; Shs: shrub savannas, Ts: tree savannas, Ws: woodland savannas, and W: woodlands.
Jaccard's similarity index and Morisita's index were computed to evaluate the similarity between vegetation communities. Both indices vary between 0 and 1:
a value close to 1 indicates a greater similarity between sites and hence low β-diversity.

TABLE 3: Structural characteristics of woody plant communities according to phytogeographical sectors and land use types (PAs and UAs).

Sectors	Vegetation communities	Land use types	Mean stems density ($N \cdot ha^{-1}$)	SE	P value	Mean basal areas ($m^2 ha^{-1}$)	SE	P value
NS	Gallery forests	PAs	551.7	184.0	0.426	19.84	11.05	0.711
		UAs	485.5	207.5		17.53	17.99	
	Shrub savannas	PAs	654.0	266.3	0.923	6.61	2.85	0.203
		UAs	666.9	360.7		5.39	1.94	
	Tree savannas	PAs	411.7	183.9	0.081	10.07	3.29	0.681
		UAs	645.0	356.1		10.94	5.56	
	Woodland savannas	PAs	468.8	159.6	0.322	13.35	7.45	0.206
		UAs	401.0	176.5		9.99	4.12	
SS	Gallery forests	PAs	703.3	172.2	0.942	81.63	48.15	0.024*
		UAs	710.0	231.0		38.22	12.93	
	Shrub savannas	PAs	931.8	190.2	0.006*	7.35	1.81	0.236
		UAs	690.0	177.8		6.39	1.89	
	Tree savannas	PAs	705.5	117.9	0.000*	13.43	6.72	0.052
		UAs	440.0	185.2		9.43	1.92	
	Woodland savannas	PAs	622.9	180.2	0.090	17.15	4.71	0.720
		UAs	496.2	193.6		16.26	7.73	
	Woodlands	PAs	765.3	188.8	0.001*	29.33	13.67	0.099
		UAs	528.0	116.0		20.99	9.05	

NS: north sudanian, SS: south sudanian, $N \cdot ha^{-1}$: number of individual per hectare, and SE: standard error.
*Significant value.

in the woodlands were the exclusive species in PAs. The mean stem density increased significantly from UAs to the PAs in the shrub savannas ($P = 0.006$), tree savannas ($P = 0.000$), and woodlands ($P = 0.001$). However, the mean basal area did not show a significant difference between both land use types (Table 3). In both management regimes, the observed SCDs for all plant communities indicated a reverse J-shaped curve with the Weibull distribution c-value smaller than or close to 1 (Figure 3(b)). This is typical of a relatively stable multispecific population. The log-linear analysis (computed for each vegetation community) indicated a good adjustment of the observed distribution to the Weibull distribution (see Table 4).

3.2. Species Diversity and Similarity between Land Use Types. Among the diversity indices, Shannon's measure of evenness and Simpson's diversity index were not significantly different between land use types for all vegetation communities (Table 1). However, according to Hill's indices all vegetation communities were relatively diverse on the two sites (NS and SS) and did not show any significant differences between the PAs and the UAs. When comparing species similarity between vegetation communities in the NS (Table 2), similarity was fairly low between tree savannas from PAs and those from UAs (33.3% and 28.8% for Jaccard's index and Morisita's index, resp.). The highest similarity (51.5% and 58.9% for Jaccard's index and Morisita's index, resp.) was observed between the shrub savannas of the PAs and the UAs in both land use types.

In the south sudanian, Shannon's measure of evenness ($P = 0.010$) and Simpson's diversity index ($P = 0.015$) were significantly different for woodlands in both land use types (Table 1).

According to Hill's indice, all vegetation communities were relatively diverse. The comparison of the species richness between vegetation types revealed a low similarity between gallery forests from PAs and those from UAs (27.5% and 37.5% for Jaccard's index and Morisita's index, resp.).

TABLE 4: Results of log-linear analysis (in SAS Inc., 1999 [22]) applied to diameter class frequency.

| Sectors | Vegetation communities | | The CATMOD procedure Maximum likelihood analysis of variance | | | | | |
| | | | Protected areas | | | Unprotected areas | | |
		Source	DF	Chi square	Pr > ChiSq	DF	Chi square	Pr > ChiSq
NS	Gallery forests	Class	9	124.23	<0.0001	9	157.98	<0.0001
		Distr.	1	0.01	0.9364	1	0.02	0.8749
		Class * Distr.	9	3.97	0.9134*	9	1.12	0.9991*
	Shrub savannas	Class	4	147.24	<0.0001	4	174.12	<0.0001
		Distr.	1	0.00	0.9634	1	0.06	0.8053
		Class * Distr.	4	0.41	0.9815*	4	0.55	0.9682*
	Tree savannas	Class	9	141.34	<0.0001	7	132.19	<0.0001
		Distr.	1	0.01	0.9400	1	0.00	1.0000
		Class * Distr.	7	0.47	0.9996*	5	1.15	0.9500*
	Woodland savannas	Class	9	136.22	<0.0001	9	117.33	<0.0001
		Distr.	1	0.32	0.5714	1	0.00	1.0000
		Class * Distr.	8	0.82	0.9992*	7	1.30	0.9885*
SS	Gallery forests	Class	9	70.19	<0.0001	9	97.12	<0.0001
		Distr.	1	0.00	0.9509	1	0.13	0.7179
		Class * Distr.	9	6.61	0.6774*	9	3.30	0.9514*
	Shrub savannas	Class	4	171.71	<0.0001	4	155.18	<0.0001
		Distr.	1	0.17	0.6828	1	0.00	0.9866
		Class * Distr.	4	0.62	0.9607*	4	0.03	0.9999*
	Tree savannas	Class	8	224.44	<0.0001	7	121.65	<0.0001
		Distr.	1	0.02	0.8998	1	0.15	0.7012
		Class * Distr.	7	1.50	0.9825*	6	1.82	0.9357*
	Woodland savannas	Class	9	142.41	<0.0001	9	120.31	<0.0001
		Distr.	1	0.00	1.0000	1	0.32	0.5714
		Class * Distr.	8	0.58	0.9998*	8	1.02	0.9981*
	Woodlands	Class	9	150.30	<0.0001	9	131.22	<0.0001
		Distr.	1	0.17	0.6758	1	0.04	0.8378
		Class * Distr.	9	0.94	0.9996*	9	2.07	0.9903*

An interaction "Class * Distr." significant ($P > 0.05$) shows that the conformation between observed structure and Weibull Distribution is accepted.
* See values.

The highest similarity (54.1% and 56.5% for Jaccard's index and Morisita's index, resp.) was observed between the shrub savannas in both land use types.

4. Discussion

The overall species richness found in this study is higher than that of native woody species reported in the country. Fontès and Guinko [11] have reported that woody flora of the country comprises 188 species. Concerning the species richness, the analysis provides nuanced results. The protection effect was evidenced by a significantly higher number of species in PAs for shrub savannas, woodland savannas, and woodlands in the south sudanian sector and tree savannas in both sectors. However, the role of protected areas in the conservation of plant diversity in savannah regions is appreciated in various ways. Although some studies demonstrated loss of plant diversity due to human pressure in exploited lands or, conversely, floristic preservation inside protected areas [23], other studies showed that phytodiversity does not differ in both land use types, and it is even higher outside protected areas [24–27]. This might be an indication that protection measures in semiarid areas ensure the conservation of the existing diversity but not always adequate for enhancing lands diversity. This idea is supported by Devineau [28] who reported an increasing recruitment of plants in fallows, favoured by moderate pasture. It is well established that the traditional long fallow system combined with agroforestry practices have a positive effect on land conservation [8]. Consequently, species richness is sometimes even higher in these areas than in the protected areas [25, 29]. The most common families were *Rubiaceae, Combretaceae, Leguminosae-Caesalpinioideae,* and *Leguminosae-Mimosoideae*; this is a typical taxonomic pattern of savanna-woodland mosaics flora in Sudano-Sahelian Africa [1, 11].

According to Shannon's and Simpson's diversity indices, woody plant populations of woodland in PAs were more diversified than in the UAs. This is most likely related to the relatively large numbers of abundant species in woodlands plots (e.g., *Cola cordifolia, Erythrophleum suaveolens, Lonchocarpus cyanescens,* and *Mimusops kummel*). Shannon's diversity index is usually found to fall between 1.5 and 3.5 and is rarely above 5.0 [15]. The values from our results are in the expected range. The similarity in species composition and abundance between the land use types was generally low, except for the shrub savannas. This plant community exhibited 51.5% and 58.9% in the north and 54.1% and 56.5% in the south sudanian sector for species composition and abundance similarity, respectively. The low values of similarity indices reflect differences in habitat and topography conditions.

Concerning the structural parameters of woody populations, a gradual increase was found for the mean stems density from UAs to the PAs only in the south Sudanian sector for shrub savannas, tree savannas, woodland savannas, and woodlands. This difference in individuals' density between both land use types might be due to the variable disturbance density [1, 27]. Generally, the survival and growth of tree species in semiarid savanna ecosystems are favoured by good moisture and edaphic conditions [30, 31] but strongly disturbed by direct and indirect human activities [32, 33]. The reverse "J" shape of SCDs for all plant communities in both zones, irrespective of land use types, is an indication of good regeneration of populations [34, 35]. A large number of individuals with dbh ≤5 cm were found in all land use types, indicating a potential renewal of the population for all plant communities [35–37].

The reverse "J" structure observed for each vegetation community may be explained by a kind of compensation between the species in the different SCDs, so that the species with slow growth are compensated by those with rapid growth. This is due to the importance of the phenomena of competition (for nutrients) between trees in the savannas [38] which make that some species are more favored than others.

5. Conclusion

Our results provide information that globally species diversity and richness are not significantly better in protected areas than in unprotected ones. However, certain plant communities are favoured by protection. This is the case of tree savannas in north and south sudanian sectors. Although they are currently disturbed, open access lands are areas of active plant materials flows (mainly by pasture and human direct activities) that favours diversification. The role of PAs, in terms of plant community conservation, is unquestionable as they ensure safety recruitment and survival of new individuals, which is primordial for population rejuvenation and consequently the plant diversity conservation. This suggests that, even if in certain cases, the PAs do not exhibit a higher phytodiversity, they play a key role in terms of warranty for the conservation of the existing diversity.

Acknowledgments

This paper was achieved within the framework of the BIOTA-West Africa Program (Project: 01LC0617D1W11) funded by the German Federal Ministry for Education and Research (BMBF). The authors are very grateful to their field assistants who helped them in data collection. Many thanks to Abel Kadéba for mapping the study site. They are also grateful to the anonymous reviewers whose pertinent comments considerably improved the quality of their paper.

References

[1] F. Bognounou, A. Thiombiano, P. Savadogo, J. I. Boussim, P. C. Odén, and S. Guinko, "Woody vegetation structure and composition at four sites along latitudinal gradient in Western Burkina Faso," *Bois et Forêts des Tropiques*, vol. 300, no. 2, pp. 29–44, 2009.

[2] M. M. Inoussa, A. Mahamane, C. Nbow, M. Saadou, and B. Yvonne, "Dynamiques spatio-temporelle des forêts claires dans le Parc national du W du Niger (Afrique de l'Ouest)," *Sécheresse*, vol. 22, pp. 108–116, 2011.

[3] B. S. Bouko, B. Sinsin, and G. B. Soulé, "Effets de la dynamique d'occupation du sol sur la structure et la diversité floristique des forêts claires et savanes au Bénin," *Tropicultura*, vol. 25, no. 4, pp. 221–227, 2007.

[4] S. L. Ariori and P. Ozer, "Évolution des ressources forestières en Afrique de l'Ouest soudano-sahélienne au cours des cinquante dernières années," *International Journal of Tropical Geology Geography and Ecology*, no. 29, pp. 61–68, 2005.

[5] H. Sawadogo, P. N. Zombré, L. Bock, and D. Lacroix, "Évolution de l'occupation du sol de Ziga dans le Yatenga (Burkina Faso) à partir de photos aériennes," *Revue de Télédétection*, vol. 8, pp. 59–73, 2008.

[6] H. Diallo, I. Bamba, Y. Sadaiou et al., "Effets combinés du climat et des pressions anthropiques sur la dynamique évolutive de la végétation d'une zone protégée du Mali (Réserve de Fina, Boucle du Baoulé)," *Sécheresse*, vol. 22, no. 2, pp. 97–107, 2011.

[7] S. Caillault, D. Delahaye, and A. Ballouche, "Des cultures temporaires face à la forêt classée, exemple des paysages à l'ouest du Burkina Faso," 13p, 2010, http://www.projetsdepaysage.fr .

[8] R. J. Buresh and P. J. M. Cooper, "The science and practice of short-term improved fallows: symposium synthesis and recommendations," *Agroforestry Systems*, vol. 47, no. 1–3, pp. 345–356, 1999.

[9] D. Zida, *Impact of forest management regimes on ligneous regeneration in the Sudanian savanna of Burkina Faso [Ph.D. thesis]*, Swedish University of Agricultural Sciences, Umeå, Sweden, 2007.

[10] P. Savadogo, *Dynamics of Sudanian savanna-woodland ecosystem in response to disturbances [Ph.D. thesis]*, Swedish University of Agricultural Sciences, Umeå, Sweden, 2007.

[11] J. Fontès and S. Guinko, "Carte de la végétation et de l'occupation du sol (Burkina Faso)," Notice Explicative, Laboratoire d'Écologie Terrestre, Institut de la Carte Internationale de la Végétation. CNRS, Université de Toulouse III, (France)/Institut du Développement Rural, Faculté des Sciences et Techniques, Université de Ouagadougou, Ouagadougou, Burkina Faso, 1995.

[12] F. Bognounou, M. Tigabu, P. Savadogo et al., "Regeneration of five Combretaceae species along a latitudinal gradient

in Sahelo-Sudanian zone of Burkina Faso," *Annals of Forest Science*, vol. 67, no. 3, 2010.

[13] U. Belemsobgo, P. Kafando, B. A. Adouabou et al., "Network of protected areas," in *Biodiversity Atlas of West Africa*, A. Thiombiano and D. Kampmann, Eds., vol. 2, pp. 354–363, Ouagadougou, Burkina Faso, 2010.

[14] A. Aubreville, "Accord à Yangambi sur la nomenclature des types africains de végétation," *Bois et Forêts des Tropiques*, no. 51, pp. 23–27, 1957.

[15] A. E. Magurran, *Measuring Biological Diversity*, Blackwell Publishing, Malden, Mass, USA, 2004.

[16] N. L. Johnson and S. Kotz, *Distributions in Statistics: Continuous Univariate Distributions*, John Wiley and Sons, New York, NY, USA, 1970.

[17] S. J. Zarnock and T. R. Dell, "An evaluation of percentile and maximum likelihood estimators of Weibull parameters," *Forest Sciences*, vol. 31, pp. 260–268, 1985.

[18] K. A. Ryniker, J. K. Bush, and O. W. Van Auken, "Structure of Quercus gambelii communities in the Lincoln National Forest, New Mexico, USA," *Forest Ecology and Management*, vol. 233, no. 1, pp. 69–77, 2006.

[19] C. G. Lorimer and A. G. Krug, "Diameter distributions in even-aged stands of shade-tolerant and midtolerant tree species," *American Midland Naturalist*, vol. 109, no. 2, pp. 331–345, 1983.

[20] P. J. Baker, S. Bunyavejchewin, C. D. Oliver, and P. S. Ashton, "Disturbance history and historical stand dynamics of a seasonal tropical forest in western Thailand," *Ecological Monographs*, vol. 75, no. 3, pp. 317–343, 2005.

[21] H. Caswell, *Matrix Population Models: Construction Analysis and Interpretation*, Sinauer Associates, 2nd edition, 2001.

[22] SAS Inc., *SAS/STAT User's Guide*, SAS Institute, Cary, NC, USA, 1999.

[23] S. Guinko, P. Ouoba, and J. Millogo-Rasolomdimby, "L'apport de l'inventaire des aires classes et protégées dans la connaissance de la diversité végétale du Burkina Faso," *Ber Sonderforschungsbereichs*, vol. 268, no. 14, pp. 257–271, 2000.

[24] C. M. Shackleton, "Comparison of plant diversity in protected and communal lands in the Bushbuckridge lowveld savanna, South Africa," *Biological Conservation*, vol. 94, no. 3, pp. 273–285, 2000.

[25] K. Hahn-Hadjali, M. Schmidt, and A. Thiombiano, "Phytodiversity dynamics in pastured and protected West African savannas," in *Taxonomy and Ecology of African Plants, Their Conservation and Sustainable Use*, S. A. Ghazanfar and H. J. Beentje, Eds., pp. 351–359, Royal Botanic Gardens, Kew, 2006.

[26] T. A. Gardner, T. Caro, E. B. Fitzherbert, T. Banda, and P. Lalbhai, "Conservation value of multiple-use areas in East Africa," *Conservation Biology*, vol. 21, no. 6, pp. 1516–1525, 2007.

[27] B. M. I. Nacoulma, K. Schumann, S. Traoré et al., "Impacts of land-use on West African savanna vegetation: a comparison between protected and communal area in Burkina Faso," *Biodiversity and Conservation*, vol. 20, no. 14, pp. 3341–3362, 2011.

[28] J. L. Devineau, "Rôle du bétail dans le cycle culture-jachère en région soudanienne : la dissémination d'espèces végétales colonisatrices d'espaces ouverts (Bondoukuy, sud-ouest du Burkina Faso)," *Revue Ecologique Terre et Vie*, vol. 54, pp. 97–121, 1999.

[29] J. L. Devineau, A. Fournier, and S. Nignan, "'Ordinary biodiversity' in western Burkina Faso (West Africa): what vegetation do the state forests conserve?" *Biodiversity and Conservation*, vol. 18, no. 8, pp. 2075–2099, 2009.

[30] R. Bellfontaine, A. Gaston, and Y. Petrucci, *Management of Natural Forest of Dry Tropical Zones*, vol. 32 of *Conservation guide*, FAO, Rome, Italy, 2000.

[31] J. J. Wiens and M. J. Donoghue, "Historical biogeography, ecology and species richness," *Trends in Ecology and Evolution*, vol. 19, no. 12, pp. 639–644, 2004.

[32] A. Ouédraogo, A. Thiombiano, K. Hahn-Adjali, and S. Guinko, "Diagnostic de l'état de dégradation des peuplements de quatre espèces ligneuses en zone soudanienne du Burkina Faso," *Sécheresse*, vol. 17, no. 4, pp. 485–491, 2006.

[33] A. Ouédraogo and A. Thiombiano, "Regeneration pattern of four threatened tree species in Sudanian savannas of Burkina Faso," *Agroforestry Systems*, vol. 861, pp. 35–48, 2012.

[34] F. E. Fongnzossie, N. Tsabang, B. A. Nkongmeneck et al., "Les peuplements d'arbres du Sanctuaire à gorilles de Mengamé au sud Cameroun," *Tropical Conservation Science*, vol. 1, no. 3, pp. 204–221, 2008.

[35] W. Bonou, R. Glèlè Kakaï, A. E. Assogbadjo, H. N. Fonton, and B. Sinsin, "Characterisation of *Afzelia Africana* Sm. habitat in the Lama forest reserve of Benin," *Forest Ecology and Management*, vol. 258, no. 7, pp. 1084–1092, 2009.

[36] H. Zegeye, D. Teketay, and E. Kelbessa, "Diversity, regeneration status and socio-economic importance of the vegetation in the islands of Lake Ziway, south-central Ethiopia," *Flora*, vol. 201, no. 6, pp. 483–498, 2006.

[37] A. Gnoumou, F. Bognounou, K. Hahn, and A. Thiombiano, "A comparison of Guibourtia copaliffera Benn. Stands in South West Burkina Faso-community structure and regeneration," *Journal of Forestry Research*, vol. 22, no. 4, pp. 551–559, 2011.

[38] J. L. Devineau, "Évolution saisonnière et taux d'accroissement des surfaces terrières des ligneux dans quelques peuplements savanicoles soudaniennes de l'ouest burkinabé," *Ecologie*, vol. 28, no. 3, pp. 217–232, 1997.

15

Integral Proteins in Plant Oil Bodies

Jason T. C. Tzen

Graduate Institute of Biotechnology, National Chung Hsing University, Taichung 402, Taiwan

Correspondence should be addressed to Jason T. C. Tzen, TCTZEN@dragon.nchu.edu.tw

Academic Editors: G. T. Maatooq and Y. Yamauchi

Hydrophobic storage neutral lipids are stably preserved in specialized organelles termed oil bodies in the aqueous cytosolic compartment of plant cells via encapsulation with surfactant molecules including phospholipids and integral proteins. To date, three classes of integral proteins, termed oleosin, caleosin, and steroleosin, have been identified in oil bodies of angiosperm seeds. Proposed structures, targeting traffic routes, and biological functions of these three integral oil-body proteins were summarized and discussed. In the viewpoint of evolution, isoforms of oleosin and caleosin are found in oil bodies of pollens as well as those of more primitive species; moreover, caleosin- and steroleosin-like proteins are also present in other subcellular locations besides oil bodies. Technically, artificial oil bodies of structural stability similar to native ones were successfully constituted and seemed to serve as a useful tool for both basic research studies and biotechnological applications.

1. Introduction

The stored energy in plant tissues is occasionally preserved in the form of proteins, yet much more commonly in the form of carbohydrates or lipids. Plant cells deposit storage resources of carbohydrates, proteins, and neutral lipids in subcellular particles termed starch granules, protein bodies, and oil bodies, respectively. In contrast with the active studies of protein bodies and starch granules [1–6], research progress on oil bodies is relatively late and slow presumably due to less research input and inevitable technical problems caused by the hydrophobic features of these lipid-storage organelles.

Oil bodies are intracellular organelles for storing neutral lipids, mainly triacylglycerols and sterol esters, and they are also referred to as lipid bodies, lipid droplets, oil globules, oleosomes, and spherosomes. These organelles have been found across a wide range of plant cells, from microalgae to the most complex angiosperms; among them, oil bodies obtained from seed cells have been studied most intensively [7]. According to the cumulative research outcome in the past three decades, it is generally assumed that an oil body is composed of a neutral lipid matrix surrounded by a mono-layer of phospholipids embedded with some unique integral proteins [8–10]. This paper focused on the integral proteins of oil bodies in terms of their proposed structures, organelle targeting, biological functions, homologous isoforms, and utilization of artificial oil bodies.

2. Identification of Integral Proteins in Oil Bodies of Angiosperm Seeds

Constituents of oil bodies in angiosperm species, particularly those in oily seeds, have been continually investigated in the past three decades [11–13]. Research approaches by using tools of molecular biology and protein chemistry were relatively active in this research area in the past two decades.

2.1. Structural Components of Seed Oil Bodies. Vegetable cooking oils commonly extracted from various oily seeds are triacylglycerol molecules that tend to segregate from aqueous solution and form a transparent layer on the top. These hydrophobic triacylglycerol molecules are originally assembled in specialized organelles termed oil bodies, and these lipid storage organelles are stably packed in aqueous environments, that is, the cytosolic compartment of seed cells, with sizes mostly ranging from 0.5 to 2 μm (Figure 1(a)) [14, 15]. Intact oil bodies isolated from oily seeds, such as sesame, form a milky layer on top of the solution after centrifugation (Figure 1(b)), and they look drastically different from the transparent vegetable cooking oils that

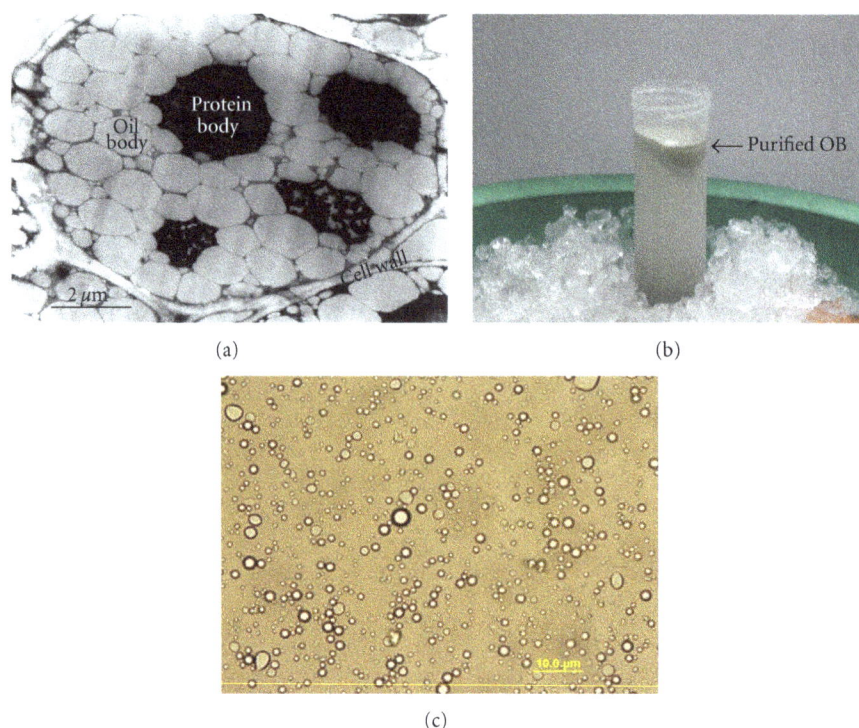

FIGURE 1: Visualizations of oil bodies. (a) Electron microscopy of a seed cell of mature sesame (adopted and modified from Figure 5 of Peng and Tzen, [14]). The most abundant gray spherical particles are oil bodies. (b) A photo of isolated sesame oil bodies. Purified oil bodies floated on the top and formed a milky layer after centrifugation. (c) Light microscopy of isolated sesame oil bodies. Isolated oil bodies were kept in 0.1 M sodium phosphate buffer, pH 7.5 for 30 min before taking the photo.

are generally extracted from seed oil bodies under relatively stringent conditions (high temperature or organic solvent). The isolated oil bodies remained maintaining their structural integrity and stability when they were suspended in an aqueous solution as observed under a light microscope (Figure 1(c)). Evidently, oil bodies are remarkably stable both in vivo and in vitro as compressed oil bodies in cells of a mature seed or in the milky layer during isolation never coalesce or aggregate. The remarkable stability of oil bodies in aqueous environments implies that surfactant molecules are present on the surface of oil bodies [16–19]. Chemical analyses suggested that phospholipids and proteins might be minor constituents (one to a few percent by weight) of oil bodies and served as surfactants to encapsulate abundant hydrophobic neutral lipids into many relatively small hydrophilic particles.

2.2. Identification of a Major Integral Protein in Seed Oil Bodies.
According to the chemical detection of phospholipids in oil bodies as well as the observation of one single boundary line on the surface of oil bodies under an electron microscope, it is generally accepted that the matrix neutral lipids of seed oil bodies are encapsulated by a monolayer of phospholipids [20–24]. In contrast, the presence of unique integral proteins, rather than nonspecifically associated contaminants of purification, in seed oil bodies was not confirmed until the striking discovery of oleosin, a major

surfactant protein in seed oil bodies of maize (*Zea mays* L.) [25]. The amino acid sequences of two maize oleosins deduced from their cDNA clones show a conservative central hydrophobic domain of approximately 70 residues that is the longest hydrophobic segment found in natural proteins so far, and apparently responsible for the anchorage of the proteins on the surface of oil bodies [26]. Oleosin was named in 1990 taking its meaning of an oil (oleo-) protein (-sin). Right after the kick-off studies on maize oleosins, homologous oleosin isoforms were subsequently identified in oil bodies of rapeseed (*Brassica napus*), soybean (*Glycine max* L.), carrot (*Daucus carota*), sunflower (*Helianthus annuus*), *Arabidopsis thaliana*, and cotton (*Gossypium hirsutum*) with their corresponding cDNA fragments cloned [27–34]. It seems that oleosin isoforms are universally present in oil bodies of angiosperm seeds including both monocotyledonous and dicotyledonous species [35–39]. Furthermore, oleosin isoforms found as the major proteins (approximately 80–90%) in seed oil bodies were demonstrated to shield the whole surface of the organelles in the company of phospholipids [40]. The structural integrity and stability of seed oil bodies were assumed to be provided by abundant oleosins via two factors, steric hindrance and electronegative repulsion [41].

2.3. Identification of Two Minor Integral Proteins in Seed Oil Bodies.
After the identification of the major protein, oleosin in seed oil bodies, whether minor integral proteins could also

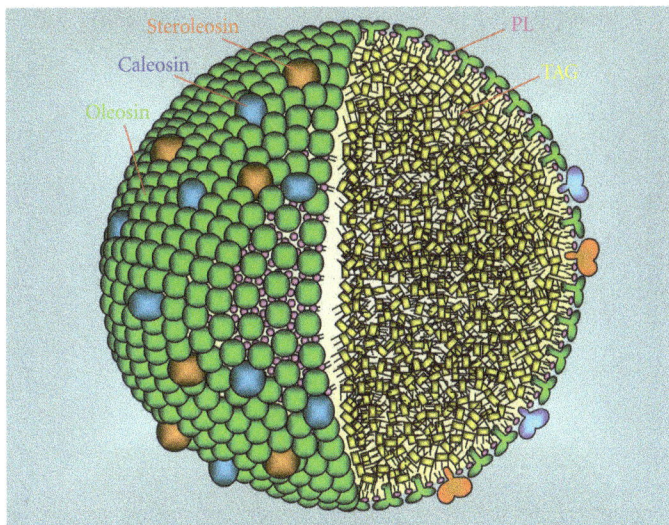

FIGURE 2: A structural model of an oil body with three types of integral oil-body proteins. An oil body is proposed to contain a triacylglycerol (TAG) matrix surrounded by a monolayer of phospholipids (PL) embedded with three classes of proteins, oleosin, caleosin, and steroleosin.

be found in these organelles became the next challenge. The approach to this challenge was severely impeded since many non-specifically associated proteins were contaminated in the preparation of oil bodies. The impediment was overcome by the development of a purification protocol that removed almost all the contaminated proteins on the surface of isolated oil bodies by washing harshly with detergent, high salt, chaotropic agent, and hexane [42]. As exemplified by seed oil bodies of sesame (*Sesamum indicum* L.), besides abundant oleosin isoforms, three minor protein bands of relatively high molecular masses were found after the harsh washing [43]. These three minor proteins were later confirmed as two classes of integral oil-body proteins named caleosin and steroleosin [44, 45]. Caleosin was named in 1999 taking its meaning of a calcium-binding (cal-) oil protein (-leosin) [46], and steroleosin was named in 2002 as a sterol-regulatory (sterol-) oil protein (-leosin) [47]. Taken together, a structural model of seed oil body was depicted in Figure 2.

2.4. Identification of Other Potential Oil-Body Proteins. Searching for more oil-body proteins has been approached by subproteomic analysis under the assistance of liquid chromatography electrospray ionization tandem mass spectrometry in the past few years [48–54]. In addition to the three known integral oil-body proteins, oleosin, caleosin, and steroleosin, several proteins were detected as potential oil-body proteins in seed oil bodies of maize and rapeseed. However, none of these newly identified proteins have been confirmed as integral oil-body proteins or peripheral proteins associated with some surface components of oil bodies for particular physiological functions. It remains to be clarified if proteins other than oleosin, caleosin, and steroleosin are embedded or peripherally associated on the surface of seed oil bodies.

3. Proposed Structures of Integral Oil-Body Proteins

Due to the insolubility of oleosin, caleosin, and steroleosin possibly caused by their hydrophobic oil-body anchoring domains, no three-dimensional structures derived from X-ray or NMR are available at the present time. Proposed structures of these three oil-body proteins are predicted based on their sequence analyses and spectrometric determination.

3.1. Proposed Oleosin Structure. An oleosin molecule is proposed to comprise three structural domains: an N-terminal amphipathic domain, a central hydrophobic oil-body anchoring domain, and a C-terminal amphipathic α-helical domain (Figure 3(a)) [55]. The N-terminus of oleosin is blocked with acetylation after the removal of the first methionine, a cotranslational modification presumably related to the enhancement of protein structural stability to fulfill the long-term storage of oil bodies within seed cells [56]. Both N- and C-terminal domains are not conserved among oleosins of diverse species, and even their lengths are quite variable in different oleosin isoforms. It is generally agreed that these two domains are putatively resided on the surface of oil bodies and stabilize the organelles via steric hindrance and electronegative repulsion [57–59]. In contrast, the central anchoring domain of oleosins is highly conserved among diverse species, particularly in a relatively hydrophilic proline knot motif at the middle of the sequence [60–63]. To date, controversial secondary structures of this domain have been proposed or determined in the past two decades, yet the contents of α-helical and β-stranded structures in these proposed models are extremely different [41, 64–72]. The controversy among these proposed structures does not seem to be receded unless a convincing three-dimensional structure of oleosin or at least its central hydrophobic domain is resolved.

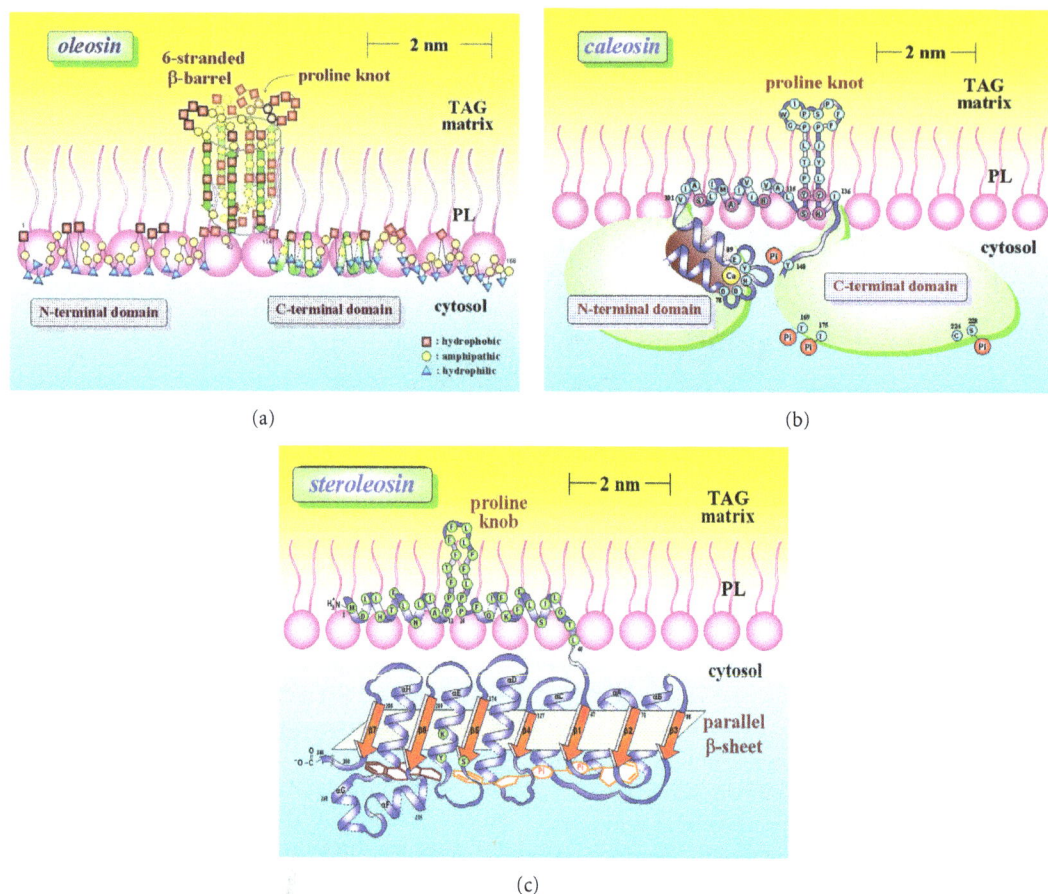

FIGURE 3: Secondary structure organizations of integral oil-body proteins on the surface of an oil body. Proposed secondary structures of oleosin (a), caleosin (b), and steroleosin (c) are adopted and modified from Figure 5 of Tzen et al., [55], Figure 7 of Chen et al., [46], and Figure 6 of Lin et al., [47], respectively.

3.2. Proposed Caleosin Structure. A caleosin molecule is also proposed to comprise three structural domains: an N-terminal hydrophilic calcium-binding domain, a central hydrophobic oil-body anchoring domain, and a C-terminal hydrophilic phosphorylation domain (Figure 3(b)) [46]. The N-terminus of caleosin is also blocked with acetylation after the removal of the first methionine, presumably modified by the same co-translational mechanism found in oleosin [56]. The N-terminal hydrophilic domain consists of an EF hand calcium-binding motif of 28 residues including an invariable glycine residue as a structural turning point and five conserved oxygen-containing residues as calcium-binding ligands [23, 46]. Owing to the presence of this calcium-binding domain, calcium ion affected electrophoretic mobility of native and recombinant caleosin on SDS-PAGE [46, 73, 74]. The calcium-binding capacity of caleosin is in agreement with the observation of calcium staining on the surface of oil bodies in electron microscopy prior to the discovery of caleosin in these lipid storage organelles [75]. The central hydrophobic domain of caleosin is relatively short in comparison with that of oleosin, and comprises an amphipathic α-helix and an anchoring region. The amphipathic α-helix is assumed to be located in the interface between hydrophobic and hydrophilic environments while

the anchoring region is predicted to comprise a pair of anti-parallel β-strands connected with a proline knot motif. The similarity of proline knot motifs of oleosin and caleosin seems to imply a significant role associated with this unique motif, such as protein folding, intermolecular assembly, or specific targeting to oil bodies [76–78]. The C-terminal hydrophilic domain of caleosin contains several potential phosphorylation sites. The native caleosin in seed oil bodies of *Arabidopsis thaliana*, but not bacterially expressed caleosin, has been found partially phosphrylated [79]. An invariable cysteine residue is present near the C-terminus of caleosin and unlikely involved in any interdisulfide linkage with another caleosin molecule or other proteins on the surface of oil bodies.

3.3. Proposed Steroleosin Structure. A steroleosin molecule is proposed to comprise a relatively small N-terminal oil-body anchoring domain and a relatively large soluble sterol-binding dehydrogenase domain (Figure 3(c)) [47]. Free N-terminus occurs in steroleosin with the translation-initiating methionine as the first residue, in contrast with the acetylation-blocked N-termini of oleosin and caleosin [56]. So far, less research investigation has been executed on the structure of steroleosin in comparison with that of oleosin or caleosin. The N-terminal anchoring segment comprises two amphipathic α-helices (12 residues in each helix) connected by a hydrophobic sequence of 14 residues bordered by 1-2 proline residues at each end. The relatively hydrophilic proline residues located in both ends of the 14-residue hydrophobic sequence are proposed to aggregate in the hydrophobic surroundings and form a unique structure, termed proline knob motif, for the integrity and stability of steroleosin anchorage on the surface of oil bodies. The soluble sterol-binding dehydrogenase domain contains an NADPH-binding subdomain, an active site region, and a sterol-binding subdomain. Three-dimensional structure of the dehydrogenase domain has been simulated by homology modeling [47]. The modeling structure reveals that the NADPH-binding region, active site, and sterol-binding region are located in the C-terminal ends of a parallel β-sheet and that the NADPH-binding region is expectably located in the crevice region, termed topological switch point, as observed in all similar α/β structures [80].

4. Possible Biological Functions of Integral Oil-Body Proteins

Except for the structural role, it is reasonable that oil-body proteins may exert some biological functions related to the synthesis or degradation of oil bodies, for example, signaling for the formation, assembly, fusion, or mobilization of these lipid storage organelles. Indeed, some physiological functions have been proposed for oil-body proteins in the past decade [81]. Further verification of these proposed physiological functions is anticipated in the come-up research progress.

4.1. Proposed Functions of Oleosin. Oleosin has been proven to play a key role in the stability of seed oil bodies via electronegative repulsion and steric hindrance [41]. This structural role prevents coalescence of oil bodies during seed desiccation and maintains them as discrete and relatively small organelles. It is demonstrated that the structural role of oleosin also protects *Arabidopsis thaliana* seeds against freeze/thaw-induced damage of their cells in vivo [82]. The contents of oleosins are found to determine sizes of seed oil bodies; presumably the ratio of oleosin over triacylglycerol is inversely proportional to the sizes of oil bodies [83–88]. The stability of isolated oil bodies could be substantially enhanced after their surface proteins were cross-linked by linker molecules, such as glutaraldehyde or genipin [87]. Being structural proteins, oleosin isoforms as well as caleosin ones are partially degraded by a special thiol-protease, thioredoxin h, after germination, and this specific degradation of oil-body structural proteins is proposed to be associated with mobilization of oil bodies in seedlings [89–92].

Furthermore, oleosin is suggested to be a bifunctional enzyme that has both monoacylglycerol acyltransferase and phospholipase activities during seed germination [93]. The regulation of these distinct dual activities seems to be controlled by the phosphorylation of oleosin presumably by a serine/threonine/tyrosine protein kinase, and the oleosin phosphorylation is also found to be activated by phosphatidylcholine and diacylglycerol, but inhibited by lysophosphatidylcholine, oleic acid, and calcium ion [94]. It will be interesting to see if these two enzymatic activities are universally detectable in oleosin isoforms of diverse species since both N- and C-terminal domains are not conserved among these oleosin isoforms. Meanwhile, most oleosins are relative small proteins of 15–20 kDa, particularly their N- and C-terminal domains are quite tiny (3 to 5 kDa for each domain; Figure 3(a)). In terms of structure-function relationship, it is a challenging task to reveal how the small structural domain(s) of oleosins construct the three dimensional active sites for the two observed enzymatic activities.

4.2. Proposed Functions of Caleosin. Having a structural organization and oil-body anchorage similar to oleosin, caleosin has been demonstrated to stabilize seed oil bodies as efficiently as oleosin [76]. The structural role of caleosin is clearly verified by the observation of stable cycad (*Cycas revoluta*) seed oil bodies that are mainly sheltered by caleosin without the presence of any oleosin isoform [95]. The investigation also invalidates a prevalent concept in this research area for two decades, declaring that oleosin is an essential constituent and can be regarded as a marker protein of plant oil bodies.

Caleosin comprises a calcium-binding motif and several potential phosphorylation sites, that is, well-known candidates involved in signal transduction, and thus may possess biological function(s) in addition to its structural role for the stability of oil bodies. According to the characterization of two independent insertion mutants lacking caleosin, it was proposed that caleosin might play a role in the degradation of storage lipids in oil bodies by inducing the interaction of oil bodies with vacuoles during germination [96]. Putative interaction between oil bodies and vacuoles was also observed in pollen cells after germination under electron microscopy; and the pollen oil bodies were presumably surrounded by tubular membrane structures and encapsulated in the vacuoles after germination [97–99]. The detailed molecular interaction between caleosin on the surface of oil bodies and its specific partner protein on the membrane of vacuoles remains to be studied.

Caleosin isoforms or caleosin-like proteins are not only localized in oil bodies but also found as membrane-bound proteins in other subcellular fractions, such as microsomal membrane; moreover, they were demonstrated to possess different biological functions, such as peroxygenase activity in biotic and abiotic stress responses in their phosphorylated forms [100–102]. Site-directed mutagenesis studies revealed that the peroxygenase catalytic activity of caleosin, an original heme-oxygenase, was dependent on two highly conserved histidines [100]. It was proposed that caleosin-like proteins might be involved in the plant-pathogen recognition, symptom development, and the basal tolerance to biotic and abiotic stresses through the salicylic acid signaling pathway [102]. In *Arabidopsis*, a stress-responsive caleosin-like protein, AtCLO4, was demonstrated to act as a negative regulator of ABA responses [103], whereas another caleosin-like protein, RD20, was involved in ABA-mediated inhibition of germination but did not response to biotic or abiotic stresses [104]. Recently, a wheat caleosin-like protein was proposed to play a role in the Ca^{2+}-triggered feedback regulation of both the canonical Gα subunit of the heterotrimeric G protein complex and phosphoinositide-specific phospholipase C [105], and a noncanonical caleosin from *Arabidopsis* was found to epoxidize unsaturated fatty acids efficiently with complete stereoselectivity [106]. Taken together, the recent research progress on the identification of caleosin functions is encouraging. However, it also raises a puzzle how the highly conserved caleosin isoforms execute several diverse functions that may require a well-structured active site for an enzymatic reaction and a well-featured binding surface for a specific protein-protein interaction.

4.3. Proposed Functions of Steroleosin. Steroleosin possesses a sterol-regulatory dehydrogenase domain that belongs to a superfamily of presignal proteins involved in signal transduction via activation of its partner receptor after binding to a regulatory sterol [45]. Besides dehydrogenase activity, no other biological functions have been experimentally proven for steroleosin so far [44, 47]. As caleosin and steroleosin are minor integral proteins of comparable contents in oil bodies of sesame seeds, it is speculated that caleosin may be regulated by a pre-signal partner, such as a sterol-activated steroleosin, to serve as a receptor or signaling molecule on the surface of oil bodies. Meanwhile, two steroleosin isoforms having distinct sterol-binding sites are found in oil bodies of angiosperm seeds, and may be involved in the activation of sterol signal transduction that regulates specialized biological functions related to the mobilization of oil bodies during seed germination.

5. Targeting of Integral Oil-Body Proteins

Targeting of oil-body proteins, particularly oleosin, has been extensively investigated via in vivo systems using transgenic techniques and in vitro systems using microsomal membranes for integration of translated proteins [107–111]. The signal segment for specific targeting of oleosin to oil bodies is apparently located within the protein itself since recombinant oleosins are able to target correctly oil bodies in different transgenic plants and yeasts [112–114]. The central hydrophobic domain, particularly the conservative proline knot motif, of oleosin or caleosin has been demonstrated to play an essential role for the protein targeting oil bodies [60–63, 78]. According to the studies with an in vitro system using microsomal membranes as targets for integration of translated proteins, it has been suggested that oleosin might target endoplasmic reticulum (ER) under the assistance of the signal recognition particle (SRP) prior to transportation to maturing oil bodies [115–119]. However, neither oleosin nor caleosin possesses a cleavable or noncleavable N-terminal signal sequence required for the SRP-dependent pathway of targeting to the ER [120, 121]. Thus, it is proposed that oleosin contains several segments that are capable of interacting with SRP to direct the protein to the ER membrane [122].

An in vitro system was established to evaluate the targeting traffic routes of oleosin, caleosin, and steroleosin by constituting artificial oil emulsions (generated by sonication of triacylglycerol and phospholipid in a buffer solution) to mimic maturing oil bodies for integration of translated oil-body proteins [9, 78]. The results suggest that steroleosin and caleosin/oleosin may be assembled to maturing oil bodies through different locations of ER membrane, that is, caleosin/oleosin directly target maturing oil bodies where ER membranes are enlarged with deposited triacylglycerol molecules, whereas steroleosin is recognized by SRP and guided to integrate into the phospholipid bilayer of ER membrane prior to its lateral migration to maturing oil bodies (Figure 4). The distinct targeting traffic routes between steroleosin and caleosin/oleosin are in agreement with the following two observations. Firstly, steroleosin and caleosin/oleosin target and anchor to oil bodies via different structural organizations. Steroleosin possesses a non-cleavable N-terminal signal sequence putatively responsible for ER targeting via SRP dependent pathway, and its anchoring to oil bodies lies mainly in the N-terminal hydrophobic domain; caleosin and oleosin, lacking an N-terminal signal sequence, target/anchor to oil bodies via their central hydrophobic domains. Secondly, steroleosin possesses a free methionine at its N-terminus while caleosin and oleosin are N-terminally blocked by acetylation after the removal of the first methionine residue [56]. Presumably, the N-terminus of steroleosin is protected by SRP complex and/or embedded in ER membrane during its synthesis and targeting while the N-termini of caleosin and oleosin are freely exposed to cytosol during their synthesis and targeting to maturing oil bodies via central hydrophobic domains.

Negatively charged phospholipids (phosphatidylserine and phosphatidylinositol) are present in a consistent amount (30–40%) in the phospholipids of oil bodies from diverse seeds [15]. According to an in vitro targeting study, inclusion of negatively charged phospholipids in artificial oil emulsions substantially enhanced the targeting efficiency of oleosin and caleosin to these emulsions [78]. It is assumed that negatively charged phospholipids in the surface area of oil bodies are involved in targeting or assembling of oil-body proteins to these organelles. Definitely, it is an important task to figure out the specific targeting interaction between the unique segments of oil-body proteins and the negatively charged phospholipids of maturing oil bodies.

6. Isoforms of Integral Oil-Body Proteins in Evolution

6.1. Oleosin Isoforms. Oleosin is an alkaline protein unique to oil bodies and has been found exclusively in plant species. Two distinct classes, H- and L- (high and low molecular weight) oleosins, are present in seed oil bodies of diverse angiosperms, and one or more isoforms may occur in each oleosin class of the same species [123].

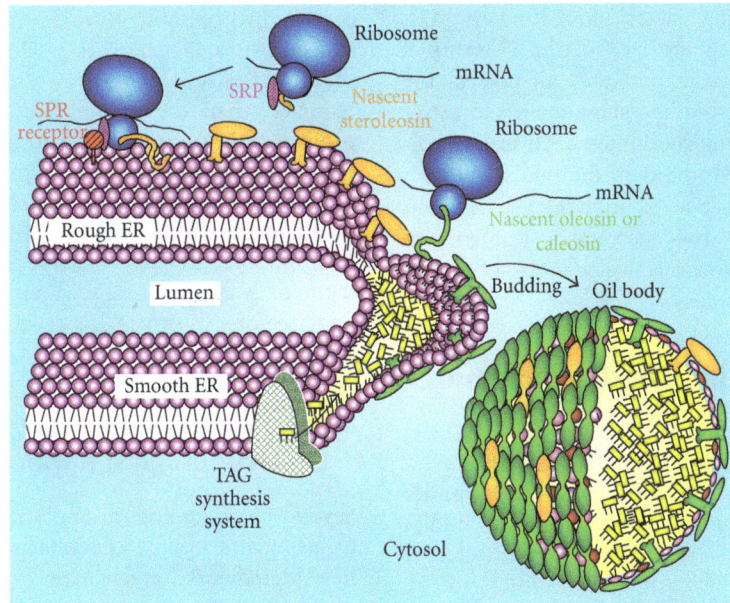

FIGURE 4: Targeting model of oleosin, caleosin, and steroleosin to a maturing oil body. Oleosin and caleosin are proposed to target directly a maturing oil body. In contrast, steroleosin is assumed to be integrated into the phospholipid bilayer of endoplasmic reticulum (ER) membrane prior to its lateral migration to a maturing oil body (adopted and modified from Figure 7 of Chen and Tzen, [78]).

It has been shown that H- and L-oleosins coexist on the surface of each oil body in seed cells of sesame [124]. The main difference between these two oleosin isoforms is an insertion of 18 residues in the C-terminal domain of H-oleosin, accounting for a 2 kDa difference in mass between the two classes found in many species [125–127]. The physiological significance of the presence of these two oleosin isoforms in oil bodies of angiosperm seeds remains to be elucidated. L-oleosin but not H-oleosin is found in megagametophytes of two gymnosperm species, pine (*Pinus koraiensis*) and ginkgo (*Ginkgo biloba*); it may imply that L-oleosin is a more primitive isoform class, with H-oleosin derived from L-oleosin before the divergence of monocot and dicot species during evolution [128]. Moreover, cDNA fragments encoding putative oleosin isoforms were found in pollen of rapeseed [129, 130]. Recently, stable oil bodies were successfully isolated from lily (*Lilium longiflorum* Thunb.) pollen, and a unique P-oleosin was found as the major integral oil-body protein [97]. Three oleosin genes, representative of early trends in evolution, were found in the model moss, *Physcomitrella* with a complex pattern of expression based on gene splicing [131]. Moreover, oleosin-like proteins, forming a distinct class termed oleopollenin, were found in tapetum and external surfaces of pollen grains [132–134].

6.2. *Caleosin Isoforms.* In angiosperm species, sequence alignment shows that caleosins in monocot seed oil bodies seem to possess an additional N-terminal appendix of approximately 40–70 residues, and thus are larger than those in dicotyledonous seed oil bodies [135–137]. Recently, a distinct P-caleosin isoform was also identified in pollen oil bodies of lily and olive (*Olea europaea* L.) [97, 98, 138].

Caleosin is also found in more primitive species, such as cycad and microalgae, and thus is assumed to be an oil-body protein more primitive than oleosin in evolution [95, 139]. Phylogenetic tree analysis supports that microalgal caleosin is the most primitive caleosin found in oil bodies to date [8, 23, 139]. The additional N-terminal appendix found in monocot caleosins is not present in pollen or cycad caleosin. Therefore, the additional N-terminal appendix found in monocot caleosins seems to be resulted from an insertion mutation in monocot seed caleosin in evolution. Of course, this hypothetic evolutionary event should be verified by further molecular evidence. Since caleosin is more primitive than oleosin, it is reasonable to speculate that caleosin is an essential integral protein of plant oil bodies [76]. However, this speculation has been invalidated as stable oil bodies located in rice aleurone layer are composed of H- and L-oleosin but not caleosin [135]. Interestingly, those rice oil bodies lacking caleosin are not mobilized after germination [39]. It remains to be studied whether caleosin is indispensable for the mobilization of oil bodies. In contrast with oleosin isoforms that are unique to oil bodies, caleosin isoforms or caleosin-like proteins are possibly present in other cellular locations, for example, ER membrane [140]. Moreover, the same caleosin isoform is possibly present in both seed oil bodies and membrane-bound fractions of other tissues [99].

6.3. *Steroleosin Isoforms.* Limited research progress has been advanced in the identification of steroleosin isoforms so far. Similar to caleosin, steroleosin isoforms or steroleosin-like proteins are possibly present not only in oil bodies but also in other subcellular locations [44]. Homologous proteins of steroleosin are presumably present in all kinds of living organisms including bacteria and humans [47]. Most steroleosin-like proteins are lack of the N-terminal hydrophobic anchoring domain, and they all possess the highly conservative NADPH-binding subdomain and active site, but diverse sterol-binding subdomains. Diverse sterol-binding subdomains are also found in the two steroleosin isoforms located in sesame oil bodies, implying that different sterols may regulate these two steroleosin isoforms to conduct distinct biological functions related to the formation or degradation of seed oil bodies [9].

7. Artificial Oil Bodies in Basic Research Studies

Artificial oil bodies of similar sizes (0.5–2 μm) and structural stability have been successfully reconstituted with triacylglycerols, phospholipids, and integral oil-body proteins under the same proportions as they are found in native oil bodies [125]. The sizes of artificial oil bodies could be controlled by changing the ratio of triacylglycerol over oil-body protein, whereas both thermostability and structural stability of artificial oil bodies decreased as their size increased, and vice versa [87]. For encapsulation of artificial oil bodies, recombinant oleosins expressed in *Escherichia coli* were found comparable to native oleosins isolated from seed oil bodies [141, 142]. It has been demonstrated that artificial oil bodies could be stabilized by oleosin or caleosin, but not steroleosin, and the average sizes (50–200 nm) of artificial oil bodies constituted with caleosin were 10-times smaller than those (0.5–2 μm) constituted with oleosin (Figure 5) [143].

Since stable artificial oil bodies could be simply generated with triacylglycerol, phospholipid, and oil-body protein, but not any two of them, it is apparent that these three constituents are essential components for the construction of oil bodies [41]. Artificial oil bodies could be stabilized by the combination of sesame oleosin isoforms or any oleosin isoform alone, that is, H1-oleosin, H2-oleosin, or L-oleosin; however, a slightly better structural stability was observed in artificial oil bodies constituted with L-oleosin than those constituted with either of the two H-oleosin isoforms [125]. Similar results were observed for the artificial oil bodies constituted with L-oleosin or H-oleosin extracted from oil bodies of rice seeds [124]. Obviously, L-oleosin is a better oil-body structural protein than H-oleosin. The relative small artificial oil bodies constituted with caleosin possessed a better thermostability (up to 70°C) than native oil bodies or artificial oil bodies stabilized with oleosin (lower than 50°C) [143]. The observation was in accordance with a later investigation on the relatively high thermostability (up to 70°C) of small cycad oil bodies that were mainly sheltered by a unique caleosin [95]. Evidently, caleosin is a better oil-body structural protein than oleosin in terms of thermal tolerance.

Artificial oil bodies constituted with truncated oleosin and caleosin have been utilized to evaluate the segments responsible for the structural stability of oil bodies. Artificial oil bodies constituted with truncated oleosins of the central hydrophobic domain longer than 36 residues were as stable as native sesame oil bodies, and those constituted with truncated oleosins lacking more than half of the original

central hydrophobic domain inclined to coalesce upon collision or aggregation [60]. Both structural stability and thermostability of artificial oil bodies were slightly or severely reduced when the amphiphatic α-helix (15 residues) or proline-knot subdomain (21 residues) of recombinant caleosin was truncated, and thus the whole central hydrophobic domain of 36 (15 + 21) residues is crucial for the stability of oil bodies [77]. Taken together, the minimal length of hydrophobic domain to serve as an oil-body anchoring segment is approximately 36 residues (mainly the proline-knot regions shown in the secondary structures of oleosin and caleosin in Figure 3).

8. Artificial Oil Bodies in Biotechnological Applications

Several biotechnological applications have been developed by using the unique characteristics of oil bodies, such as new ingredients for flavoring or emulsifying agents, affinity matrices for enzyme fixation/purification, and expression/purification systems for producing recombinant proteins via transgenic plants [144–154]. Many applications related to the utilization of oil bodies have been patented [155–163]. These applications of seed oil bodies can be authentically applied to the utilization of artificial oil bodies. Moreover, some novel usages of artificial particles have also been developed in the past decade.

8.1. Protein Expression/Purification System. A bacterial expression/purification system to produce recombinant proteins was developed by using artificial oil bodies [141, 164–169]. In this system, a target protein was first overexpressed as an insoluble oleosin-fused polypeptide, collected from the pellet of cell lysate simply by centrifugation, assembled into artificial oil bodies, separated from oleosin, and then harvested by concentrating the ultimate supernatant. This technique offers a powerful and competitive option to replace the conventional affinity chromatography used for protein purification. However, the requirement of using a relatively expensive endopeptidase, for example, factor Xa, for specific release of the target protein from the recombinant oleosin-fused polypeptide raises the processing cost substantially, and thus severely restricts its potential applications. To cost down this process, an improved system was developed by replacing the specific proteolytic cleavage sequence between oleosin and the target protein with an intein (an inducible self-splicing polypeptide) linker. In this revised system, the target protein was released from artificial oil bodies via self-splicing of the intein linker, induced by temperature alteration or dithiothreitol supplement, without using the expensive endopeptidase.

8.2. Matrix for Enzyme Immobilization. A new technique of enzyme fixation was designed to achieve, in one step, protein refolding and immobilization by linking a target enzyme, for example, D-hydantoinase, to oleosin on the surface of artificial oil bodies [170, 171]. The immobilized enzyme remained stable for at least 15 days when stored at

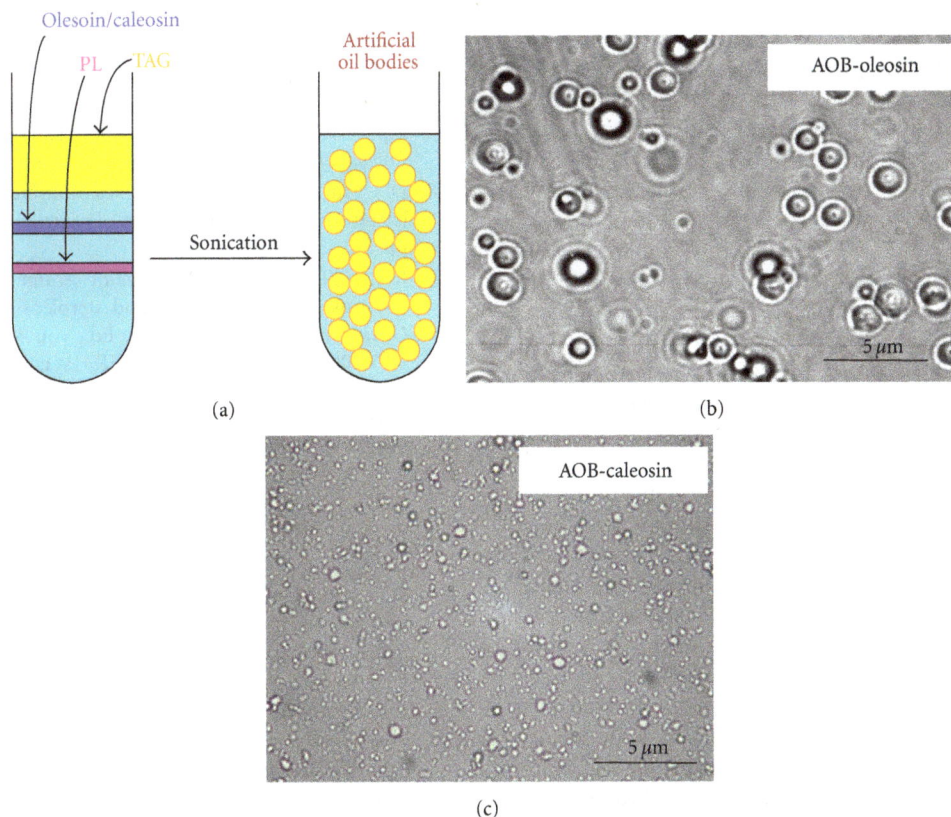

FIGURE 5: Constitution of artificial oil bodies. A cartoon diagram showed the preparation of artificial oil bodies by sonication (a). Artificial oil bodies (AOB) were generated with triacylglycerol (TAG) and phospholipid (PL) in the presence of oil-body protein, oleosin (b), or caleosin (c). Photos of (b) and (c) are adopted from Figure 3 of Chen et al., [143].

$4°C$, and its conversion yield exceeded 80% after 7 cycles of repeated use. Apparently, the simple and effective system by fixing target enzymes on the surface of artificial oil bodies is practical and useful for the routine operation of industrial enzymatic reactions.

8.3. Formula for Encapsulation of Bacteria. Numerous healthy and nutritional benefits have been ascribed to probiotics, such as lactic acid bacteria. Since probiotics may not survive in sufficient number to retain their functionality in human gastrointestinal tract, many approaches have been explored to increase their viability when used as food supplements. A technique was developed to protect lactic acid bacteria against simulated gastrointestinal conditions by encapsulation of bacterial cells within artificial oil bodies [172]. Compared with nonencapsulated cells, the entrapped bacteria demonstrated a significant increase (approximately 10,000 times) in survival rate in the presence of simulated high acid gastric or bile salt conditions. It is recommended that artificial oil bodies may represent a suitable formula of biocapsule to encapsulate bacteria for commercial utilization in dairy products.

8.4. Carrier for Drug Delivery. Relatively small artificial oil bodies stabilized with caleosin have been used to develop an oral delivery system for hydrophobic drugs, for example,

cyclosporine A, a drug commonly utilized as a clinical immunosuppressant to prevent transplant rejection and to treat several autoimmune diseases [173]. Cyclosporine A efficiently encapsulated in artificial oil bodies stabilized with caleosin could be stably stored for weeks at $4°C$. An oral delivery formulation with cyclosporine A in artificial oil bodies was demonstrated to exhibit satisfactory bioavailability in an animal test [173]. This drug delivery system or its improved formula may also be used as an adequate carrier for many other hydrophobic drugs, such as antitumor drugs [174–177].

8.5. Antibody Generation System. Recently, a system of generating antibodies against small molecules (haptens) was established under the assistance of artificial oil bodies [178]. To develop this system, a series of recombinant caleosins were engineered with more Lys residues to link and render small molecules on the surface of artificial oil bodies for antibody production. In this design, covalently conjugated haptens were anticipated to cover the whole surface of artificial oil bodies constituted with hapten-charged caleosins. The results indicate that engineered Lys-rich caleosins are suitable carrier proteins for the production of monospecific antibodies against small molecules, such as drug, herbal compounds, pesticides, herbicides, antibiotics, and hormones.

9. Perspective

In the past three decades, continual research advancement has confirmed the presence of three classes of integral proteins, oleosin caleosin, and steroleosin on the surface of oil bodies. A lot of potential oil-body proteins have been recently screened by the subproteomic approaches under the assistance of mass spectrometry; though a few of them seem to be contaminants apparently, some candidate proteins are waiting for further verification to see if they are real integral oil-body proteins or peripheral proteins associated with some surface components of oil bodies for particular physiological functions. Controversial structures have been proposed for oleosin, and the controversy cannot be receded until a convincing three-dimensional structure of oleosin is determined. Several physiological functions other than structural role have been actively demonstrated or proposed for oleosin and caleosin in the past decade. Taken together, it seems unlikely that these two relatively small proteins are capable of executing several diverse biological functions jointly, and thus some of the proposed functions may not be correct and should be ruled out in the follow-up researches. Oleosin- and caleosin-stabilized artificial oil bodies have been successfully constituted and used to develop various systems for biotechnological applications. Further investigation and technical improvement will create novel artificial oil bodies as versatile vehicles to fulfill many other requirements for specialized applications.

Acknowledgments

The work was supported by Grants from the National Science Council, Taiwan, ROC (NSC 100-2313-B-005-012-MY3 and NSC 100-2313-B-005-015-MY3 to JTC Tzen). The author cordially expresses appreciation to all the colleagues, particularly graduate students, contributing to this research area synergistically in the past two decades.

References

[1] V. Ibl and E. Stoger, "The formation, function and fate of protein storage compartments in seeds," *Protoplasma*, vol. 249, no. 2, pp. 379–392, 2012.

[2] M. R. Tandang-Silvas, E. M. Tecson-Mendoza, B. Mikami, S. Utsumi, and N. Maruyama, "Molecular design of seed storage proteins for enhanced food physicochemical properties," *Annual Review of Food Science and Technology*, vol. 2, pp. 59–73, 2011.

[3] T. Kawakatsu and F. Takaiwa, "Cereal seed storage protein synthesis: fundamental processes for recombinant protein production in cereal grains," *Plant Biotechnology Journal*, vol. 8, no. 9, pp. 939–953, 2010.

[4] S. Pérez and E. Bertoft, "The molecular structures of starch components and their contribution to the architecture of starch granules: a comprehensive review," *Starch/Staerke*, vol. 62, no. 8, pp. 389–420, 2010.

[5] S. G. Ball and M. K. Morell, "From bacterial glycogen to starch: understanding the biogenesis of the plant starch granule," *Annual Review of Plant Biology*, vol. 54, pp. 207–233, 2003.

[6] A. M. Smith, "The biosynthesis of starch granules," *Biomacromolecules*, vol. 2, no. 2, pp. 335–341, 2001.

[7] K. D. Chapman, J. M. Dyer, and R. T. Mullen, "Biogenesis and functions of lipid droplets in plants: Thematic Review Series: Lipid Droplet Synthesis and Metabolism: from Yeast to Man," *Journal of Lipid Research*, vol. 53, no. 2, pp. 215–226, 2012.

[8] D. J. Murphy, "The dynamic roles of intracellular lipid droplets: from archaea to mammals," *Protoplasma*, vol. 249, no. 3, pp. 541–585, 2011.

[9] J. T. C. Tzen, "Seed oil bodies of sesame and their surface proteins, oleosin, caleosin, and steroleosin," in *Sesame, the Genus Sesamum*, D. Bedigian, Ed., vol. 48, chapter 10, pp. 187–200, CRC Press, London, UK, 1st edition, 2011.

[10] A. H. C. Huang, "Oleosins and oil bodies in seeds and other organs," *Plant Physiology*, vol. 110, no. 4, pp. 1055–1061, 1996.

[11] D. L. Brasaemle and N. E. Wolins, "Packaging of fat: an evolving model of lipid droplet assembly and expansion," *Journal of Biological Chemistry*, vol. 287, no. 4, pp. 2273–2279, 2012.

[12] D. Zweytick, K. Athenstaedt, and G. Daum, "Intracellular lipid particles of eukaryotic cells," *Biochimica et Biophysica Acta*, vol. 1469, no. 2, pp. 101–120, 2000.

[13] J. A. Napier, A. K. Stobart, and P. R. Shewry, "The structure and biogenesis of plant oil bodies: the role of the ER membrane and the oleosin class of proteins," *Plant Molecular Biology*, vol. 31, no. 5, pp. 945–956, 1996.

[14] C. C. Peng and J. T. C. Tzen, "Analysis of the three essential constituents of oil bodies in developing sesame seeds," *Plant and Cell Physiology*, vol. 39, no. 1, pp. 35–42, 1998.

[15] J. T. C. Tzen, Y. Z. Cao, P. Laurent, C. Ratnayake, and A. H. C. Huang, "Lipids, proteins, and structure of seed oil bodies from diverse species," *Plant Physiology*, vol. 101, no. 1, pp. 267–276, 1993.

[16] C. R. Slack, W. S. Bertaud, B. D. Shaw, R. Holland, J. Browse, and H. Wright, "Some studies on the composition and surface properties of oil bodies from the seed cotyledons of safflower (*Carthamus tinctorius*) and linseed (*Linum ustatissimum*)," *Biochemical Journal*, vol. 190, no. 3, pp. 551–561, 1980.

[17] R. Qu, S. M. Wang, Y. H. Lin, V. B. Vance, and A. H. Huang, "Characteristics and biosynthesis of membrane proteins of lipid bodies in the scutella of maize (*Zea mays* L.)," *The Biochemical Journal*, vol. 235, no. 1, pp. 57–65, 1986.

[18] E. M. Herman, "Immunogold-localization and synthesis of an oil-body membrane protein in developing soybean seeds," *Planta*, vol. 172, no. 3, pp. 336–345, 1987.

[19] D. J. Murphy, "Storage lipid bodies in plants and other organisms," *Progress in Lipid Research*, vol. 29, no. 4, pp. 299–324, 1990.

[20] L. Y. Yatsu and T. J. Jacks, "Spherosome membranes: half unit-membranes," *Plant Physiology*, vol. 49, no. 6, pp. 937–943, 1972.

[21] A. H. C. Huang, "Oil bodies and oleosins in seeds," *Annual Review of Plant Physiology and Plant Molecular Biology*, vol. 43, no. 1, pp. 177–200, 1992.

[22] D. J. Murphy, "Structure, function and biogenesis of storage lipid bodies and oleosins in plants," *Progress in Lipid Research*, vol. 32, no. 3, pp. 247–280, 1993.

[23] G. I. Frandsen, J. Mundy, and J. T. C. Tzen, "Oil bodies and their associated proteins, oleosin and caleosin," *Physiologia Plantarum*, vol. 112, no. 3, pp. 301–307, 2001.

[24] Z. Purkrtova, P. Jolivet, M. Miquel, and T. Chardot, "Structure and function of seed lipid body-associated proteins," *Comptes Rendus Biologies*, vol. 331, no. 10, pp. 746–754, 2008.

[25] V. B. Vance and A. H. Huang, "The major protein from lipid bodies of maize. Characterization and structure based on cDNA cloning," *Journal of Biological Chemistry*, vol. 262, no. 23, pp. 11275–11279, 1987.

[26] R. Qu and A. H. C. Huang, "Oleosin KD 18 on the surface of oil bodies in maize. Genomic and cDNA sequences and the deduced protein structure," *Journal of Biological Chemistry*, vol. 265, no. 4, pp. 2238–2243, 1990.

[27] D. J. Murphy and D. M. Y. Au, "A new class of highly abundant apolipoproteins involved in lipid storage in oilseeds," *Biochemical Society Transactions*, vol. 117, no. 4, pp. 682–683, 1989.

[28] D. J. Murphy, J. N. Keen, J. N. O'Sullivan et al., "A class of amphipathic proteins associated with lipid storage bodies in plants. Possible similarities with animal serum apolipoproteins," *Biochimica et Biophysica Acta*, vol. 1088, no. 1, pp. 86–94, 1991.

[29] J. S. Keddie, G. Hübner, S. P. Slocombe et al., "Cloning and characterisation of an oleosin gene from *Brassica napus*," *Plant Molecular Biology*, vol. 19, no. 3, pp. 443–453, 1992.

[30] A. Kalinski, D. S. Loer, J. M. Weisemann, B. F. Matthews, and E. M. Herman, "Isoforms of soybean seed oil body membrane protein 24 kDa oleosin are encoded by closely related cDNAs," *Plant Molecular Biology*, vol. 17, no. 5, pp. 1095–1098, 1991.

[31] P. Hatzopoulos, G. Franz, L. Choy, and R. Z. Sung, "Interaction of nuclear factors with upstream sequences of a lipid body membrane protein gene from carrot," *Plant Cell*, vol. 2, no. 5, pp. 457–467, 1990.

[32] I. Cummins and D. J. Murphy, "cDNA sequence of a sunflower oleosin and transcript tissue specificity," *Plant Molecular Biology*, vol. 19, no. 5, pp. 873–876, 1992.

[33] G. J. H. van Rooijen, L. I. Terning, and M. M. Moloney, "Nucleotide sequence of an *Arabidopsis thaliana* oleosin gene," *Plant Molecular Biology*, vol. 18, no. 6, pp. 1177–1179, 1992.

[34] D. W. Hughes, H. Y. Wang, and G. A. Galau, "Cotton (*Gossypium hirsutum*) MatP6 and MatP7 oleosin genes," *Plant Physiology*, vol. 101, no. 2, pp. 697–698, 1993.

[35] Q. Liu, Y. Sun, W. Su et al., "Species-specific size expansion and molecular evolution of the oleosins in angiosperms," *Gene*, vol. 509, no. 2, pp. 247–257, 2012.

[36] R. B. Aalen, "The transcripts encoding two oleosin isoforms are both present in the aleurone and in the embryo of barley (*Hordeum vulgare* L.) seeds," *Plant Molecular Biology*, vol. 28, no. 3, pp. 583–588, 1995.

[37] R. L. C. Chuang, J. C. F. Chen, J. Chu, and J. T. C. Tzen, "Characterization of seed oil bodies and their surface oleosin isoforms from rice embryos," *Journal of Biochemistry*, vol. 120, no. 1, pp. 74–81, 1996.

[38] J. C. F. Chen, R. H. Lin, H. C. Huang, and J. T. C. Tzen, "Cloning, expression and isoform classification of a minor oleosin in sesame oil bodies," *Journal of Biochemistry*, vol. 122, no. 4, pp. 819–824, 1997.

[39] L. S. H. Wu, L. D. Wang, P. W. Chen, L. J. Chen, and J. T. C. Tzen, "Genomic cloning of 18 kDa oleosin and detection of triacylglycerols and oleosin isoforms in maturing rice and postgerminative seedlings," *Journal of Biochemistry*, vol. 123, no. 3, pp. 386–391, 1998.

[40] J. T. C. Tzen and A. H. C. Huang, "Surface structure and properties of plant seed oil bodies," *Journal of Cell Biology*, vol. 117, no. 2, pp. 327–335, 1992.

[41] J. T. C. Tzen, G. C. Lie, and A. H. C. Huang, "Characterization of the charged components and their topology on the surface of plant seed oil bodies," *Journal of Biological Chemistry*, vol. 267, no. 22, pp. 15626–15634, 1992.

[42] J. T. C. Tzen, C. C. Peng, D. J. Cheng, E. C. F. Chen, and J. M. H. Chiu, "A new method for seed oil body purification and examination of oil body integrity following germination," *Journal of Biochemistry*, vol. 121, no. 4, pp. 762–768, 1997.

[43] E. C. F. Chen, S. S. K. Tai, C. C. Peng, and J. T. C. Tzen, "Identification of three novel unique proteins in seed oil bodies of sesame," *Plant and Cell Physiology*, vol. 39, no. 9, pp. 935–941, 1998.

[44] L. J. Lin and J. T. C. Tzen, "Two distinct steroleosins are present in seed oil bodies," *Plant Physiology and Biochemistry*, vol. 42, no. 7-8, pp. 601–608, 2004.

[45] J. T. C. Tzen, M. M. Wang, J. C. F. Chen, L. J. Lin, and M. C. M. Chen, "Seed oil body proteins: oleosin, caleosin, and steroleosin," *Current Topic in Biochemical Reseaech*, vol. 5, pp. 133–139, 2003.

[46] J. C. F. Chen, C. C. Y. Tsai, and J. T. C. Tzen, "Cloning and secondary structure analysis of caleosin, a unique calcium-binding protein in oil bodies of plant seeds," *Plant and Cell Physiology*, vol. 40, no. 10, pp. 1079–1086, 1999.

[47] L. J. Lin, S. S. K. Tai, C. C. Peng, and J. T. C. Tzen, "Steroleosin, a sterol-binding dehydrogenase in seed oil bodies," *Plant Physiology*, vol. 128, no. 4, pp. 1200–1211, 2002.

[48] H. Tnani, I. López, T. Jouenne, and C. M. Vicient, "Quantitative subproteomic analysis of germinating related changes in the scutellum oil bodies of *Zea mays*," *Plant Science*, vol. 191-192, pp. 1–7, 2012.

[49] H. Tnani, I. López, T. Jouenne, and C. M. Vicient, "Protein composition analysis of oil bodies from maize embryos during germination," *Journal of Plant Physiology*, vol. 168, no. 5, pp. 510–513, 2011.

[50] S. Popluechai, M. Froissard, P. Jolivet et al., "*Jatropha curcas* oil body proteome and oleosins: L-form *JcOle3* as a potential phylogenetic marker," *Plant Physiology and Biochemistry*, vol. 49, no. 3, pp. 352–356, 2011.

[51] F. Capuano, N. J. Bond, L. Gatto et al., "LC-MS/MS methods for absolute quantification and identification of proteins associated with chimeric plant oil bodies," *Analytical Chemistry*, vol. 83, no. 24, pp. 9267–9272, 2011.

[52] P. Jolivet, C. Boulard, A. Bellamy et al., "Protein composition of oil bodies from mature *Brassica napus* seeds," *Proteomics*, vol. 9, no. 12, pp. 3268–3284, 2009.

[53] V. Katavic, G. K. Agrawal, M. Hajduch, S. L. Harris, and J. J. Thelen, "Protein and lipid composition analysis of oil bodies from two *Brassica napus* cultivars," *Proteomics*, vol. 6, no. 16, pp. 4586–4598, 2006.

[54] P. Jolivet, E. Roux, S. D'Andrea et al., "Protein composition of oil bodies in *Arabidopsis thaliana* ecotype WS," *Plant Physiology and Biochemistry*, vol. 42, no. 6, pp. 501–509, 2004.

[55] J. T. C. Tzen, M. M. C. Wang, S. S. K. Tai, T. T. T. Lee, and C. C. Peng, "The abundant proteins in sesame seed: storage proteins in protein bodies and oleosins in oil bodies," *Advances in Plant Physiology*, vol. 6, pp. 93–105, 2003.

[56] L. J. Lin, P. C. Liao, H. H. Yang, and J. T. C. Tzen, "Determination and analyses of the N-termini of oil-body proteins, steroleosin, caleosin and oleosin," *Plant Physiology and Biochemistry*, vol. 43, no. 8, pp. 770–776, 2005.

[57] D. J. Murphy, "The biogenesis and functions of lipid bodies in animals, plants and microorganisms," *Progress in Lipid Research*, vol. 40, no. 5, pp. 325–438, 2001.

[58] F. Capuano, F. Beaudoin, J. A. Napier, and P. R. Shewry, "Properties and exploitation of oleosins," *Biotechnology Advances*, vol. 25, no. 2, pp. 203–206, 2007.

[59] T. L. Shimada and I. Hara-Nishimura, "Oil-body-membrane proteins and their physiological functions in plants," *Biological and Pharmaceutical Bulletin*, vol. 33, no. 3, pp. 360–363, 2010.

[60] C. C. Peng, V. S. Y. Lee, M. Y. Lin, H. Y. Huang, and J. T. C. Tzen, "Minimizing the central hydrophobic domain in oleosin for the constitution of artificial oil bodies," *Journal of Agricultural and Food Chemistry*, vol. 55, no. 14, pp. 5604–5610, 2007.

[61] B. M. Abell, M. Hahn, L. A. Holbrook, and M. M. Moloney, "Membrane topology and sequence requirements for oil body targeting of oleosin," *Plant Journal*, vol. 37, no. 4, pp. 461–470, 2004.

[62] K. Giannoulia, G. Banilas, and P. Hatzopoulos, "Oleosin gene expression in olive," *Journal of Plant Physiology*, vol. 164, no. 1, pp. 104–107, 2007.

[63] B. M. Abell, L. A. Holbrook, M. Abenes, D. J. Murphy, M. J. Hills, and M. M. Moloney, "Role of the proline knot motif in oleosin endoplasmic reticulum topology and oil body targeting," *Plant Cell*, vol. 9, no. 8, pp. 1481–1493, 1997.

[64] K. B. Vargo, R. Parthasarathy, and D. A. Hammer, "Self-assembly of tunable protein suprastructures from recombinant oleosin," *Proceedings of the National Academy of Sciences of the United States of America*, vol. 109, no. 29, pp. 11657–11662, 2012.

[65] Y. Gohon, J. D. Vindigni, A. Pallier et al., "High water solubility and fold in amphipols of proteins with large hydrophobic regions: oleosins and caleosin from seed lipid bodies," *Biochimica et Biophysica Acta*, vol. 1808, no. 3, pp. 706–716, 2011.

[66] L. G. Alexander, R. B. Sessions, A. R. Clarke, A. S. Tatham, P. R. Shewry, and J. A. Napier, "Characterization and modelling of the hydrophobic domain of a sunflower oleosin," *Planta*, vol. 214, no. 4, pp. 546–551, 2002.

[67] M. Li, D. J. Murphy, K. H. K. Lee et al., "Purification and structural characterization of the central hydrophobic domain of oleosin," *Journal of Biological Chemistry*, vol. 277, no. 40, pp. 37888–37895, 2002.

[68] B. M. Abell, S. High, and M. M. Moloney, "Membrane protein topology of oleosin is constrained by its long hydrophobic domain," *Journal of Biological Chemistry*, vol. 277, no. 10, pp. 8602–8610, 2002.

[69] D. J. Lacey, N. Wellner, F. Beaudoin, J. A. Napier, and P. R. Shewry, "Secondary structure of oleosins in oil bodies isolated from seeds of safflower (*Carthamus tinctorius* L.) and sunflower (*Helianthus annuus* L.)," *Biochemical Journal*, vol. 334, no. 2, pp. 469–477, 1998.

[70] M. Millichip, A. S. Tatham, F. Jackson, G. Griffiths, P. R. Shewry, and A. K. Stobart, "Purification and characterization of oil-bodies (oleosomes) and oil-body boundary proteins (oleosins) from the developing cotyledons of sunflower (*Helianthus annuus* L.)," *Biochemical Journal*, vol. 314, no. 1, pp. 333–337, 1996.

[71] M. Li, J. S. Keddie, L. J. Smith, D. C. Clark, and D. J. Murphy, "Expression and characterization of the N-terminal domain of an oleosin protein from sunflower," *Journal of Biological Chemistry*, vol. 268, no. 23, pp. 17504–17512, 1993.

[72] M. Li, L. J. Smith, D. C. Clark, R. Wilson, and D. J. Murphy, "Secondary structures of a new class of lipid body proteins from oilseeds," *Journal of Biological Chemistry*, vol. 267, no. 12, pp. 8245–8253, 1992.

[73] S. Takahashi, T. Katagiri, K. Yamaguchi-Shinozaki, and K. Shinozaki, "An *Arabidopsis* gene encoding a Ca^{2+} -binding protein is induced by abscisic acid during dehydration," *Plant and Cell Physiology*, vol. 41, no. 7, pp. 898–903, 2000.

[74] Z. Purkrtova, C. Le Bon, B. Kralova, M. H. Ropers, M. Anton, and T. Chardot, "Caleosin of *Arabidopsis thaliana*: effect of calcium on functional and structural properties," *Journal of Agricultural and Food Chemistry*, vol. 56, no. 23, pp. 11217–11224, 2008.

[75] M. B. Busch, K. H. Kortje, H. Rahmann, and A. Sievers, "Characteristic and differential calcium signals from cell structures of the root cap detected by energy-filtering electron microscopy (EELS/ESI)," *European Journal of Cell Biology*, vol. 60, no. 1, pp. 88–100, 1993.

[76] P. L. Jiang and J. T. C. Tzen, "Caleosin serves as the major structural protein as efficient as oleosin on the surface of seed oil bodies," *Plant Signaling and Behavior*, vol. 5, no. 4, pp. 447–449, 2010.

[77] T. H. Liu, C. L. Chyan, F. Y. Li, and J. T. C. Tzen, "Stability of artificial oil bodies constituted with recombinant caleosins," *Journal of Agricultural and Food Chemistry*, vol. 57, no. 6, pp. 2308–2313, 2009.

[78] J. C. F. Chen and J. T. C. Tzen, "An in vitro system to examine the effective phospholipids and structural domain for protein targeting to seed oil bodies," *Plant and Cell Physiology*, vol. 42, no. 11, pp. 1245–1252, 2001.

[79] Z. Purkrtova, S. d'Andrea, P. Jolivet et al., "Structural properties of caleosin: a MS and CD study," *Archives of Biochemistry and Biophysics*, vol. 464, no. 2, pp. 335–343, 2007.

[80] C. I. Brändeén, "Relation between structure and function of alpha/beta-proteins," *Quarterly Reviews of Biophysics*, vol. 13, no. 3, pp. 317–338, 1980.

[81] C. van der Schoot, L. K. Paul, S. B. Paul, and P. L. Rinne, "Plant lipid bodies and cell-cell signaling: a new role for an old organelle?" *Plant Signaling and Behavior*, vol. 6, no. 11, pp. 1732–1738, 2011.

[82] T. L. Shimada, T. Shimada, H. Takahashi, Y. Fukao, and I. Hara-Nishimura, "A novel role for oleosins in freezing tolerance of oilseeds in *Arabidopsis thaliana*," *Plant Journal*, vol. 55, no. 5, pp. 798–809, 2008.

[83] Y. Y. Wu, Y. R. Chou, C. S. Wang, T. H. Tseng, L. J. Chen, and J. T. C. Tzen, "Different effects on triacylglycerol packaging to oil bodies in transgenic rice seeds by specifically eliminating one of their two oleosin isoforms," *Plant Physiology and Biochemistry*, vol. 48, no. 2-3, pp. 81–89, 2010.

[84] M. A. Schmidt and E. M. Herman, "Suppression of soybean oleosin produces micro-oil bodies that aggregate into oil body/ER complexes," *Molecular Plant*, vol. 1, no. 6, pp. 910–924, 2008.

[85] R. M. P. Siloto, K. Findlay, A. Lopez-Villalobos, E. C. Yeung, C. L. Nykiforuk, and M. M. Moloney, "The accumulation of oleosins determines the size of seed oilbodies in *Arabidopsis*," *Plant Cell*, vol. 18, no. 8, pp. 1961–1974, 2006.

[86] E. Roux, S. Baumberger, M. A. V. Axelos, and T. Chardot, "Oleosins of *Arabidopsis thaliana*: expression in *Escherichia coli*, purification, and functional properties," *Journal of Agricultural and Food Chemistry*, vol. 52, no. 16, pp. 5245–5249, 2004.

[87] C. C. Peng, I. P. Lin, C. K. Lin, and J. T. C. Tzen, "Size and stability of reconstituted sesame oil bodies," *Biotechnology Progress*, vol. 19, no. 5, pp. 1623–1626, 2003.

[88] J. T. L. Ting, K. Lee, C. Ratnayake, K. A. Platt, R. A. Balsamo, and A. H. C. Huang, "Oleosin genes in maize kernels having diverse oil contents are constitutively expressed independent of oil contents: size and shape of intracellular oil bodies are determined by the oleosins/oils ratio," *Planta*, vol. 199, no. 1, pp. 158–165, 1996.

[89] N. Babazadeh, M. Poursaadat, H. R. Sadeghipour, and A. Hossein Zadeh Colagar, "Oil body mobilization in sunflower seedlings is potentially regulated by thioredoxin h," *Plant Physiology and Biochemistry*, vol. 57, pp. 134–142, 2012.

[90] M. Rudolph, A. Schlereth, M. Körner et al., "The lipoxygenase-dependent oxygenation of lipid body membranes is promoted by a patatin-type phospholipase in cucumber cotyledons," *Journal of Experimental Botany*, vol. 62, no. 2, pp. 749–760, 2011.

[91] E. S. L. Hsiao and J. T. C. Tzen, "Ubiquitination of oleosin-H and caleosin in sesame oil bodies after seed germination," *Plant Physiology and Biochemistry*, vol. 49, no. 1, pp. 77–81, 2011.

[92] S. Vandana and S. C. Bhatla, "Evidence for the probable oil body association of a thiol-protease, leading to oleosin degradation in sunflower seedling cotyledons," *Plant Physiology and Biochemistry*, vol. 44, no. 11-12, pp. 714–723, 2006.

[93] V. Parthibane, S. Rajakumari, V. Venkateshwari, R. Iyappan, and R. Rajasekharan, "Oleosin is bifunctional enzyme that has both monoacylglycerol acyltransferase and phospholipase activities," *The Journal of Biological Chemistry*, vol. 287, no. 3, pp. 1946–1954, 2012.

[94] V. Parthibane, R. Iyappan, A. Vijayakumar, V. Venkateshwari, and R. Rajasekharan, "Serine/threonine/tyrosine protein kinase phosphorylates oleosin, a regulator of lipid metabolic functions," *Plant Physiology*, vol. 159, no. 1, pp. 95–104, 2012.

[95] P. L. Jiang, J. C. F. Chen, S. T. Chiu, and J. T. C. Tzen, "Stable oil bodies sheltered by a unique caleosin in cycad megagametophytes," *Plant Physiology and Biochemistry*, vol. 47, no. 11-12, pp. 1009–1016, 2009.

[96] M. Poxleitner, S. W. Rogers, A. Lacey Samuels, J. Browse, and J. C. Rogers, "A role for caleosin in degradation of oil-body storage lipid during seed germination," *Plant Journal*, vol. 47, no. 6, pp. 917–933, 2006.

[97] P. L. Jiang, C. S. Wang, C. M. Hsu, G. Y. Jauh, and J. T. C. Tzen, "Stable oil bodies sheltered by a unique oleosin in lily pollen," *Plant and Cell Physiology*, vol. 48, no. 6, pp. 812–821, 2007.

[98] P. L. Jiang, G. Y. Jauh, C. S. Wang, and J. T. C. Tzen, "A unique caleosin in oil bodies of lily pollen," *Plant and Cell Physiology*, vol. 49, no. 9, pp. 1390–1395, 2008.

[99] K. Zienkiewicz, A. J. Castro, J. D. D. Alché, A. Zienkiewicz, C. Suárez, and M. I. Rodríguez-García, "Identification and localization of a caleosin in olive (*Olea europaea* L.) pollen during in vitro germination," *Journal of Experimental Botany*, vol. 61, no. 5, pp. 1537–1546, 2010.

[100] A. Hanano, M. Burcklen, M. Flenet et al., "Plant seed peroxygenase is an original heme-oxygenase with an EF-hand calcium binding motif," *Journal of Biological Chemistry*, vol. 281, no. 44, pp. 33140–33151, 2006.

[101] M. Partridge and D. J. Murphy, "Roles of a membrane-bound caleosin and putative peroxygenase in biotic and abiotic stress responses in *Arabidopsis*," *Plant Physiology and Biochemistry*, vol. 47, no. 9, pp. 796–806, 2009.

[102] H. Feng, X. Wang, Y. Sun et al., "Cloning and characterization of a calcium binding EF-hand protein gene *TaCab1* from wheat and its expression in response to *Puccinia striiformis* f. sp. *tritici* and abiotic stresses," *Molecular Biology Reports*, vol. 38, no. 6, pp. 3857–3866, 2011.

[103] Y. Y. Kim, K. W. Jung, K. S. Yoo, J. U. Jeung, and J. S. Shin, "A stress-responsive caleosin-like protein, AtCLO4, Acts as a Negative Regulator of ABA Responses in *Arabidopsis*," *Plant and Cell Physiology*, vol. 52, no. 5, pp. 874–884, 2011.

[104] Y. Aubert, L. Leba, C. Cheval et al., "Involvement of RD20, a member of caleosin family, in ABA-mediated regulation of germination in *Arabidopsis thaliana*," *Plant Signaling and Behavior*, vol. 6, no. 4, pp. 538–540, 2011.

[105] H. B. Khalil, Z. Wang, J. A. Wright et al., "Heterotrimeric Gα subunit from wheat (*Triticum aestivum*), GA3, interacts with the calcium-binding protein, Clo3, and the phosphoinositide-specific phospholipase C, PI-PLC1," *Plant Molecular Biology*, vol. 77, pp. 145–158, 2011.

[106] E. Blée, M. Flenet, B. Boachon, and M. L. Fauconnier, "A non-canonical caleosin from *Arabidopsis* efficiently epoxidizes physiological unsaturated fatty acids with complete stereoselectivity," *The FEBS Journal*, vol. 279, no. 20, pp. 3981–3995, 2012.

[107] S. De Domenico, S. Bonsegna, M. S. Lenucci et al., "Localization of seed oil body proteins in tobacco protoplasts reveals specific mechanisms of protein targeting to leaf lipid droplets," *Journal of Integrative Plant Biology*, vol. 53, no. 11, pp. 858–868, 2011.

[108] J. Hänisch, M. Wältermann, H. Robenek, and A. Steinbüchel, "Eukaryotic lipid body proteins in oleogenous actinomycetes and their targeting to intracellular triacylglycerol inclusions: impact on models of lipid body biogenesis," *Applied and Environmental Microbiology*, vol. 72, no. 10, pp. 6743–6750, 2006.

[109] W. Li, L. G. Li, X. F. Sun, and K. X. Tang, "An oleosin-fusion protein driven by the CaMV35S promoter is accumulated in *Arabidopsis* (Brassicaceae) seeds and correctly targeted to oil bodies," *Genetics and Molecular Research*, vol. 11, no. 3, pp. 2138–2146, 2012.

[110] G. J. Van Rooijen and M. M. Moloney, "Structural requirements of oleosin domains for subcellular targeting to the oil body," *Plant Physiology*, vol. 109, no. 4, pp. 1353–1361, 1995.

[111] I. Cummins, M. J. Hills, J. H. E. Ross, D. H. Hobbs, M. D. Watson, and D. J. Murphy, "Differential, temporal and spatial expression of genes involved in storage oil and oleosin accumulation in developing rapeseed embryos: implications for the role of oleosins and the mechanisms of oil-body formation," *Plant Molecular Biology*, vol. 23, no. 5, pp. 1015–1027, 1993.

[112] W. S. Lee, J. T. C. Tzen, J. C. Kridl, S. E. Radke, and A. H. C. Huang, "Maize oleosin is correctly targeted to seed oil bodies in *Brassica napus* transformed with the maize oleosin gene," *Proceedings of the National Academy of Sciences of the United States of America*, vol. 88, no. 14, pp. 6181–6185, 1991.

[113] T. Wahlroos, J. Soukka, A. Denesyuk, R. Wahlroos, T. Korpela, and N. J. Kilby, "Oleosin expression and trafficking during oil body biogenesis in tobacco leaf cells," *Genesis*, vol. 35, no. 2, pp. 125–132, 2003.

[114] J. T. L. Ting, R. A. Balsamo, C. Ratnayake, and A. H. C. Huang, "Oleosin of plant seed oil bodies is correctly targeted to the lipid bodies in transformed yeast," *Journal of Biological Chemistry*, vol. 272, no. 6, pp. 3699–3706, 1997.

[115] F. Beaudoin, B. M. Wilkinson, C. J. Stirling, and J. A. Napier, "In vivo targeting of a sunflower oil body protein in yeast

secretary (*sec*) mutants," *Plant Journal*, vol. 23, no. 2, pp. 159–170, 2000.

[116] F. Beaudoin and J. A. Napier, "The targeting and accumulation of ectopically expressed oleosin in non-seed tissues of *Arabidopsis thaliana*," *Planta*, vol. 210, no. 3, pp. 439–445, 2000.

[117] C. Sarmiento, J. H. E. Ross, E. Herman, and D. J. Murphy, "Expression and subcellular targeting of a soybean oleosin in transgenic rapeseed. Implications for the mechanism of oil-body formation in seeds," *Plant Journal*, vol. 11, no. 4, pp. 783–796, 1997.

[118] P. J. Thoyts, M. I. Millichip, A. K. Stobart, W. T. Griffiths, P. R. Shewry, and J. A. Napier, "Expression and in vitro targeting of a sunflower oleosin," *Plant Molecular Biology*, vol. 29, no. 2, pp. 403–410, 1995.

[119] D. S. Loer and E. M. Herman, "Cotranslational integration of soybean (*Glycine max*) oil body membrane protein oleosin into microsomal membranes," *Plant Physiology*, vol. 101, no. 3, pp. 993–998, 1993.

[120] M. J. Hills, M. D. Watson, and D. J. Murphy, "Targeting of oleosins to the oil bodies of oilseed rape (*Brassica napus* L.)," *Planta*, vol. 189, no. 1, pp. 24–29, 1993.

[121] M. Froissard, S. D'Andréa, C. Boulard, and T. Chardot, "Heterologous expression of AtClo1, a plant oil body protein, induces lipid accumulation in yeast," *FEMS Yeast Research*, vol. 9, no. 3, pp. 428–438, 2009.

[122] F. Beaudoin and J. A. Napier, "Targeting and membrane-insertion of a sunflower oleosin in vitro and in *Saccharomyces cerevisiae*: the central hydrophobic domain contains more than one signal sequence, and directs oleosin insertion into the endoplasmic reticulum membrane using a signal anchor sequence mechanism," *Planta*, vol. 215, no. 2, pp. 293–303, 2002.

[123] J. T. C. Tzen, Y. K. Lai, K. L. Chan, and A. H. C. Huang, "Oleosin isoforms of high and low molecular weights are present in the oil bodies of diverse seed species," *Plant Physiology*, vol. 94, no. 3, pp. 1282–1289, 1990.

[124] J. T. C. Tzen, R. L. C. Chuang, J. C. F. Chen, and L. S. H. Wu, "Coexistence of both oleosin isoforms on the surface of seed oil bodies and their individual stabilization to the organelles," *Journal of Biochemistry*, vol. 123, no. 2, pp. 318–323, 1998.

[125] S. S. K. Tai, M. C. M. Chen, C. C. Peng, and J. T. C. Tzen, "Gene family of oleosin isoforms and their structural stabilization in sesame seed oil bodies," *Bioscience, Biotechnology and Biochemistry*, vol. 66, no. 10, pp. 2146–2153, 2002.

[126] A. C. N. Chua, P. L. Jiang, L. S. Shi, W. M. Chou, and J. T. C. Tzen, "Characterization of oil bodies in jelly fig achenes," *Plant Physiology and Biochemistry*, vol. 46, no. 5-6, pp. 525–532, 2008.

[127] A. J. Simkin, T. Qian, V. Caillet et al., "Oleosin gene family of *Coffea canephora*: quantitative expression analysis of five oleosin genes in developing and germinating coffee grain," *Journal of Plant Physiology*, vol. 163, no. 7, pp. 691–708, 2006.

[128] L. S. H. Wu, G. H. H. Hong, R. F. Hou, and J. T. C. Tzen, "Classification of the single oleosin isoform and characterization of seed oil bodies in gymnosperms," *Plant and Cell Physiology*, vol. 40, no. 3, pp. 326–334, 1999.

[129] M. R. Roberts, R. Hodge, and R. Scott, "*Brassica napus* pollen oleosins possess a characteristic C-terminal domain," *Planta*, vol. 195, no. 3, pp. 469–470, 1995.

[130] D. J. Murphy and J. H. E. Ross, "Biosynthesis, targeting and processing of oleosin-like proteins, which are major pollen coat components in *Brassica napus*," *Plant Journal*, vol. 13, no. 1, pp. 1–16, 1998.

[131] C. Y. Huang, C. I. Chung, Y. C. Lin, Y. I. C. Hsing, and A. H. C. Huang, "Oil bodies and oleosins in *Physcomitrella* possess characteristics representative of early trends in evolution," *Plant Physiology*, vol. 150, no. 3, pp. 1192–1203, 2009.

[132] L. S. Robert, J. Gerster, S. Allard, L. Cass, and J. Simmonds, "Molecular characterization of two *Brassica napus* genes related to oleosins which are highly expressed in the tapetum," *Plant Journal*, vol. 6, no. 6, pp. 927–933, 1994.

[133] J. H. E. Ross and D. J. Murphy, "Characterization of anther-expressed genes encoding a major class of extracellular oleosin-like proteins in the pollen coat of Brassicaceae," *Plant Journal*, vol. 9, no. 5, pp. 625–637, 1996.

[134] L. O. Franco, C. L. De, S. Hamdi, G. Sachetto-Martins, and D. E. De Oliveira, "Distal regulatory regions restrict the expression of *cis*-linked genes to the tapetal cells," *The FEBS Letters*, vol. 517, no. 1–3, pp. 13–18, 2002.

[135] D. H. Chen, C. L. Chyan, P. L. Jiang, C. S. Chen, and J. T. C. Tzen, "The same oleosin isoforms are present in oil bodies of rice embryo and aleurone layer while caleosin exists only in those of the embryo," *Plant Physiology and Biochemistry*, vol. 60, pp. 18–24, 2012.

[136] H. C. Lu, P. L. Jiang, L. R. C. Hsu, C. L. Chyan, and J. T. C. Tzen, "Characterization of Oil bodies in adlay (*Coix lachryma-jobi* L.)," *Bioscience, Biotechnology and Biochemistry*, vol. 74, no. 9, pp. 1841–1847, 2010.

[137] H. Liu, P. Hedley, L. Cardle et al., "Characterisation and functional analysis of two barley caleosins expressed during barley caryopsis development," *Planta*, vol. 221, no. 4, pp. 513–522, 2005.

[138] K. Zienkiewicz, A. Zienkiewicz, M. I. Rodríguez-García, and A. J. Castro, "Characterization of a caleosin expressed during olive (*Olea europaea* L.) pollen ontogeny," *BMC Plant Biology*, vol. 11, article 122, 2011.

[139] I. P. Lin, P. L. Jiang, C. S. Chen, and J. T. C. Tzen, "A unique caleosin serving as the major integral protein in oil bodies isolated from *Chlorella* sp. cells cultured with limited nitrogen," *Plant Physiology and Biochemistry*, vol. 44, no. 61, pp. 80–87, 2012.

[140] H. Næsted, G. I. Frandsen, G. Y. Jauh et al., "Caleosins: Ca^{2+}-binding proteins associated with lipid bodies," *Plant Molecular Biology*, vol. 44, no. 4, pp. 463–476, 2000.

[141] C. C. Peng, J. C. F. Chen, D. J. H. Shyu, M. J. Chen, and J. T. C. Tzen, "A system for purification of recombinant proteins in *Escherichia coli* via artificial oil bodies constituted with their oleosin-fused polypeptides," *Journal of Biotechnology*, vol. 111, no. 1, pp. 51–57, 2004.

[142] C. C. Peng, D. J. H. Shyu, W. M. Chou, M. J. Chen, and J. T. C. Tzen, "Method for bacterial expression and purification of sesame cystatin via artificial oil bodies," *Journal of Agricultural and Food Chemistry*, vol. 52, no. 10, pp. 3115–3119, 2004.

[143] M. C. M. Chen, C. L. Chyan, T. T. T. Lee, S. H. Huang, and J. T. C. Tzen, "Constitution of stable artificial oil bodies with triacylglycerol, phospholipid, and caleosin," *Journal of Agricultural and Food Chemistry*, vol. 52, no. 12, pp. 3982–3987, 2004.

[144] G. J. H. Van Rooijen and M. M. Moloney, "Plant seed oil-bodies as carriers for foreign proteins," *Bio/Technology*, vol. 13, no. 1, pp. 72–77, 1995.

[145] J. G. Boothe, J. A. Saponja, and D. L. Parmenter, "Molecular farming in plants: oilseeds as vehicles for the production of pharmaceutical proteins," *Drug Development Research*, vol. 42, no. 3-4, pp. 172–181, 1997.

[146] N. Markley, C. Nykiforuk, J. Boothe, and M. Moloney, "Producing proteins using transgenic oilbody-oleosin technology," *BioPharm International*, vol. 19, no. 6, pp. 34–57, 2006.

[147] S. C. Bhatla, V. Kaushik, and M. K. Yadav, "Use of oil bodies and oleosins in recombinant protein production and other biotechnological applications," *Biotechnology Advances*, vol. 28, no. 3, pp. 293–300, 2010.

[148] M. D. McLean, R. Chen, D. Yu et al., "Purification of the therapeutic antibody trastuzumab from genetically modified plants using safflower Protein A-oleosin oilbody technology," *Transgenic Research*. In press.

[149] G. Banilas, G. Daras, S. Rigas, M. M. Moloney, and P. Hatzopoulos, "Oleosin di-or tri-meric fusions with GFP undergo correct targeting and provide advantages for recombinant protein production," *Plant Physiology and Biochemistry*, vol. 49, no. 2, pp. 216–222, 2011.

[150] C. Y. Yang, S. Y. Chen, and G. C. Duan, "Transgenic peanut (*Arachis hypogaea* L.) expressing the urease subunit B gene of *Helicobacter pylori*," *Current Microbiology*, vol. 63, no. 4, pp. 387–391, 2011.

[151] W. Li, L. Li, K. Li, J. Lin, X. Sun, and K. Tang, "Expression of biologically active human insulin-like growth factor 1 in *Arabidopsis thaliana* seeds via oleosin fusion technology," *Biotechnology and Applied Biochemistry*, vol. 58, no. 3, pp. 139–146, 2011.

[152] A. Ahmad, E. O. Pereira, A. J. Conley, A. S. Richman, and R. Menassa, "Green biofactories: recombinant protein production in plants," *Recent Patents on Biotechnology*, vol. 4, no. 3, pp. 242–259, 2010.

[153] C. E. Orozco-Barrios, S. F. Battaglia-Hsu, M. L. Arango-Rodriguez et al., "Vitamin B12-impaired metabolism produces apoptosis and Parkinson phenotype in rats expressing the transcobalamin-oleosin chimera in substantia nigra," *PLoS ONE*, vol. 4, no. 12, Article ID e8268, 2009.

[154] L. Pons, S. F. Battaglia-Hsu, C. E. Orozco-Barrios et al., "Anchoring secreted proteins in endoplasmic reticulum by plant oleosin: the example of vitamin B12 cellular sequestration by transcobalamin," *PLoS ONE*, vol. 4, no. 7, Article ID e6325, 2009.

[155] M. M. Moloney, "Oil-body proteins as carriers of high-value peptides in plants," Patent US, 659835, 1991.

[156] M. M. Moloney and G. van Rooijen Sembiosys, "Expression of epidermal growth factor in plant seeds," Patent US, 7091401, 2006.

[157] M. M. Moloney, "Oil-body proteins as carriers of high-value peptides in plants," Patent US, 5650554, 1997.

[158] M. M. Moloney, J. Boothe, and G. van Rooijen, "Oil bodies and associated proteins as affinity matrices," Patent US, 6509453, 2003.

[159] S. Szarka, G. van Rooijen, and M. M. Moloney, "Methods for the production of multimeric immunoglobulins, and related compositions," Patent US, 7098383, 2006.

[160] G. van Rooijen, S. Zaplachinski, P. B. Heifetz et al., "Methods for the production of multimeric protein complexes, and related compositions," Patent US, 2006/0179514, 2006.

[161] J. McCarthy and S. A. Nestec, "Recombinant oleosins from cacao and their use as flavoring or emulsifying agents," Patent US, 7126042, 2006.

[162] T. Harada, K. Kashihara, and N. Nio, "Oleosin/phospholipid complex and process for producing the same," Patent WO, 2002/026788, 2002.

[163] H. M. Deckers, G. van Rooijen, J. Boothe et al., "Uses of oil bodies," Patent US, 6210742, 2001.

[164] C. J. Chiang, H. C. Chen, Y. P. Chad, and J. T. C. Tzen, "Efficient system of artificial oil bodies for functional expression and purification of recombinant nattokinase in *Escherichia coli*," *Journal of Agricultural and Food Chemistry*, vol. 53, no. 12, pp. 4799–4804, 2005.

[165] J. R. Liu, C. H. Duan, X. Zhao, J. T. C. Tzen, K. J. Cheng, and C. K. Pai, "Cloning of a rumen fungal xylanase gene and purification of the recombinant enzyme via artificial oil bodies," *Applied Microbiology and Biotechnology*, vol. 79, no. 2, pp. 225–233, 2008.

[166] Y. J. Hung, C. C. Peng, J. T. C. Tzen, M. J. Chen, and J. R. Liu, "Immobilization of *Neocallimastix patriciarum* xylanase on artificial oil bodies and statistical optimization of enzyme activity," *Bioresource Technology*, vol. 99, no. 18, pp. 8662–8666, 2008.

[167] R. W. Scott, S. Winichayakul, M. Roldan et al., "Elevation of oil body integrity and emulsion stability by polyoleosins, multiple oleosin units joined in tandem head-to-tail fusions," *Plant Biotechnology Journal*, vol. 8, no. 8, pp. 912–927, 2010.

[168] J. M. Tseng, J. R. Huang, H. C. Huang, J. T. C. Tzen, W. M. Chou, and C. C. Peng, "Facilitative production of an antimicrobial peptide royalisin and its antibody via an artificial oilbody system," *Biotechnology Progress*, vol. 27, no. 1, pp. 153–161, 2011.

[169] S. Winichayakul, A. Pernthaner, S. Livingston et al., "Production of active single-chain antibodies in seeds using trimeric polyoleosin fusion," *Journal of Biotechnology*, vol. 161, no. 4, pp. 407–413, 2012.

[170] C. J. Chiang, H. C. Chen, H. F. Kuo, Y. P. Chao, and J. T. C. Tzen, "A simple and effective method to prepare immobilized enzymes using artificial oil bodies," *Enzyme and Microbial Technology*, vol. 39, no. 5, pp. 1152–1158, 2006.

[171] C. J. Chiang, H. C. Chen, Y. P. Chao, and J. T. C. Tzen, "One-step purification of insoluble hydantoinase overproduced in *Escherichia coli*," *Protein Expression and Purification*, vol. 52, no. 1, pp. 14–18, 2007.

[172] R. C. W. Hou, M. Y. Lin, M. M. C. Wang, and J. T. C. Tzen, "Increase of viability of entrapped cells of *Lactobacillus delbrueckii* ssp. bulgaricus in artificial sesame oil emulsions," *Journal of Dairy Science*, vol. 86, no. 2, pp. 424–428, 2003.

[173] M. C. M. Chen, J. L. Wang, and J. T. C. Tzen, "Elevating bioavailability of cyclosporine A via encapsulation in artificial oil bodies stabilized by caleosin," *Biotechnology Progress*, vol. 21, no. 4, pp. 1297–1301, 2005.

[174] C. J. Chiang, C. C. Lin, T. L. Lu, and H. F. Wang, "Functionalized nanoscale oil bodies for targeted delivery of a hydrophobic drug," *Nanotechnology*, vol. 22, no. 41, Article ID 415102, 2011.

[175] C. J. Chiang, C. J. Chen, L. J. Lin, C. H. Chang, and Y. P. Chao, "Selective delivery of cargo entities to tumor cells by nanoscale artificial oil bodies," *Journal of Agricultural and Food Chemistry*, vol. 58, no. 22, pp. 11695–11702, 2010.

[176] C. J. Chiang, L. J. Lin, C. C. Lin, C. H. Chang, and Y. P. Chao, "Selective internalization of self-assembled artificial oil bodies by HER2/neu-positive cells," *Nanotechnology*, vol. 22, no. 1, Article ID 015102, 2011.

[177] M. T. Chang, C. R. Chen, T. H. Liu, C. P. Lee, and J. T. C. Tzen, "Development of a protocol to solidify native and artificial oil bodies for long-term storage at room temperature," *Journal of the Science of Food and Agriculture*. In press.

[178] T. H. Liu, C. L. Chyan, F. Y. Li, Y. J. Chen, and J. T. C. Tzen, "Engineering lysine-rich caleosins as carrier proteins torenderbiotin as a hapten on artificial oil bodies for antibody production," *Biotechnology Progress*, vol. 27, no. 6, pp. 1760–1767, 2011.

16

No Reproductive Interference from an Alien to a Native Species in *Cerastium* (Caryophyllaceae) at the Stage of Seed Production

Koh-Ichi Takakura

Division of Urban Environment, Osaka City Institute of Public Health and Environmental Sciences, 8-34 Tojo-cho, Tennoji, Osaka 543-0026, Japan

Correspondence should be addressed to Koh-Ichi Takakura, takakura@nature.email.ne.jp

Academic Editors: F. A. Culianez-Macia and T. L. Weir

Reproductive interference, adverse interspecific interaction during the mating process, has been regarded as a powerful driver of species displacement between species. Recent empirical reports have described its importance in biological invasions. This study was undertaken to test whether a rare herbaceous plant species indigenous to Japan suffered reproductive interference from an alien species of the genus *Cerastium*. Field observations and a transplanting experiment were conducted to ascertain the effects of coexistence with an alien species on the seed production of the native species. Results show that coexistence with the alien species did not significantly decrease seed numbers, but it significantly affected the seed weight only in field observations. In this study, the reproductive process of the native species was examined only at or before the seed production stage. Because the interspecific pollen transfer might produce hybrids with low viability or fertility, reproductive interference cannot be denied in this study. To test reproductive interference at such latter stages, additional studies should be conducted. Consequently, detection of reproductive interference demands high costs in some species. Based on these results and suggestions, the necessity of narrowing down the target species for testing of reproductive interference is discussed to elucidate the universality of reproductive interference.

1. Introduction

In recent years, reproductive interference has gathered much attention as a powerful force driving displacement between species. Reproductive interference refers to any interspecific interaction during the mating process that decreases the reproductive success of either or both interacting species [1]. Typical interactions causing reproductive interference are interspecific courtship in animals and interspecific pollen transfer in plants. Reproductive interference can easily eliminate species of smaller abundance or ones with high sensitivity to reproductive interference [2, 3] because of its frequency dependence [3]. Consequently, reproductive interference has become an important key to determining the community structure of both animals and plants [4, 5].

Reproductive interference has been regarded as important also in terms of conservation biology. Recent studies have demonstrated that it is involved with displacement of the native by the alien species. For example, the native dandelion in Japan has been displaced by the alien congener, especially in urban areas. Our previous reports have described that the reproductive interference played a crucial role in displacement [6, 7]. Similarly, reproductive interference has been suspected as an important factor of the extinction of the Japanese cocklebur, *Xanthium japonica*, in western Japan [8]. More examples of reproductive interference have been reported in animals [9–11]. Consequently, reproductive interference has often been found in the pair of the alien species and its allied native species and has been recognized as an important factor related to the invasiveness of alien species.

It is a fact that considerable reproductive interference has been detected in pairs of alien species and allied native species, but is this truly universal? To assess the true universality of the phenomenon, it is necessary to consider publication bias [12], which occurs because researchers might not seek to report negative results and editors might not desire to accept such reports for publication. In fact, most published studies of reproductive interference have reported positive results only ([6, 7, 13], but see [14]). To discuss the universality of the reproductive interference fairly, it is necessary to review not only the detection of the reproductive interference but also reports of its absence. Furthermore, the pattern of the significant reproductive interference should be discussed to reveal what factors and conditions are involved with the intensity and direction of the reproductive interference.

This study tested the reproductive interference between the native and the alien species pair of mouse-ear chickweeds of the genus *Cerastium*. The native species was the Japanese mouse-ear chickweed, which is rare to the present day. However, the alien species, sticky mouse-ear chickweed, is abundant throughout the world, including Japan. The reproductive interference that the native species suffered from the alien species was tested using field observations and a transplanting experiment. Based on these results, the universality of reproductive interference in plants and how it should be addressed are discussed.

2. Materials and Methods

2.1. Study Species. The Japanese mouse-ear chickweed, *Cerastium holosteoides* Fries var. *hallaisanense* (Nakai) Mizushima, a herbaceous biennial plant endemic in Japan [15] (Figure 1), was a common weed a century ago [16], but it is a rare species in residential and commercial areas in the present day [17]. Its present habitats are limited to roadsides and orchards in mountainous regions [17]. It cannot be observed in lowland regions of western Japan today (Takakura KI personal obs.).

However, the alien congeneric species is the sticky mouse-ear chickweed, *C. glomeratum* Thuill (Figure 1). This species was reported to establish not only in Japan but also in many other countries [18]: it is a cosmopolitan alien weed and is quite common also in Japan [17]. This species appeared to displace *C. h. hallaosanense* in developed areas of Japan [17], but the direct adverse effect on the growth and reproduction of the native species has not been studied. The pollinator fauna on these two species remains unknown, but these two plant species probably share pollinators because their flower morphologies are simple and mutually similar (Figure 1). Moreover, the flowering seasons are early spring in both species [15].

2.2. Field Observation. In these observations, the *C. h.* var. *hallaisanense* seed production of individuals coexisting with the alien *C. glomeratum* and those without the alien species in close vicinity were compared to test the effects of the alien species on the seed production in the native species. This study site was a rice field on the periphery of Kawachi-Nagano, Osaka, Japan (34°23′32″N, 135°34′37″E) on 20 April 2007. The area containing the site was an agricultural landscape surrounded by mountains. I collected fruits that

(a)

(b)

FIGURE 1: Flowers of the native species, *C. h.* var. *hallaisanense* (a), and the alien species, *C. glomeratum* (b).

had grown sufficiently to count seeds in them and immediately before dehiscence from 22 *C. h.* var. *hallaisanense* individuals. Ten individuals grew within a 30 cm radius from the closest individuals of *C. glomeratum* (coexisting group). The remaining 12 individuals occurred 1 m or farther apart from the closest individuals of *C. glomeratum* (control).

Collected fruits were dissected in the laboratory to count their seeds. Furthermore, after drying for two weeks at room temperature, the seed weight was measured in a mass for every fruit with an electric balance (Mettler Toledo AG 285; Mettler Instrumente AG, Zurich, Switzerland) with 0.1 mg accuracy.

2.3. Transplanting Experiment. Plants were reared in pots and used for this experiment. Seeds of *C. h.* var. *hallaisanense* were collected at the site of Kawachi-Nagano (see above) during May 2006. Seeds were put individually on soil in plastic pots (75 mm diameter, 75 mm depth) in July 2006. Pots were put in the laboratory yard of the Osaka City Institute of Public Health and Environmental Sciences, Osaka, Japan (34°39′47″N, 135°31′42″E) and were irrigated arbitrarily. Individuals that had flowered in April 2007 were used for the experiment.

No Reproductive Interference from an Alien to a Native Species in Cerastium (Caryophyllaceae) at
the Stage of Seed Production

143

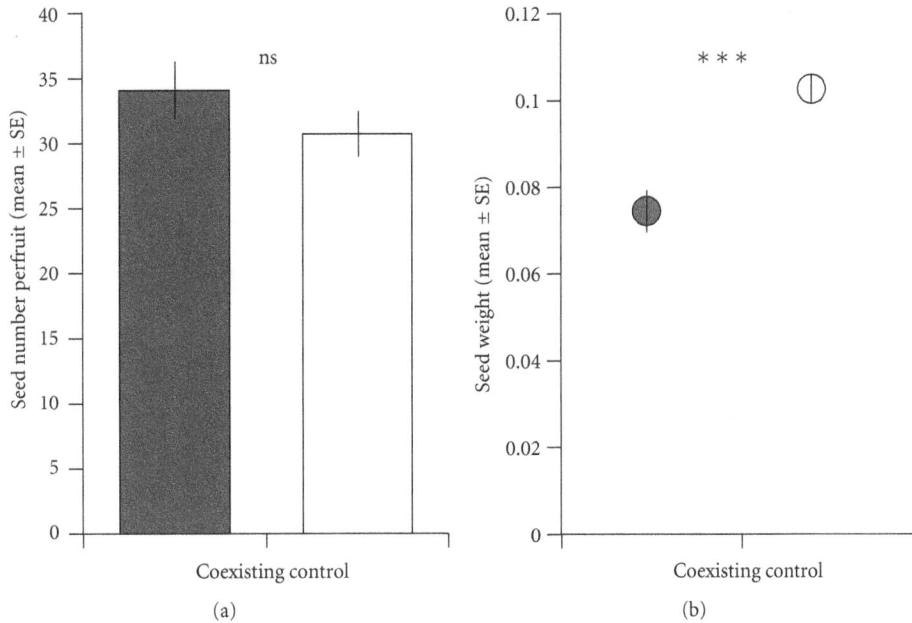

FIGURE 2: Seed numbers per fruit (a) and seed weights (b) in fruits produced from *C. h.* var. *hallaisanense* individuals that coexisted with the alien individuals (filled bar) and those of the control (open bar). Error bars represent 95% confidence interval. Symbols denote statistical significance (GLMMs; $* * *$, $P < 0.001$; ns, $P > 0.05$).

The experimental site was Tsurumi Park, Osaka, Japan ($34°42'32''$N, $135°34'55''$E). Eighteen pots, each with an individual of *C. h.* var. *hallaisanense*, were put in the vicinity of different patches of *C. glomeratum* in the site on 16 April 2007. These were designated as the coexisting group below as above. The 18 different individuals (pots) were left in the laboratory yard (control). Before the experiment, all flowers and fruits were removed with precision scissors for all individuals to prevent seeds that had been produced before the experiment from being counted afterward. Ten days after transplanting, all pots were brought to the laboratory yard again and were irrigated arbitrarily. Fruits borne during the transplantation were cropped immediately before dehiscence. Their number and weight were measured as the observation described above.

2.4. Statistical Analyses.

2.4. Statistical Analyses. Data from the observations and the experiment were similarly analyzed statistically. The seed numbers were compared between treatments: the coexisting groups and controls. Generalized linear mixed models (GLMMs, [19]) were applied for all datasets. In GLMMs for seed numbers, Poisson distributions were assumed as error structures. The response was the number of seeds in each fruit and the factor was each treatment: coexisting with the alien species or not. The individual of *C. h.* var. *hallaisanense* was treated as a random effect that would absorb the interindividual variance. For the seed weight, first, the weight per seed was calculated for each fruit. Subsequently, it was compared between groups. In these analyses, the Gaussian distributions were assumed as error structures. The factor and the random effect was the same as in analyses of seed numbers. All analyses were conducted using R 2.15.1 [20].

3. Results

3.1. Field Observation. The seed number per fruit was 34.10 ± 2.19 (mean $\pm$ SE) in the mixed group and 30.7 ± 1.75 in the control of *C. h.* var. *hallaisanense* (Figure 2(a)). The difference between groups was not significant (GLMM, coefficient of the coexisting group $\pm$ SE $= 0.222 \pm 0.184$, $Z = 1.21$, $P = 0.227$). The seed weight was 0.075 ± 0.005 mg in the coexisting group and smaller than that in the control, 0.103 ± 0.003 mg (Figure 2(b)). The difference between groups was significant (GLMM, coefficient of the coexisting group $\pm$ SE $= -0.029 \pm 0.006$, $t = 4.52$, $P < 0.001$).

3.2. Transplanting Experiment. The seed number per fruit was 34.8 ± 1.78 (mean $\pm$ SE) in the mixed group and 30.9 ± 1.71 in the control of *C. h.* var. *hallaisanense* (Figure 3(a)). The difference between groups was not significant (GLMM, coefficient of mixed group $\pm$ SE $= -0.102 \pm 0.126$, $Z = -0.82$, $P = 0.415$). The seed weight was 0.100 ± 0.004 in the mixed group and 0.106 ± 0.006 in the control (Figure 3(b)). The difference between groups was not significant (GLMM, coefficient of mixed group $\pm$ SE $= -0.005 \pm 0.012$, $t = 0.375$, $P = 0.710$).

4. Discussion

In this study, both the field observations and the transplanting experiment demonstrated that coexistence with the alien congener did not decrease seed production of the native species, *C. h.* var. *hallaisanense*, which appeared to have out-competed the alien species. The seed weight was significantly lower in field observations but the difference

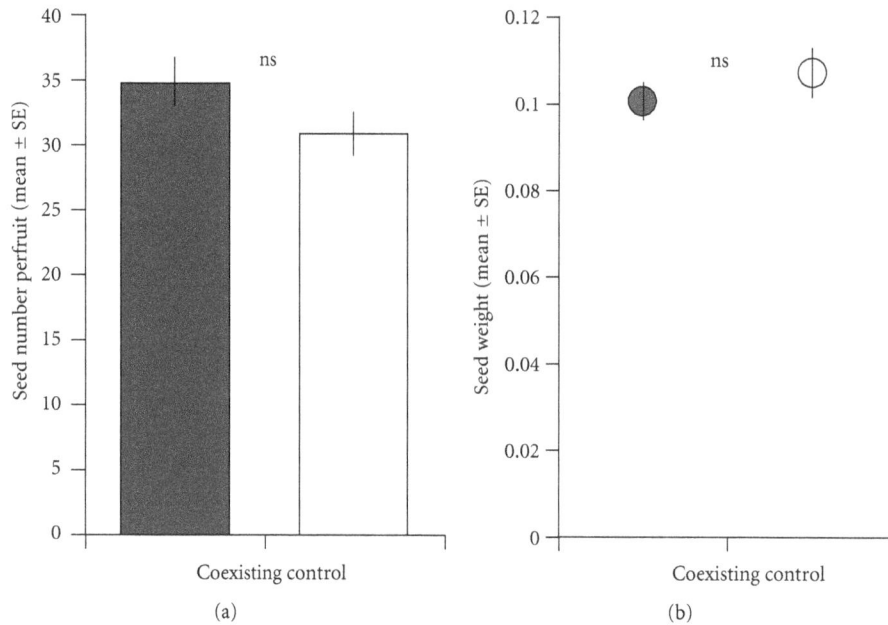

FIGURE 3: Seed numbers per fruit (a) and seed weights (b) in fruits produced from *C. h.* var. *hallaisanense* individuals that transplanted in the vicinity of the patches of the alien individuals (filled bar) and those of the control (open bar). Error bars represent a 95% confidence interval. Symbols denote statistical significance (GLMMs; ns, $P > 0.05$).

was not significant in the transplanting experiment. These results demonstrate that the native species rarely suffered from competition with the congeneric alien species at or before the seed production stage.

Defects of the reproductive interference in plants can be expressed in various stages through reproductive processes, not only at or before the seed production, but also thereafter. Among studies of plants, reproductive interference has been detected predominantly at or before the seed production stage [6, 8, 13]. However, this predominance might be a publication bias because such a form of reproductive interference is easily detected and is therefore a positive result that is easily obtained. Some reproductive interference is expressed in subsequent developmental stages, although the possibility has not been discussed to date. Indeed, many past reports have described sterile hybrids between allied species pairs [21–23]. The production of sterile hybrids must directly engender the reproductive interference because such hybrids are produced at the sacrifice of fertile offspring. This study detected no definite reproductive interference at or before the seed production stage in *C. h.* var. *hallaisanense*. However, this failure does not necessarily indicate the absence of the reproductive interference from *C. glomeratum* to *C. h.* var. *hallaisanense*. For example, the lesser weight of seeds produced by *C. h.* var. *hallaisanense* thatcoexisted with *C. glomeratum* in the field might mean less viability of seedlings although it should be tested in future studies. For complete testing for the reproductive interference, the seed and seedling viability and fertility after growth should be examined.

Reproductive interference has gathered much attention as a novel hypothetical power driving the extinction of native species. In fact, recent studies have detected intense reproductive interference in some alien species [6, 8, 10]. However, complete testing of reproductive interference sometimes entails high costs for experiments because the reproductive interference can occur in several stages of reproduction: pollen transfer, attachment of pollen to the stigma, pollen tube growth, seed development, seedling viability, and offspring fertility. For efficient testing of the universality of reproductive interference in plants, we must narrow down target species by reference to past studies. Theoretical studies have predicted that the species pair could not coexist if there were reproductive interference between them [3, 24]. Empirical studies have confirmed [6, 8–11, 13], also suggested that reproductive interference tends to occur between congeneric species pairs. Based on these results, we should specifically examine congeneric species pairs without sympatric coexistence. The empirical testing of the reproductive interference in plants is just beginning. It is necessary to continue exploration of the universality and the importance of reproductive interference efficiently in plant community structures.

5. Conclusion

This study investigated a case in which no significant reproductive interference was detected, even in the species pair of the native and alien species between which displacement was suspected. However, the possibilities of reproductive interference expressed in stages of germination, seedling, and reproduction stages were not denied by results of this study. Additional studies must be undertaken to assess such

No Reproductive Interference from an Alien to a Native Species in Cerastium (Caryophyllaceae) at the Stage of Seed Production

145

possibilities. Furthermore, we must narrow down the target systems by reference to past theoretical and empirical studies.

Acknowledgment

This work was partly supported by a Grant-in-Aid for Young Scientists (B, no. 22710240) from the Ministry of Education, Culture, Sports, Science and Technology, Japan.

References

[1] J. Gröning and A. Hochkirch, "Reproductive interference between animal species," *The Quarterly Review of Biology*, vol. 83, no. 3, pp. 257–282, 2008.

[2] J. M. C. Ribeiro and A. Spielman, "The Satyr Effect: a model predicting parapatry and species extinction," *The American Naturalist*, vol. 128, no. 4, pp. 513–528, 1986.

[3] E. Kuno, "Competitive exclusion through reproductive interference," *Researches on Population Ecology*, vol. 34, no. 2, pp. 275–284, 1992.

[4] A. Hochkirch, J. Gröning, and A. Bücker, "Sympatry with the devil: reproductive interference could hamper species coexistence," *Journal of Animal Ecology*, vol. 76, no. 4, pp. 633–642, 2007.

[5] D. A. R. Eaton, C. B. Fenster, J. Hereford, S. Q. Huang, and R. H. Ree, "Floral diversity and community structure in *Pedicularis* (Orobanchaceae)," *Ecology*, vol. 93, no. 8, supplement, pp. S182–S194, 2012.

[6] K. I. Takakura, T. Nishida, T. Matsumoto, and S. Nishida, "Alien dandelion reduces the seed-set of a native congener through frequency-dependent and one-sided effects," *Biological Invasions*, vol. 11, no. 4, pp. 973–981, 2009.

[7] T. Matsumoto, K. I. Takakura, and T. Nishida, "Alien pollen grains interfere with the reproductive success of native congener," *Biological Invasions*, vol. 12, no. 6, pp. 1617–1626, 2010.

[8] K. I. Takakura and S. Fujii, "Reproductive interference and salinity tolerance differentiate habitat use between two alien cockleburs: *Xanthium occidentale* and *X. italicum* (Compositae)," *Plant Ecology*, vol. 206, no. 2, pp. 309–319, 2009.

[9] E. A. Dame and K. Petren, "Behavioural mechanisms of invasion and displacement in Pacific island geckos (Hemidactylus)," *Animal Behaviour*, vol. 71, no. 5, pp. 1165–1173, 2006.

[10] A. D'Amore, E. Kirby, and V. Hemingway, "Reproductive interference by an invasive species: an evolutionary trap?" *Herpetological Conservation and Biology*, vol. 124, no. 4, pp. 325–330, 2009.

[11] D. W. Crowder, M. I. Sitvarin, and Y. Carrière, "Plasticity in mating behaviour drives asymmetric reproductive interference in whiteflies," *Animal Behaviour*, vol. 79, no. 3, pp. 579–587, 2010.

[12] R. Rosenthal, "The "file-drawer problem" and tolerance for null results," *Psychological Bulletin*, vol. 86, no. 3, pp. 85–97, 1979.

[13] R. B. Runquist and M. L. Stanton, "Asymmetric and frequency-dependent pollinator-mediated interactions may influence competitive displacement in two vernal pool plants," *Ecology Letters*. In press.

[14] S. Nishida, K. I. Takakura, T. Nishida, T. Matsumoto, and M. M. Kanaoka, "Differential effects of reproductive interference by an alien congener on native Taraxacum species," *Biological Invasions*, vol. 14, no. 2, pp. 439–447, 2012.

[15] M. Kitagawa, "Caryophyllaceae," in *Wildflowers of Japan*, Y. Satake, T. Ooi, S. Kitamura, T. Watari, and T. Tominari, Eds., pp. 39–45, Heibonsha, Tokyo, Japan, 1982.

[16] T. Makino, *Researches on Wild Plants*, Sanbunsha, Tokyo, Japan, 1907.

[17] H. Fukuhara, R. Murata, H. Noda, and H. Igarashi, "Comparative ecology of the alien species, *Cerastium glomeratum* and native species, *C. holosteoides* var. *hallaisanense*: (1) distribution and life history," *Memoirs of the Faculty of Education. Natural Sciences*, vol. 10, no. 1, pp. 23–37, 2007 (Japanese).

[18] USDA, ARS, National Genetic Resources Program, "*Cerastium glomeratum* Thuill," in *Germplasm Resources Information Network [Online Database]*, Beltsville, Md, USA, 2012.

[19] B. M. Bolker, M. E. Brooks, C. J. Clark et al., "Generalized linear mixed models: a practical guide for ecology and evolution," *Trends in Ecology and Evolution*, vol. 24, no. 3, pp. 127–135, 2009.

[20] R Core Team, *R: A Language and Environment for Statistical Computing*, R Foundation for Statistical Computing, Vienna, Austria, 2012.

[21] M. Ownbey, "Natural hybridization and amphiploidy in the genus *Tragopogon*," *American Journal of Botany*, vol. 37, no. 5, pp. 487–499, 1950.

[22] P. J. Brownsey, "*Asplenium* hybrids in the New Zealand flora," *New Zealand Journal of Botany*, vol. 15, no. 3, pp. 601–637, 1977.

[23] A. Aparicio, "Fitness components of the hybrid *Phlomis x margaritae* aparicio and silvestre (Lamiaceae)," *Botanical Journal of the Linnean Society*, vol. 124, no. 4, pp. 331–343, 1997.

[24] J. Yoshimura and C. W. Clark, "Population dynamics of sexual and resource competition," *Theoretical Population Biology*, vol. 45, no. 2, pp. 121–131, 1994.

17

Female Short Shoot and Ovule Development in *Ginkgo biloba* L. with Emphasis on Structures Associated with Wind Pollination

Biao Jin,[1] Di Wang,[1] Yan Lu,[1] Xiao Xue Jiang,[1] Min Zhang,[1] Lei Zhang,[1] and Li Wang[1,2]

[1] *College of Horticulture and Plant Protection, Yangzhou University, Yangzhou 225009, China*
[2] *Key Laboratory of Plant Molecular Physiology, Institute of Botany, Chinese Academy of Sciences, Beijing 100093, China*

Correspondence should be addressed to Li Wang, liwang@yzu.edu.cn

Academic Editors: C. Bolle, A. M. Rashotte, and S. Sawa

The orientation and morphology of the female cone are important for wind pollination in gymnosperms. To examine the role of female reproductive structures associated with wind pollination in *Ginkgo biloba*, we used scanning electron microscopy and semithin section techniques to observe the development of female short shoots and ovules in *G. biloba* before and during the pollination period. The ovule differentiation process was divided into six stages: undifferentiated, general stalk differentiation, integument differentiation, nucellus differentiation, collar differentiation, and mature stage. Before the pollination period, the integument tip generated the micropylar canal and the micropyle, while the nucellus tip cells degenerated to form the pollen chamber. During pollination, the micropylar canal surface became smooth, the micropyle split into several pieces and bore a pollination drop, and the pollen chamber directly faced the straight micropylar canal. The leaves and ovules were spirally arranged on the female short shoot, with the ovules erect and the fan-shaped leaves bent outwards and downwards. The ovules of *G. biloba* have differentiated some special architectural features adapted for pollen capture and transport. Together, these structures constitute a reproductive structural unit that may improve wind pollination efficiency at the female level.

1. Introduction

Pollination is an important process in the life cycle of seed plants. During the evolution of angiosperms, diverse forms of pollination with complex mechanisms have evolved. Among the gymnosperm species, the majorities are wind pollinated (anemophilous) and their pollen grains are carried from their initial location to female parts via air flow [1]. Thus, wind pollination in gymnosperms is a random and inefficient process with high pollen losses. There have been many studies documenting how gymnosperms achieve pollination via such relatively passive processes. For example, the pollination process has been studied in several conifer families, such as the Podocarpaceae, Cupressaceae, Taxaceae, and Pinaceae, and the proposed evolution of different pollination mechanisms has been widely discussed [2–4]. Based on the results of such studies, it was concluded that adaptive characteristics to wind pollination that contribute to the pollination mechanisms of gymnosperms involve (1) orientation of the ovulate cone, as well as ovule structure and position at the time of pollination [2, 5]; (2) pollen characteristics, especially the presence or absence of sacci [6, 7]; (3) whether a pollination drop appears at pollination or not [8, 9]. Among these characteristics, the ovule as the female reproductive organ varies widely in its morphology among gymnosperms. Studies on ovule development in gymnosperms have revealed various ovule structures that have evolved to increase the chances of pollination success. For instance, the tip of the integument in the ovule shows a variety of shapes for pollen collection. Owens et al. [2] reported that in the Cupressaceae, Taxodiaceae, and Cephalotaxaceae, the ovules are flask-shaped, variable in number, and attached in the axil of the bract scale, and the integument tip has a narrow neck and a small, unspecialized micropyle. However, in *Abies* species, the integument tip forms a short funnel, often with fluted edges, around a large micropyle [10]. In addition, the orientation and morphology of the megastrobilus (female cone) are important features for wind pollination. For example, in some conifer species, ovulate cones may enhance the probability of pollen entrapment

Female Short Shoot and Ovule Development in Ginkgo biloba L. with Emphasis on Structures Associated with Wind Pollination

147

by aerodynamically predetermining airflow patterns around scale-bract complexes [11]. Together, the results of such studies showed that the diverse morphologies and structures of female reproductive organs among gymnosperms are an evolutionary consequence of adaptation for effective pollen reception. Thus, it is important to study ovule development of gymnosperms to further clarify the roles of various ovule characteristics in the wind pollination mechanism.

Ginkgo biloba L. is the unique living species of the ancient lineage of Ginkgophyta, which originated in the Paleozoic Carboniferous period; thus, it is regarded as a "living fossil" [12]. The female reproductive organ of G. biloba, particularly the ovule, has been described and studied extensively because of several unique and primitive characteristics [13, 14]. The ovule differentiates into the integument, nucellus, and collar [15]. Primarily, the integument develops in a circumferential manner and becomes slightly lobed during the formation of the micropyle [16]. After that, the nucellar tissues are derived from periclinal divisions in the hypodermal layer, and the tip cells of the nucellus die, resulting in formation of the pollen chamber [17]. Almost simultaneously, one cell in the sporangial tissues differentiates to form a spore mother cell, which further develops to form the female gametophyte, while the other cells differentiate to form the tapetum [18]. During the pollination period, the ovule produces a pollination drop [19, 20]. Although these studies have described the ovule developmental process in G. biloba, most have focused on either the morphological structures or the floristic features. Since G. biloba is a wind-pollinated dioecious plant and this lineage has existed for more than 200 million years, the Ginkgo ovule is obviously a survivor of evolutionary consequences related to wind pollination. However, so far, there have been no systematic studies of how ovule structure has differentiated to adapt to wind pollination. Moreover, while previous studies have provided information about the formation of the female gametophyte, the developmental conditions of the female gametophyte before and during the pollination period remain unclear [21, 22]. In addition, the erect ovules surrounding by several fan-shaped leaves constitute the overall female short shoot system, which clearly works as a whole during wind pollination. However, the specific spatial arrangement and architecture of the short shoot have received little attention, despite the fact that these are important features for comprehensively understanding the female reproductive system associated with wind pollination in G. biloba.

Based on our previous studies [21, 23], in this paper we investigated the morphological and structural development of female short shoots and ovules in G. biloba, focusing on ontogenetic features and their relationships with pollen capture. We used continuous image acquisition, semithin sectioning, and SEM observation techniques to observe the morphological and anatomical changes. We present morphological data that clarify the relationships between female reproductive structures and wind pollination in G. biloba, providing further insights into the wind pollination mechanisms in gymnosperms.

2. Materials and Methods

2.1. Plant Material. Healthy female G. biloba L. trees were selected from the Ginkgo Experimental Station at Yangzhou University, Yangzhou, China (32°20′N, 119°30′E). Ovules were collected weekly from August 2008 to April 2009. Between mid-March and early April, the period for female gametophyte and pollen chamber development, ovules were collected daily. Ovules in bud were dissected out using a razor blade and a steel needle, and the nucellus together with the pollen chamber was manually isolated from ovules under a dissecting microscope.

2.2. Digital Camera Observation Studies. Potential female buds were randomly sampled and tagged to follow their development. Photographs were taken against a black background each week from August 2008 to mid March 2009 using a DSC-H7 Sony digital camera. Between mid-March and late April 2009, female buds and ovules were photographed daily until pollination ended.

2.3. Semithin Section Studies. The material was prepared for semi-thin sectioning as follows: at each sampling, 30 fresh ovules were prefixed in 2.5% (v/v) glutaraldehyde in 0.1 mol/L phosphate buffer (pH 7.2) for 20 h at room temperature, postfixed in 1% (w/v) osmium tetroxide for 6 h at 4°C, rinsed twice, dehydrated in an alcohol series (30, 50, 70, 80, 90, 95, and 100%, 15 min each step), treated twice for 10 min with propylene oxide, infiltrated with 1 : 1 propylene oxide/resin in embedding capsules overnight, and then finally embedded in Spurr's resin. Semi-thin sections (1 μm) were cut with an Leica EM UC6 ultramicrotome (Leica Microsystems GmbH, Wetzlar, Germany) and collected onto glass slides, then stained with 1% (w/v) toluidine blue O dissolved in 1% (w/v) sodium borate prior to examination [24]. Observations and photography were performed under a microscope (Zeiss Axioskop 40: Carl Zeiss Shanghai Company Ltd., Shanghai, China).

2.4. SEM Observation Studies. Samples were prepared for SEM observations as follows: at each developmental stage, 30 intact ovules were dissected from buds under a dissecting microscope. The samples were fixed in 2.5% glutaraldehyde (in 0.1 mol/L phosphate buffer, pH 7.0) for 4 h at room temperature and then kept for 12 h at 4°C. Then, they were dehydrated in a graded ethanol series (30, 50, 70, 80, 90, 95 and 100%, 15 min each step). After drying at the critical point, the specimens were coated with gold using an EMITECH K550 sputter coater for 150 s, and observed under a Hitachi H-300 scanning electron microscope (Tokyo, Japan) at 15 kv accelerating voltage [25].

3. Results

3.1. Female Cone Initiation and Sprouting Process. The female buds of G. biloba, borne at the apex of the shortshoots, were conical-shaped with the apex slightly acuminate and nearly enclosed by seven to nine thick-bud scales from September to the next February (Figure 1(a)). In late March,

FIGURE 1: Female cone development of *G. biloba*. (a) Female cone enclosed by thick bud scales during winter. ((b)–(d)) Bud scales opened gradually in late March. (e) Ovules began to appear in early April. ((f)–(h)) Ovular stalks elongated and leaves grew rapidly. (i) Leaves and ovules spirally arranged on the short shoot apex just before pollination. (j) Leaves and ovules during pollination stage (mid-April). (k) Long shoot with several female short shoots during pollination. (l) Pollination drop secreted from micropyle. (m) Ovules changed from yellow to green after pollination. (n) Ovules attached directly to general stalk. (o) Ovule with ovular stalks. bs, bud scale; gs, general stalk; le, leaf; o, ovule; os, ovular stalk; pd, pollination drop; ss, short shoot. Bars: ((a)–(f)) = 50 mm; ((i)–(k)) = 1 cm; ((l)–(m)) = 5 mm; ((n)–(o)) = 2 mm.

the bud scales began to unfold, and over several days, the outer brown scales opened to expose the inner green ones (Figure 1(b)). Continued unfolding of the bud scales resulted in the appearance of the leaves and ovules (Figures 1(c)–1(e)). Most of the ovules were concealed in the bud because both the leaf stalks and ovular stalks were very short. At this stage, the leaves were young and curved inward from the two lateral sides (Figures 1(d) and 1(e)). As the

leaves developed, all of the ovules in the buds became visible (Figures 1(f) and 1(g)). At this stage, the ovules were yellow. Afterwards, the leaves began to spread gradually and the leaf stalks rapidly elongated. At the same time, the ovular stalks began to extend rapidly, pushing the ovules outward from the bud (Figure 1(h)). This resulted in a spiral arrangement of the ovules and leaf tufts on the female short shoot. The fan-shaped leaves spread out and bent

Female Short Shoot and Ovule Development in Ginkgo biloba L. with Emphasis on Structures Associated with
Wind Pollination

149

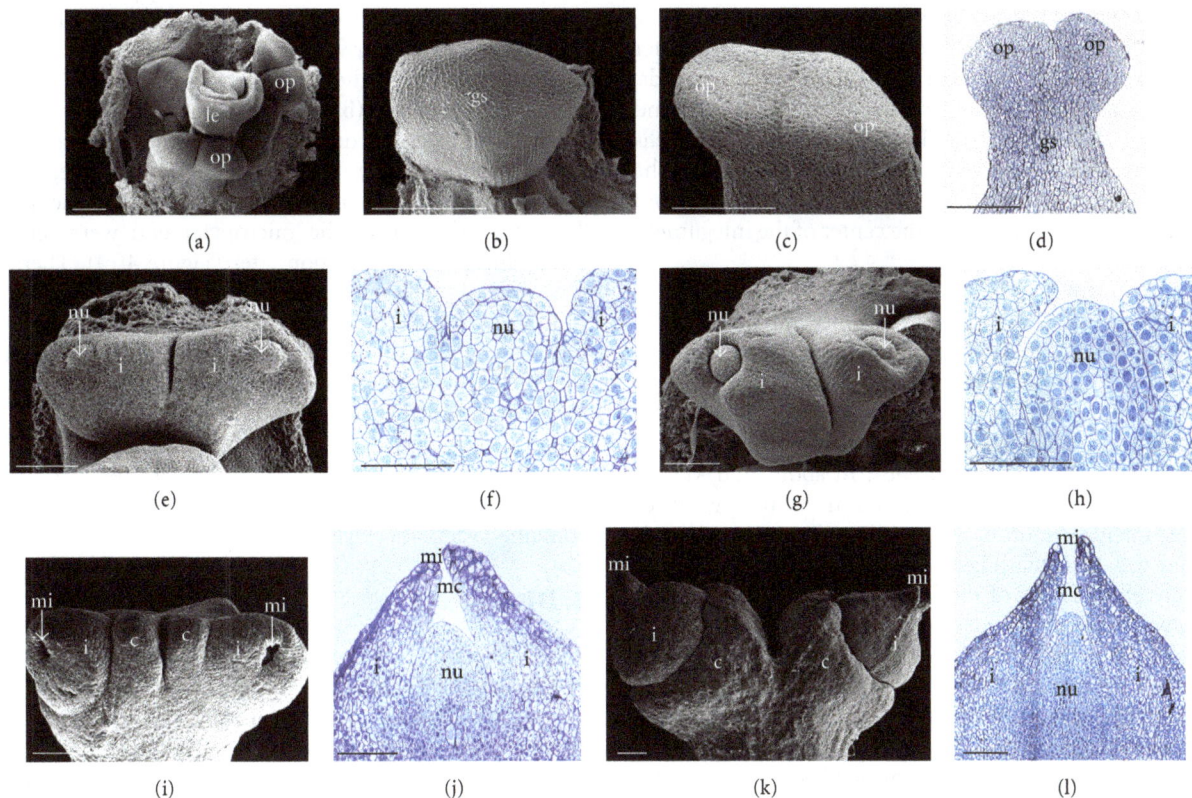

FIGURE 2: Tissues during ovule differentiation process in *G. biloba*. (a) Position of ovules and leaf in female bud. (b) General stalk differentiated and its tip extended and swelled. ((c)-(d)) Formation of integument primordium on tip general stalk. ((e)-(f)) Center of integument differentiated into nucellar tissue. ((g)–(j)) Integument tissue grew to encircle nucellus. ((k)-(l)) Mature ovules with well-developed collar, integument, nucellus, micropyle, and micropyle tunnel. co, collar; gs, general stalk; i, integument; le, leaf; mi, micropyle; nu, nucellus; op, ovular primordium. Bars: ((a)–(e)) = 100 μm; ((f), (h), (j), (l)) = 100 μm; ((g), (i), (k)) = 200 μm.

outwards and downwards (Figure 1(i)). During this period, the ovules were still yellow, erect, and inclined obliquely at various angles. During the pollination period, in mid-April, almost all of the female short shoots were open with several ovules in the center (Figures 1(j) and 1(k)), and each ovule produced a pollination drop on the micropyle (Figure 1(l)). After pollination, the color of ovules changed from yellow to green as the pollination drop disappeared (Figure 1(m)). Usually, a single general stalk bore two ovules that were directly attached to the stalk via ovular collars (Figure 1(n)). However, the general stalk of ovules occasionally differentiated into dichotomous separated stalks (ovule stalk), which connected the two ovular collars with the general stalk (Figure 1(o)).

3.2. Differentiation of Ovule Tissues.

The ovule differentiation process was observed using scanning electron microscopy (SEM) and semi-thin sections. Our materials collected from Yangzhou in January 2009 had ovules in the early stage of development (Figure 2(a)), although we did not know the exact time of ovule initiation (maybe in December or earlier). The general stalk primordia of ovules were initiated at and around the boundary between bud scales and leaves in the first year, and they remained in this stage for a long time (about 1 month). Until the next

January, the tip of the general stalk was swollen, giving the stalk an ellipse-like appearance (Figure 2(b)). Afterwards, the swollen primordium stretched on both lateral sides to differentiate into the integument primordium (Figure 2(c)). The cells of the integument primordium were very small and tightly arranged, and the nuclei could not be distinguished easily (Figure 2(d)). The integument tissues showed active cell division, in contrast to the center part of the general stalk. The cell division between the integument tissues led to the formation of a dorsal raphe between them (Figures 2(c) and 2(d)). Further differentiation became visible in mid-March. As the integument developed rapidly, the dorsal raphe developed further, causing the integument to separate into two independent parts. Then, the nucellus was initiated on the convex region of each integument, forming a faint ridge (Figure 2(e)). At this stage, the nucellar tissue was rounded with a flat tip, and its cells were similar in size and appearance to those in the integument (Figure 2(f)). Afterwards, the nucellus continued to differentiate and became free from the integument, causing the latter to show a cup-like shape (Figure 2(g)). Then, rapid cell division resulted in rapid growth of the integument, which grew bigger than, and eventually engulfed, the nucellus (Figure 2(h)). The integument completely surrounded the nucellus in late March (Figure 2(i)). Meanwhile, the ovule collar differentiated from cells between the base of the integument and the top of the ovular stalk (Figure 2(i)). After that, the integument proceeded to differentiate upwards, instead of ceasing growth, giving rise to the micropyle and micropylar canal (Figure 2(j)). The mature ovule had a well-defined integument and ovular collar (Figure 2(k)). The micropyle was situated on the opposite side of the collar, and the column-like nucellus was in the center of the integument (Figure 2(l)).

3.3. Formation of Pollen Capturing and Transporting Tissues. In *G. biloba*, the pollen capturing and transporting tissues include the micropyle, micropylar canal, and pollen chamber. Among them, the micropyle and micropylar canal are formed by the integument, while the pollen chamber is formed by the tip of the nucellus. In about mid-March, after the nucellus had differentiated from the center of the integument primordium, it continued to grow upwards by rapid cell division (Figure 3(a)). Because of the more frequent cell divisions of the integument, the nucellus was gradually engulfed by the surrounding integument (Figure 3(b)). When the nucellus was buried deeply in the ovule, the tip of the integument fused toward the very center at the top of the nucellus (Figure 3(c)). Then, the nucellus ceased to extend and became completely concealed by the integument (Figure 3(d)), while the latter continued to grow upwards (Figures 3(e)–3(h)). Therefore, the integument encircled the top of the nucellus to emerge as the micropyle (Figure 3(i)). Some of the cells under the micropyle formed a tunnel-like structure, the micropylar canal (Figure 3(j)). During its differentiation process, the micropyle showed a wavy shape, rather than a ring-shape (Figures 3(f)–3(i)). This was probably because of the differential growth rate of different regions of the upper integument. The micropylar canal cells were almost rectangular and had smooth surfaces (Figures 3(k) and 3(k-1)). As the micropyle and micropylar canal further developed, the nucellus became column-shaped (Figure 3(l)) and slightly acute on the tip side (Figures 3(m) and 3(n)). Some cells near the micropyle elongated longitudinally and collapsed (Figure 2(l)), resulting in a cavity (the pollen chamber) at the nucellus tip (Figures 3(o) and 3(p)). During the pollination period, all tissues were specialized for pollen capture and transport. The micropyle slightly curled outwards to form a funnel-like shape, and had a diameter of approximately 200 μm (Figures 3(q) and 3(r)). The micropylar canal was straight and smooth, and was approximately 100 μm wide and 500 μm long (Figures 3(k) and 3(q)). The pollen chamber was well developed with its orifice directly oriented towards the micropyle and micropylar canal, and the diameter was approximately 100 μm (Figures 3(p) and 3(r)).

3.4. Formation of the Female Gametophyte. After the micropyle was enclosed by the integument in late March (Figure 2(j)), the megasporocyte (megaspore mother cell) initiated from the center of the nucellus, and was tightly surrounded by spongy tissue. The megasporocyte was very large relative to surrounding cells with a central nucleus (Figure 4(a)). In early April, when the tip of the nucellus

began to degenerate to form the pollen chamber, the megasporocyte underwent meiosis. It divided along the longitudinal axis of the ovule, giving rise to the two cells of the dyad (Figure 4(b)). Upon formation of the dyad, the two cells immediately divided again to yield the linear tetrad of megaspores (Figure 4(c)). Among all the megaspores, the one near the chalazal end was typically the largest. The other three near the micropylar end were relatively small, and degenerated soon after (Figure 4(c)). Therefore, the chalazal megaspore enlarged and developed into the functional megaspore (Figure 4(d)). During the pollination period, when the pollen chamber reached its largest volume, the nucleus of the functional megaspore began to undergo series of karyokinetic divisions, forming many free nuclei (Figures 4(e) and 4(g)), indicating the beginning of the female gametophyte phase. At this stage, the surrounding spongy tissue was distinguishable from other cells as they contained vacuoles (Figure 4(f)).

4. Discussion

4.1. Differential Stages of the G. biloba Ovule. The structure of the mature *G. biloba* ovule is similar to that of many other gymnosperms. The mature ovule has a general stalk bearing two ovules at its tip, a single integument with the micropyle positioned opposite the stalk, some pollen grains in the pollen chamber, and many free nuclei [26]. However, some of the developmental stages of the ovule between initiation and maturation remained unclear. In our study, we observed the ovule differentiation process in *G. biloba* and divided it into six stages: (1) undifferentiated stage (from July to November), the leaf primordium had differentiated, but the ovule primordium was not visible; (2) general stalk differentiation stage (in December), the general stalk primordium appeared in axils of leaves and/or bracts; (3) integument differentiation stage (from January to early March), the tip of the general stalk was extended and swollen (Figure 2(b)), and later, the shoulder of the tip protruded upwards to form the integument primordium (Figures 2(c) and 2(d)), whose upper end then developed into the micropyle and the micropylar canal; (4) nucellus differentiation stage (from mid-March to early April), the nucellus differentiated from the center of the integument (Figure 2(e)); (5) collar differentiation stage (in late March), the ovulate collar differentiated quickly at the base of the integument (Figure 2(i)); and (6) mature stage (in mid-April), a pollination drop appeared on each micropyle (Figure 1(l)), signaling that the ovule was well prepared, morphologically and structurally, for pollen entrapment. The nucellus differentiation stage was the key stage in ovule development, because the nucellus gives rise to the pollen chamber, and later, the female gametophyte.

4.2. Functions of Leaves during Pollination. Many plant lineages show a fossil history of consolidating individual reproductive organs into compound structures with emergent aerodynamic properties [26]. This appears to have increased the probability of pollination of the reproductive units. For example, wind-pollinated plants tend to show

Female Short Shoot and Ovule Development in Ginkgo biloba L. with Emphasis on Structures Associated with Wind Pollination

151

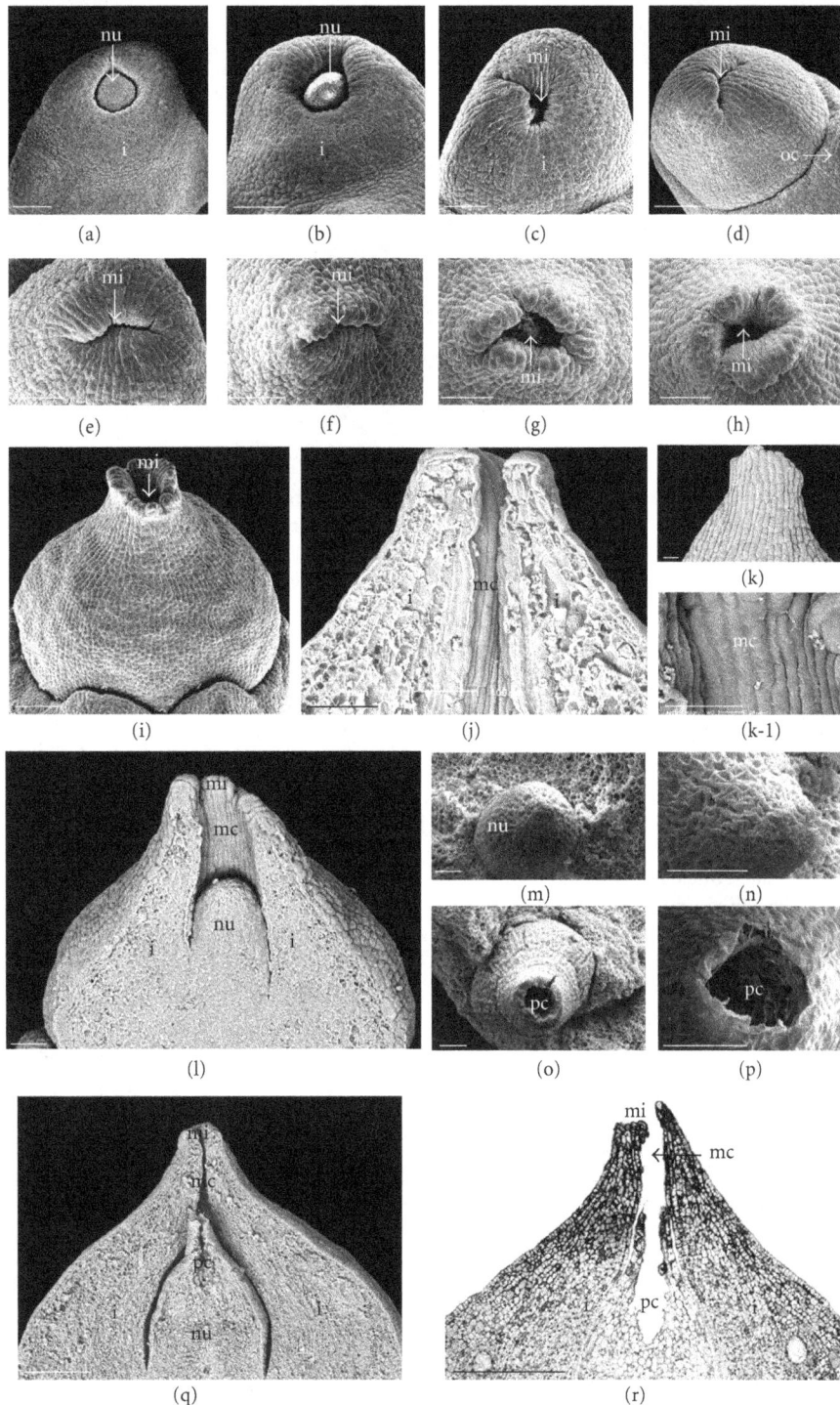

FIGURE 3: Development of tissues for pollen capture and transport in *G. biloba*. ((a)-(b)) After nucellus differentiation, integument continued to grow upward, surrounding nucellus. ((c)-(d)) Tip of integument gradually fused toward the center on top of nucellus. ((e)–(h)) Tip of integument continued growing upwards. (i) Formation of micropyle. (j) Micropylar canal. ((k), (k-1)) Cells of micropylar canal with smooth surfaces. (l) Column-shaped nucellus and developing micropylar canal. ((m)-(n)) Tip of nucellus became slightly acute. ((o)-(p)) Emergence of pollen chamber. ((q)-(r)) Ovules during pollination period with well-developed micropyle, micropylar canal, and pollen chamber. i, integument; mi, micropyle; mc, micropylar canal; nu, nucellus; oc, ovular collar; pc, pollen chamber. Bars: ((a)–(j), (l)–(p)) = 100 μm; (k) = 50 μm; ((q)-(r)) = 500 μm.

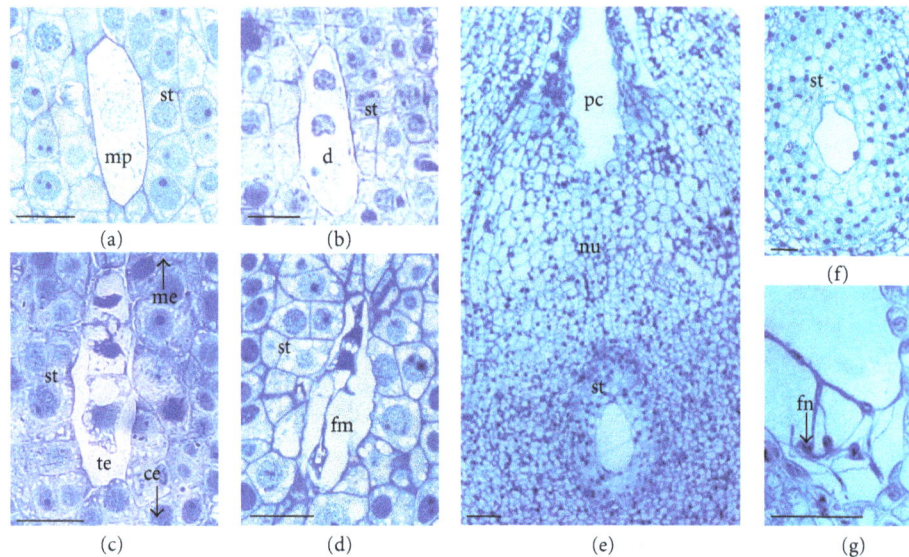

FIGURE 4: Formation of female gametophyte in *G. biloba*. (a) Megasporocyte differentiated from inner nucellar tissue in late March. (b) Dyad. (c) Tetrad. (d) Functional megaspore. ((e)–(g)) Approaching pollination stage, pollen chamber formed at apex of nucellus, and free nuclear stage of female gametocyte occurred. ce, chalazal end; d, dyad; fm, functional megaspore; fn, free nucleus; me, micropylar end; mp, megasporocyte; nu, nucellus; pc, pollen chamber; st, spongy tissue; te, tetrad. Bars: ((a)–(d), (g)) = 50 μm; ((e)-(f)) = 100 μm.

a reduction in the size of branches, leaves, and other structures surrounding pollen-receptive spaces that could catch pollen grains, thereby reducing the ultimate number of pollen grains reaching receptive surfaces [27]. Moreover, the reduction in the size of these structures has been associated with modification of air currents towards receptive surfaces [27]. In addition, in some, but not all species of pine, leaf fascicles emerge around ovulate cones before pollination. As a result, pollen capture by the cone involves the aerodynamics of subtending leaf clusters [28]. In our study, we examined the development of the female short shoot and the orientation of leaves and ovules of *G. biloba* during pollination. We found before the bud scales unfolded, and ovules and leaves were borne at the apex of the short shoots and were surrounded by thick bud scales (Figure 1(a)). The ovules began to appear as the bud scales and leaves continued to unfold (Figures 1(b)–1(h)). Eventually, all scales, leaves, and ovules formed a spiral arrangement on the short shoot. The ovules were upright, and closer to the vertical axis of the shoot, while the leaves were fan-shaped, and extended further from the center, bending outwards and downwards (Figure 1(i)). Considering this architecture of the leaves and ovules during pollination, we supposed that the shapes and angles of leaves of *G. biloba* changed the aerodynamics around the ovule, which could increase the probability of pollen capture by the ovule. However, further experiments are required to determine whether or not the leaves of the female short shoot in *G. biloba* can produce favorable aerodynamics for pollen deposition.

4.3. The Ovulate Stalk. The general evolutionary trends of ovules in *G. biloba* include a reduction in the number of ovules, an increase in ovule size [29], and shortening or disappearance of the ovulate stalk [17]. Zhou [30] insisted

that under the k-strategy, to balance weight and sustainability, the ovulate stalk became shorter, and then ultimately disappeared. Indeed, the single living species of *Ginkgo* has no ovulate stalk between the collar and general stalk. However, in our study, we found a low frequency of ovulate stalks that had differentiated on the top of the general stalk (Figure 1(o)). Considering that this phenomenon appeared more frequently in Ginkgo fossils [31], we believe the emergence of the ovulate stalk in living *Ginkgo* might reflect the vestigial expression of an underlying ancestral trait. From the view of the spatial arrangement of the ovules and leaves, the ovulate stalk represents advantageous architecture for ovules to trap pollen grains from the air. However, on the other hand, a long stalk also increases the risk of the pollination drop falling from the micropyle as a result of wind action during the pollination period. In addition, it is unfavorable to have a long developmental period after pollination, owing to the longer distance for nutrient transport and the increased energy demands of ovule bearing. These factors may have contributed to the disappearance of the furcated ovulate stalk during long-term evolution.

4.4. Ovule Morphology and Structures Associated with Wind Pollination. Ovule orientation and morphology are vital features for wind pollination in gymnosperms. Traditionally, wind pollination is viewed as a simple matter of "chance," a mode of reproduction dominated by stochastic processes that predominate in the biology of long-distance dispersal [28]. However, a certain number of morphological features of reproductive units have evolved in divergent land plants that may contribute to the success of wind pollination [7]. For instance, female reproductive organs of gymnosperms showed considerable developmental flexibility, which would significantly influence the behavior of pollen grains in the air

Female Short Shoot and Ovule Development in Ginkgo biloba L. with Emphasis on Structures Associated with Wind Pollination

153

around the ovulate cones [11]. Furthermore, ovule geometry and position in relation to floral parts seem to play an essential role in affecting pollination efficiency [28]. In our study, we observed the *G. biloba* female cone and ovule structure and found many morphological and structural characteristics adapted to wind pollination. First, different from conifers in which ovules are often engulfed by other structures, such as bracts in *Pinus monticola* or lobes in *Cunninghamia* [3, 32], the organization of ovulate structures in *G. biloba* is very simple: the erect ovule is fully naked and is not enclosed by seed scales (Figure 1(j)). In addition, there is no distinctly organized lobe surrounding the ovule; this reduces boundary layer effects near ovules and increases the pollen-scavenging area (Figure 2(k)). Therefore, during pollination, the micropyle orifice enlarges and forms a funnel-like shape, and a pollination drop is produced on the micropyle (Figures 1(l) and 3(q)). It is interesting that the integument tip of the ovule, which includes the micropyle, does not emerge as strictly ring-shaped, but splits to several pieces (Figures 3(i) and 3(p)). Therefore, we propose that the architecture of the micropyle in *G. biloba* is adapted to produce a pollination drop with its size optimized for attachment of pollen grains, similar to other modifications of the integument tip in some conifers [3, 33–35]. Also, these adaptations could increase the contact area between the ovule and the pollination drop, decreasing the possibility of the pollination drop falling from the micropyle owing to the unexpected motion of the long shoot. Third, before pollination, the ovule differentiates into tissues adapted for pollen transport, including the micropyle, the micropylar canal, and the pollen chamber, of which the micropylar canal is straight and the cells on the inner surface are smooth (Figures 3(j) and 3(k)). The pollen chamber is well-developed, and the orifice reaches its largest size as it faces the micropylar canal (Figure 3(r)). Thus, we assume that, at this moment, the structure of the ovule enables it to transport and accept pollen grains, and facilitates the movement of the pollen from the pollination drop to the pollen chamber.

4.5. Ovule Developmental State during Pollination. At the time of pollination, the developmental state of ovules varies among species, owing to wide variations in the period between pollination and fertilization, or the progamic phase [36]. Ordinarily, the angiosperm ovule, which contains an egg cell, two synergids, and a binucleated central cell at the micropylar pole, is already present during the pollination period, because angiosperms often have a relatively short progamic phase [37–39]. In contrast, since the time interval from pollination to fertilization varies among families, the ovules of gymnosperms have a variable developmental state at the time of pollination. For instance, pollination in *Podocarpus totara* [40] and *Larix decidua* [41] occurred when the functional megaspore inside the ovule developed and became the mother cell of the female gametophyte, whereas the female gametophyte of interior spruce was cellular at pollination [42]. In the present study, the female gametophyte of the *G. biloba* ovule was in the free nuclei stage (Figures 4(e) and 4(g)) during the pollination period. According to Wang et al. [22], there were approximately 130 d between

the initiation of karyomitosis in the female gametophyte and the formation of the mature archegonium. Meanwhile, there were also approximately 130 d between entry of the pollen grain into the pollen chamber at the time of pollination and development of the male gametophyte into the mature spermatozoid at the time of fertilization. Thus, we can conclude that the female gametophyte phase occurring inside the ovule during the pollination period might be in preparation for fertilization at the later period. Therefore, the developmental processes of the male and female gametophyte are coordinated after pollination, thereby ensuring that the male and female gametes meet at the correct developmental stage during the short fertilization period.

Author's Contribution

D. Wang and Y. Lu contributed equally to this work.

Acknowledgments

The authors thank Prof. Peng Chen and Prof. Zhong Wang of Yangzhou University, China, for technical advice. This work was supported by the Natural Science Fund of Jiangsu Province (no. BK2011444), Science and Technology Innovation Fund of Xuzhou City (no. XF11C001), and the National Natural Science Foundation of China (no. 30870436).

References

[1] M. Proctor, P. Yeo, and A. Lack, *The Natural History of Pollination*, Timber Press, Portland, Ore, USA, 1996.

[2] J. N. Owens, T. Takaso, and C. John Runions, "Pollination in conifers," *Trends in Plant Science*, vol. 3, no. 12, pp. 479–485, 1998.

[3] J. N. Owens, G. Catalano, and J. S. Bennett, "The pollination mechanism of western white pine," *Canadian Journal of Forest Research*, vol. 31, no. 10, pp. 1731–1741, 2001.

[4] L. M. Chandler and J. N. Owens, "The pollination mechanism of Abies amabilis," *Canadian Journal of Forest Research*, vol. 34, no. 5, pp. 1071–1080, 2004.

[5] C. J. Runions, K. H. Rensing, T. Takaso, and J. N. Owens, "Pollination of *Picea orientalis* (Pinaceae): saccus morphology governs pollen buoyancy," *American Journal of Botany*, vol. 86, no. 2, pp. 190–197, 1999.

[6] A. B. Schwendemann, G. Wang, M. L. Mertz, R. T. McWilliams, S. L. Thatcher, and J. M. Osborn, "Aerodynamics of saccate pollen and its implications for wind pollination," *American Journal of Botany*, vol. 94, no. 8, pp. 1371–1381, 2007.

[7] Y. Lu, B. Jin, L. Wang et al., "Adaptation of male reproductive structures to wind pollination in gymnosperms: cones and pollen grains," *Canadian Journal of Plant Science*, vol. 91, no. 5, pp. 897–906, 2011.

[8] G. Gelbart and P. Von Aderkas, "Ovular secretions as part of pollination mechanisms in conifers," *Annals of Forest Science*, vol. 59, no. 4, pp. 345–357, 2002.

[9] S. Mugnaini, M. Nepi, M. Guarnieri, B. Piotto, and E. Pacini, "Pollination drop in *Juniperus communis*: response to deposited material," *Annals of Botany*, vol. 100, no. 7, pp. 1475–1481, 2007.

[10] H. Singh and J. N. Owens, "Sexual reproduction in grand fir (*Abies grandis*)," *Canadian Journal of Botany*, vol. 60, no. 11, pp. 2197–2214, 1982.

[11] K. J. Niklas and K. T. P. U, "Pollination and airflow patterns around conifer ovulate cones," *Science*, vol. 217, no. 4558, pp. 442–444, 1982.

[12] L. Wang, D. Wang, M.-M. Lin, Y. Lu, X.-X. Jiang, and B. Jin, "An embryological study and systematic significance of the primitive gymnosperm *Ginkgo biloba*," *Journal of Systematics and Evolution*, vol. 49, no. 4, pp. 353–361, 2011.

[13] M. Favre-Duchatre, "Ginkgo, an oviparous plant," *Phytomorphology*, vol. 8, pp. 377–396, 1958.

[14] Z. L. Lee, "Recent advances (1949–1959) in morphology, anatomy and cytology of *Ginkgo Biloba* L," *Acta Botanica Sinica*, vol. 8, no. 4, pp. 262–270.

[15] A. W. Douglas, D. W. Stevenson, and D. P. Little, "Ovule development in *Ginkgo biloba* L., with emphasis on the collar and nucellus," *International Journal of Plant Sciences*, vol. 168, no. 9, pp. 1207–1236, 2007.

[16] L. Wang, Y. P. Wang, Q. Wang et al., "Anatomical study of development of ovule in *Ginkgo biloba* L," *Acta Botanica Boreali-Occidentalia Sinica*, vol. 27, no. 7, pp. 1349–1356, 2007.

[17] D. H. Li, X. Yang, X. Cui, K. M. Cui, Z. L. Li, and C. L. Lee, "Early development of pollen chamber in *Ginkgo biloba* Ovule," *Acta Botanica Sinica*, vol. 44, no. 7, pp. 757–763, 2002.

[18] C. J. Ji, X. Yang, and Z. L. Li, "Morphological studies on megaspore formation in *Ginkgo biloba*," *Acta Botanica Sinica*, vol. 41, no. 7, pp. 219–221, 1999.

[19] W. E. Friedman, "Growth and development of the male gametophyte of *Ginkgo biloba* within the ovule (in Vivo)," *American Journal of Botany*, vol. 74, no. 12, pp. 1797–1815.

[20] B. Jin, L. Zhang, Y. Lu et al., "The mechanism of pollination drop withdrawal in *Ginkgo biloba* L," *BMC Plant Biology*, vol. 12, no. 59, 2012.

[21] L. Wang, B. Jin, M. M. Lin, Y. Lu, N. J. Teng, and P. Chen, "Studies of the development of female reproductive organs in *Ginkgo biloba* L," *Chinese Bulletin of Botany*, vol. 44, no. 6, pp. 673–681, 2009.

[22] L. Wang, Y. Lu, B. Jin, M. M. Lin, and P. Chen, "Gametophyte development and embryogenesis in *Ginkgo biloba*: a current view," *Chinese Bulletin of Botany*, vol. 45, no. 1, pp. 119–127, 2010.

[23] L. Wang, B. Jin, Y. Lu, X. X. Jin, N. J. Teng, and P. Chen, "Developmental characteristics of the ovule and its biological significance in *Ginkgo biloba* L," *Journal of Beijing Forestry University*, vol. 32, no. 2, pp. 79–85, 2010.

[24] B. Jin, L. Wang, J. Wang et al., "The effect of experimental warming on leaf functional traits, leaf structure and leaf biochemistry in *Arabidopsis thaliana*," *BMC Plant Biology*, vol. 11, article 35, 2011.

[25] B. Jin, L. Wang, J. Wang et al., "The structure and roles of sterile flowers in *Viburnum macrocephalum* f. *keteleeri* (Adoxaceae)," *Plant Biology*, vol. 12, no. 6, pp. 853–862, 2010.

[26] M. W. Frohlich and M. W. Chase, "After a dozen years of progress the origin of angiosperms is still a great mystery," *Nature*, vol. 450, no. 7173, pp. 1184–1189, 2007.

[27] K. J. Niklas, "Wind pollination-a study in controlled chaos," *American Scientist*, vol. 73, no. 5, pp. 462–470, 1985.

[28] K. J. Niklas, "The aerodynamics of wind pollination," *The Botanical Review*, vol. 51, no. 3, pp. 328–386, 1985.

[29] S. Zheng and Z. Zhou, "A new Mesozoic *Ginkgo* from western Liaoning, China and its evolutionary significance," *Review of Palaeobotany and Palynology*, vol. 131, no. 1-2, pp. 91–103, 2004.

[30] Z. Y. Zhou, "Heterochronic origin of *Ginkgo biloba*-type organs," *Acta Palaeontologica Sinica*, vol. 33, no. 2, pp. 131–139, 1994.

[31] Z. Y. Zhou, "An overview of fossil Ginkgoales," *Palaeoworld*, vol. 18, no. 1, pp. 1–22, 2009.

[32] A. Farjon and S. O. Garcia, "Cone and ovule development in *Cunninghamia* and *Taiwania* (Cupressaceae sensu lato) and its significance for conifer evolution," *American Journal of Botany*, vol. 90, no. 1, pp. 8–16, 2003.

[33] J. R. McWilliam, "The role of the micropyle in the pollination of Pinus," *Botanical Gazette*, vol. 120, no. 2, pp. 109–117, 1995.

[34] J. N. Owens, S. J. Simpson, and M. Molder, "The pollination mechanism and the optimal time of pollination in Douglas-fir (*Pseudotsuga menziesii*)," *Canadian Journal of Forest Research*, vol. 11, no. 1, pp. 36–50, 1981.

[35] J. N. Owens, S. J. Simpson, and G. Caron, "The pollination mechanism of Engelmann spruce (*Picea engelmannii* Parry)," *Canadian Journal of Botany*, vol. 65, no. 7, pp. 1439–1450, 1987.

[36] J. H. Williams, "Novelties of the flowering plant pollen tube underlie diversification of a key life history stage," *Proceedings of the National Academy of Sciences of the United States of America*, vol. 105, no. 32, pp. 11259–11263, 2008.

[37] W. E. Friedman and K. C. Ryerson, "Reconstructing the ancestral female gametophyte of angiosperms: insights from *Amborella* and other ancient lineages of flowering plants," *American Journal of Botany*, vol. 96, no. 1, pp. 129–143, 2009.

[38] V. Olmedo-Monfil, N. Durán-Figueroa, M. Arteaga-Vázquez et al., "Control of female gamete formation by a small RNA pathway in *Arabidopsis*," *Nature*, vol. 464, no. 7288, pp. 628–632, 2010.

[39] A. Cisneros, R. Benega Garcia, and N. Tel-Zur, "Ovule morphology, embryogenesis and seed development in three *Hylocereus* species (Cactaceae)," *Flora*, vol. 206, pp. 1076–1084, 2011.

[40] V. R. Wilson and J. N. Owens, "The reproductive biology of totara (*Podocarpus totara*) (Podocarpaceae)," *Annals of Botany*, vol. 83, no. 4, pp. 401–411, 1999.

[41] K. Rafińska and E. Bednarska, "Localisation pattern of homogalacturonan and arabinogalactan proteins in developing ovules of the gymnosperm plant *Larix decidua* Mill," *Sexual Plant Reproduction*, vol. 24, no. 1, pp. 75–87, 2011.

[42] J. J. Runions and J. N. Owens, "Sexual reproduction of interior spruce (Pinaceae). I. Pollen germination to archegonial maturation," *International Journal of Plant Sciences*, vol. 160, no. 4, pp. 631–640, 1999.

18

Phytoliths as a Tool for the Identification of Some Chloridoideae Grasses in Kerala

P. I. Jattisha and M. Sabu

Department of Botany, University of Calicut, Kerala 673 635, India

Correspondence should be addressed to P. I. Jattisha, jattisha.p@gmail.com

Academic Editors: T. Berberich, F. A. Culianez-Macia, and S.-W. Park

The phytoliths of eight genera including fifteen species of grasses under the subfamily Chloridoideae in Kerala were studied. Phytoliths were studied after chemical isolation. Every species was found to produce a diverse array of phytoliths. However the frequency assemblages of phytoliths, their size, and orientation in the epidermal layer appear to vary among the different species and hence can be used for the delimitation of the taxa. Consequently, an identification key following the International Code for Phytolith Nomenclature was developed for all the species studied.

1. Introduction

Silicon is deposited in plants as hydrated amorphous silica ($SiO_2 \cdot nH_2O$) through the polymerization of monosilicic acid ($Si(OH)_4$) absorbed by roots from soil [1]. Silica deposited as inclusions within the cells are usually termed as phytoliths or silica bodies. Phytoliths have proved to be a potential tool in palaeoecological studies as they remain stable in the soil for even millions of years after plant tissues decay [2–12].

Depending upon the species of the plant, silica is deposited between the cells, within the cell walls, or even sometimes completely infilling the cells themselves. Silica bodies in the silica cells of grasses assume characteristic forms when the leaf is mature. Silica deposits in grasses have been well documented [13–17]. According to Metcalfe and Kaufman et al. [13, 18] silica deposits in grasses occur primarily in epidermal long cells, trichomes (hairs), specialized silica short cells, and as fillings within the bulliform cells of plant leaves. Phytolith shapes are found to be consistent within a species and hence they can provide significant taxonomic information [6, 19–21].

Metcalfe [13] considered the structure of silica bodies as one of the characteristics useful in plant identification. Since the publication of Metcalfe's work, the utility of phytoliths in plant taxonomy has been demonstrated in other publications. Studies made by Ollendorf et al. [14] have shown that phytolith characters can be used to distinguish *Arundo donax* from *Phragmites communis*, two giant reed grasses that is otherwise difficult to identify from the field. Various studies have been made to distinguish phytoliths in cultivated crops from those of wild relatives of crop plants [22–24].

The present study is concerned with the identification of some Chloridoideae grasses in the vegetative stage based on the size, shape, frequency, and orientation of phytoliths in the leaf epidermis.

2. Materials and Methods

Plants included in the present study belong to the subfamily Chloridoideae of the family Poaceae, according to the classification made by Clayton and Renvoize [25]. In particular, a total of fifteen species of grasses under eight genera were selected for the present study (Table 1). Specimens were collected fresh from different parts of Kerala. Voucher specimens of collected plants are deposited in Calicut University Herbarium (CALI). Herbarium specimens collected from CALI were also utilized for the study.

Mature leaf blades were selected for the present study. The leaf blades were first washed in water and wiped dry. They were then cut into $1\,cm^2$ pieces and soaked in acid mixture (con. HNO_3 + con. H_2SO_4 in $1:1$ proportion) for 15 days [15]. The mixture was then centrifuged and the

TABLE 1: Grasses included in the present study.

Si. no.	Latin name	Collection no.
(1)	*Chloris barbata* Sw.	127309, 127380, 6177
(2)	*Cynadon arcuatus* J.S. Presl. ex C.B. Presl.	127304
(3)	*C. dactylon* (Linn.) Pers.	113370, 11276
(4)	*Dactyloctenium aegyptium* (Linn.) P. Beauv.	113373, 8951, 3580
(5)	*Eleusine corocana* (Linn.) Gaertn.	124901, 521, 3819
(6)	*E. indica* (Linn.) Gaertn.	113343, 9479, 2667
(7)	*Eragrostis nutans* (Retz.) Steud.	127390
(8)	*E. tenella* (Linn.) Roem. and Schult	113375, 34419
(9)	*E. unioloides* (Retz.) Steud.	113383, 3587, 2565
(10)	*E. viscosa* (Retz.) Trin.	113351, 12406
(11)	*Leptochloa chinensis* (Linn.) Nees	127355
(12)	*Perotis indica* (Linn.) O. Kuntze	113381, 5803, 3986
(13)	*Sporobolus diander* (Retz.) P. Beauv.	113377, 1337
(14)	*S. piliferus* (Trin.) Kunth	127307, 2891, 4650
(15)	*S. tenuissimus* (Schrank) O. Kuntze	113357, 113375

TABLE 2: Different types of phytoliths observed and their abbreviations.

Si. no.	Type	Abbreviation
(1)	Associated epidermal polygonal cells	AEP
(2)	Associated long cells	ALC
(3)	Associated papilla	AP
(4)	Associated short cells	AS
(5)	Bulliform elements	BE
(6)	Stomatal complex	SC
(7)	Subepidermal elements	SEE
(8)	Simple bilobate	SB
(9)	Complex bilobate	CB
(10)	Nodular bilobate	NB
(11)	Quadralobate	Q
(12)	Normal saddle	S
(13)	Elongate	E
(14)	Fan	F
(15)	Horned towers	HT
(16)	Macrohairs	LH
(17)	Microhairs	MH
(18)	Prickle hairs	PR
(19)	Papilla	P

Abbreviations for lobate phytoliths are followed by C, F, or X denoting concave, fattened, or convex margins.

acid decanted. The residue was then suspended in distilled water and again centrifuged. This process was repeated upto 4 cycles inorder to completely remove the acid from the residue. The final residue was then suspended in rectified spirit and stored in storage vials for further study.

A drop of the phytolith alcohol mixture was placed on a clean slide and warmed under the spirit lamp. Then a drop of crystal violet stain was applied and again warmed. Excess stain was removed by suddenly dipping the slide and taking out from water. This was warmed again and the slide was mounted permanently in DPX (Distyrene Plasticizer Xylene) for further observations and photography.

For observations and photography Motic Digital Microscope with image analyser was used. Observations and photography were taken under oil immersion objective (100x). Measurements were taken from surface view of phytoliths using Motic Image Analyzer software. Various features of phytoliths noted include length, width, nature of outer margins, shape, and so forth.

In addition to measurement, frequency of phytolith assemblages was also noted. About 1000 phytoliths from each species were counted and frequency determined. For both frequency and measurement range, average and standard error were calculated.

Photographs of different types of phytoliths observed for different species are also provided. Species taken for the present study and their collection number are shown in Table 1. The serial numbers of the species shown in this table are used to denote the name of the species in Tables 3 and 4.

3. Results

Data obtained from the frequency of phytolith assemblages as well as measurements were utilized for the preparation of an identification key for grasses upto species level. Length and width dimensions and frequency of about 1000 phytoliths were measured from each species. The present study on phytoliths applies the rules of International Code for Phytolith Nomenclature [26]. Silica bodies were found in two forms—articulated (associated with other cells) and isolated. Various types of phytoliths observed and their abbreviations used are given in Table 2.

3.1. Articulated Forms. These are the associated forms of phytoliths found joined either with the same type or different types of cells. These include the following subcategories.

TABLE 3: Measurements (in μm) of phytoliths.

Shape	Dimension		1	2	3	4	5	6	7	8	9	10	11	12	13	14	15
*	L	Avge ± SE	—	—	—	—	—	—	—	—	—	—	17.5 ± 0.6	14.3 ± 0.5	20.9 ± 1	12.1 ± 0.3	12.6 ± 0.6
		Range	—	—	—	—	—	—	—	—	—	—	10–21.1	8.3–20.9	11.9–29	8.6–15.4	9.1–17
	W	Avge ± SE	—	—	—	—	—	—	—	—	—	—	8.1 ± 0.2	6.2 ± 0.2	8.1 ± 0.2	7.2 ± 0.2	5.3 ± 0.3
		Range	—	—	—	—	—	—	—	—	—	—	5.8–10.9	4.3–9	6.7–10.4	4.8–8.7	4.4–9.1
SB	L	Avge ± SE	—	—	—	—	—	—	—	13.2 ± 0.5	—	—	—	—	—	—	—
		Range	—	—	—	—	—	—	—	10.2–14.5	—	—	—	—	—	—	—
	W	Avge ± SE	—	—	—	—	—	—	—	5.8 ± 0.31	—	—	—	—	—	—	—
		Range	—	—	—	—	—	—	—	4.5–7.6	—	—	—	—	—	—	—
E	L	Avge ± SE	—	—	48.5 ± 3.6	—	—	60.6 ± 7.2	—	59.4 ± 5.7	—	—	—	—	—	—	—
		Range	—	—	36.1–66.8	—	—	44.8–89.7	—	58.2–72.4	—	—	—	—	—	—	—
	W	Avge ± SE	—	—	11 ± 0.6	—	—	11.8 ± 1.1	—	11.9 ± 0.7	—	—	—	—	—	—	—
		Range	—	—	8.6–15.7	—	—	11.2–16	—	11.5–13.6	—	—	—	—	—	—	—
S	L	Avge ± SE	11.9 ± 0.5	7.98 ± 0.2	8.03 ± 0.2	11.69 ± 0.2	8.4 ± 0.7	7.9 ± 0.3	7.2 ± 0.3	—	—	11.7 ± 0.9	—	—	9.5 ± 0.2	—	5.7 ± 0.3
		Range	7.5–18.1	5.7–9.9	4.8–10.5	8.7–13.4	5.9–10.5	6.3–10.7	5.9–10	—	—	9.5–13	—	—	8.5–10.8	—	3.1–7.9
	W	Avge ± SE	12.98 ± 0.4	9.29 ± 0.36	8.58 ± 0.17	10.6 ± 0.1	6.5 ± 0.5	10.3 ± 0.4	7.9 ± 0.8	—	—	7.8 ± 0.7	—	—	4.8 ± 0.3	—	7.3 ± 0.3
		Range	8.8–17.3	5.4–16.2	5–11.3	8.9–15.8	5.3–9.9	8–13.4	2.8–14.4	—	—	6.3–9.6	—	—	3.4–6.2	—	5–10.1
SC	L	Avge ± SE	—	—	—	—	—	28.6 ± 0.6	23.4 ± 0.5	19.3 ± 0.3	22.7 ± 0.6	26.3 ± 0.7	—	—	29.01 ± 2.7	—	—
		Range	—	—	—	—	—	25.8–31	21.7–26.6	16.7–21.9	19.1–32.1	20.7–30.9	—	—	17.3–35.8	—	—
	W	Avge ± SE	—	—	—	—	—	21.1 ± 1.2	16.1 ± 0.5	12.5 ± 0.3	16.4 ± 0.6	17.98 ± 0.4	—	—	17.7 ± 1.7	—	—
		Range	—	—	—	—	—	16.4–26	14.5–17.9	10.5–15.5	12.5–24.5	15–21.8	—	—	10.4–23	—	—
PR	L	Avge ± SE	—	—	43.4 ± 4.3	—	—	—	—	29.8 ± 1.3	—	—	—	—	—	—	43.7 ± 2.1
		Range	—	—	24.2–67.5	—	—	—	—	24.6–32.5	—	—	—	—	—	—	30.1–58
	W	Avge ± SE	—	—	18.8 ± 2.7	—	—	—	—	21.7 ± 1.4	—	—	—	—	—	—	23.7 ± 2.01
		Range	—	—	7.4–34.5	—	—	—	—	16.5–25.7	—	—	—	—	—	—	11.7–36.2

TABLE 4: Frequency (in percentage) of phytoliths.

Phytolith type	1	2	3	4	5	6	7	8	9	10	11	12	13	14	15
AEP	—	—	—	—	—	—	0.4 ± 0.2	—	—	—	—	—	—	—	—
ALC	5.6 ± 0.3	3.6 ± 0.5	4.3 ± 1.1	1.8 ± 0.5	10.7 ± 0.9	15.7 ± 0.8	28.5 ± 1.4	64 ± 1	67.2 ± 1.4	31.2 ± 1.8	0.1 ± 0.1	0.5 ± 0.2	6.4 ± 0.9	—	2.3 ± 0.4
AP	0.4 ± 0.2	—	1.4 ± 0.4	—	—	—	—	—	—	—	—	—	—	—	—
AS	—	—	1.4 ± 0.6	0.5 ± 0.3	—	—	—	—	—	—	—	0.5 ± 0.2	0.2 ± 0.1	—	0.3 ± 0.2
BE	—	—	—	0.3 ± 0.2	—	—	—	2.7 ± 0.2	—	—	—	—	3.1 ± 0.8	—	—
SC	0.4 ± 0.2	—	—	—	11.9 ± 0.9	3.6 ± 0.3	22.8 ± 0.6	12.3 ± 0.6	16.4 ± 0.8	40.7 ± 1.3	—	—	9.6 ± 0.9	—	1 ± 0.2
SEE	—	—	0.3 ± 0.2	0.7 ± 0.2	0.8 ± 0.3	—	0.3 ± 0.2	—	—	0.4 ± 0.2	1 ± 0.3	—	0.6 ± 0.2	—	—
SBC	—	—	—	—	—	4.7 ± 0.5	5.2 ± 0.5	2.8 ± 0.3	2 ± 0.3	1.4 ± 0.4	—	12.6 ± 1.5	6 ± 0.9	51.4 ± 0.8	6.5 ± 0.3
SBF	—	—	—	0.1 ± 0.1	—	0.5 ± 0.2	6 ± 0.8	1.1 ± 0.3	2.9 ± 0.7	0.3 ± 0.2	79.7 ± 0.7	54.1 ± 1.8	4.9 ± 0.6	15.8 ± 0.5	4.5 ± 0.6
SBF/C	—	—	—	—	—	1.3 ± 0.3	1.2 ± 0.3	—	—	1.3 ± 0.4	3.5 ± 0.5	8.5 ± 0.7	1.2 ± 0.4	16.4 ± 0.7	3.3 ± 0.4
SBF/X	—	—	—	—	—	—	—	—	—	1 ± 0.2	6.7 ± 0.3	11.9 ± 0.7	—	0.6 ± 0.3	0.6 ± 0.2
SBX	—	—	—	—	—	—	—	—	1.2 ± 0.1	—	—	6.8 ± 0.4	—	—	—
SBX/C	—	—	—	—	—	—	—	0.5 ± 0.2	0.6 ± 0.2	0.8 ± 0.2	—	1.5 ± 0.2	—	1.5 ± 0.2	2 ± 0.2
CBC	—	—	—	—	—	—	—	—	—	—	0.6 ± 0.2	1 ± 0.4	—	—	—
CBF	—	—	—	—	—	—	—	—	—	—	—	0.5 ± 0.2	—	—	—
CBF/C	—	—	—	—	—	—	—	—	—	—	—	—	—	—	—
E	0.2 ± 0.1	—	8.8 ± 0.7	2.1 ± 0.5	2.9 ± 0.3	6.9 ± 0.4	0.8 ± 0.3	7 ± 0.5	5.5 ± 0.6	4 ± 0.6	2.1 ± 0.5	0.4 ± 0.2	4.3 ± 0.5	—	3.1 ± 0.3
F	—	—	—	—	—	—	—	—	—	0.2 ± 0.1	—	—	0.4 ± 0.4	0.5 ± 0.2	—
HT	—	—	—	0.6 ± 0.3	0.3 ± 0.2	—	—	—	—	1 ± 0.4	—	—	—	0.7 ± 0.3	—
LH	—	—	0.1 ± 0.1	—	—	3.7 ± 0.5	—	0.5 ± 0.2	0.7 ± 0.3	0.1 ± 0.1	—	—	20.1 ± 1.8	—	0.4 ± 0.2
MH	—	—	—	—	—	—	—	—	—	0.1 ± 0.1	—	—	—	—	—
NBF	—	—	—	—	—	—	—	—	—	0.4 ± 0.2	—	0.4 ± 0.2	—	0.9 ± 0.3	—
NBF/C	—	—	—	0.1 ± 0.1	—	—	—	—	—	—	—	—	—	—	—
P	—	—	2.4 ± 1	0.1 ± 0.1	0.1 ± 0.1	—	—	—	—	—	—	—	—	—	—
Q	—	—	—	0.1 ± 0.1	—	60.7 ± 0.9	0.4 ± 0.2	1.5 ± 0.2	1.5 ± 1	10.4 ± 1	—	—	—	1.3 ± 0.4	—
S	89.4 ± 0.7	94.8 ± 0.5	75.5 ± 2.6	93.2 ± 0.8	57.2 ± 1.2	—	18 ± 1.1	—	—	10.4 ± 1	—	0.4 ± 0.2	39.5 ± 1.3	10.2 ± 1.2	71.3 ± 1.1
PR	4 ± 0.5	1.6 ± 0.2	5.8 ± 0.9	0.5 ± 0.3	15.8 ± 0.6	15.9 ± 1.1	15.9 ± 1.1	7.5 ± 0.3	3.5 ± 0.4	6.2 ± 0.8	6.3 ± 0.6	0.6 ± 0.2	3.6 ± 0.7	0.7 ± 0.2	5 ± 0.3
ST	—	—	—	—	0.4 ± 0.2	—	—	—	—	0.2 ± 0.1	—	—	—	—	—

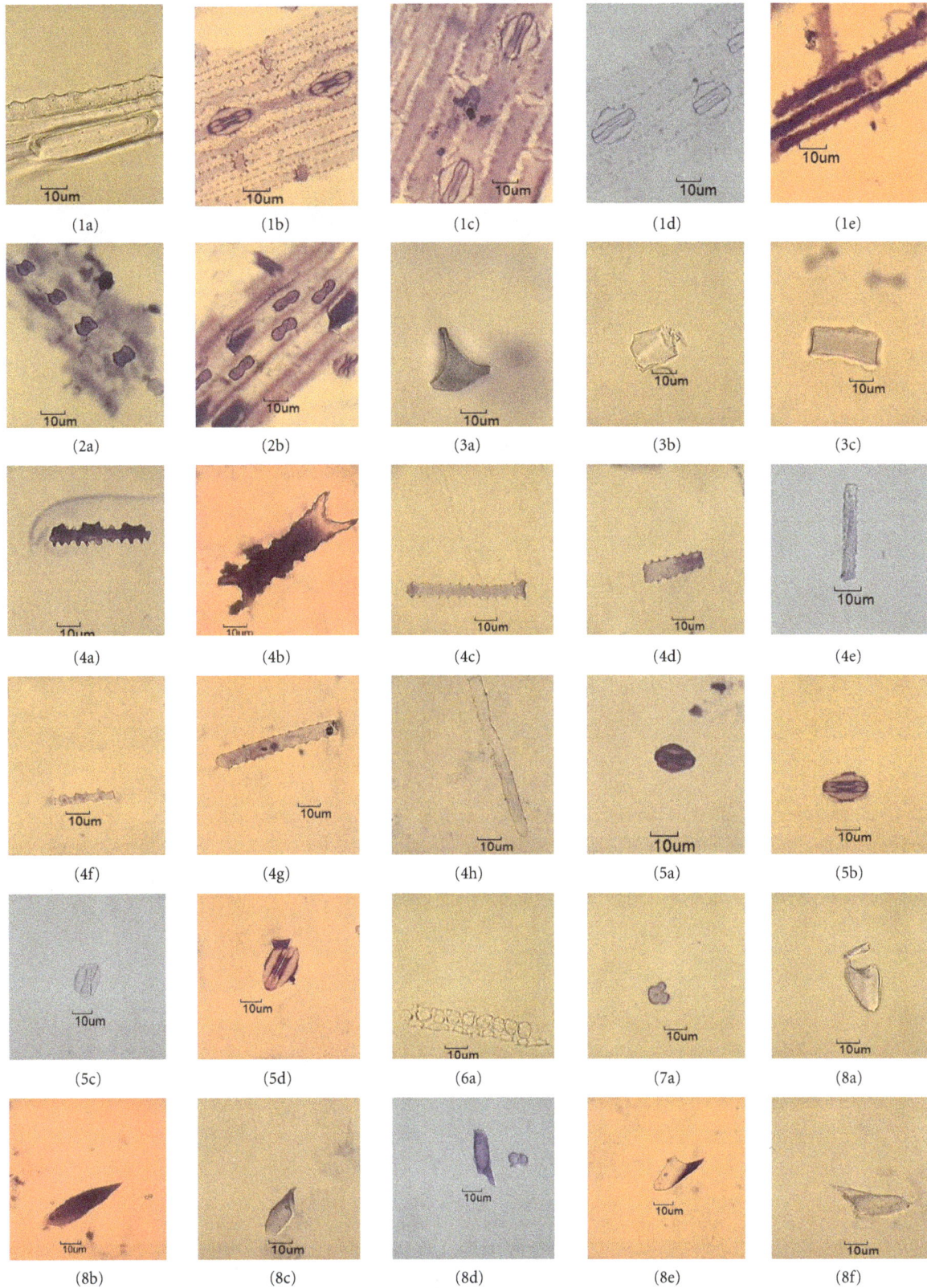

FIGURE 1: Associated long cells of (1a) *Dactyloctenium aegyptium,* (1b) *Eragrostis tenella,* (1c) *E. unioloides,* (1d) *E. viscosa,* (1e) *Sporobolus diander;* (2) associated short cells of (2a) *Cynadon arcuatus,* (2b) *Eragrostis tenella;* (3) fan cells of (3a) *Cynadon arcuatus,* (3b) *C. dactylon,* (3c) *Perotis indica;* (4) elongate cells of (4a) *Cynadon dactylon,* (4b) *Eleusine corocana,* (4c) *Eragrostis tenella,* (4d) *E. unoloides,* (4e) *E. viscosa,* (4f) *Perotis indica,* (4g) *Sporobolus diander,* (4h) *S. tenuissimus;* (5) stomatal complex of (5a) *Cynadon dactylon,* (5b), (5c) *E. viscosa,* (5d) *Sporobolus diander;* (6) (6a) subepidermal elements of *Dactyloctenium aegyptium;* (7) (7a) horned tower of Sporobolus piliferus (8) prickle hairs of (8a) *Dactyloctenium aegyptium,* (8b) *Eleusine corocana,* (8c) *Eragrostis tenella,* (8d) *E. viscosa,* (8e) *Sporobolus tenuissimus,* (8f) *S. diander.*

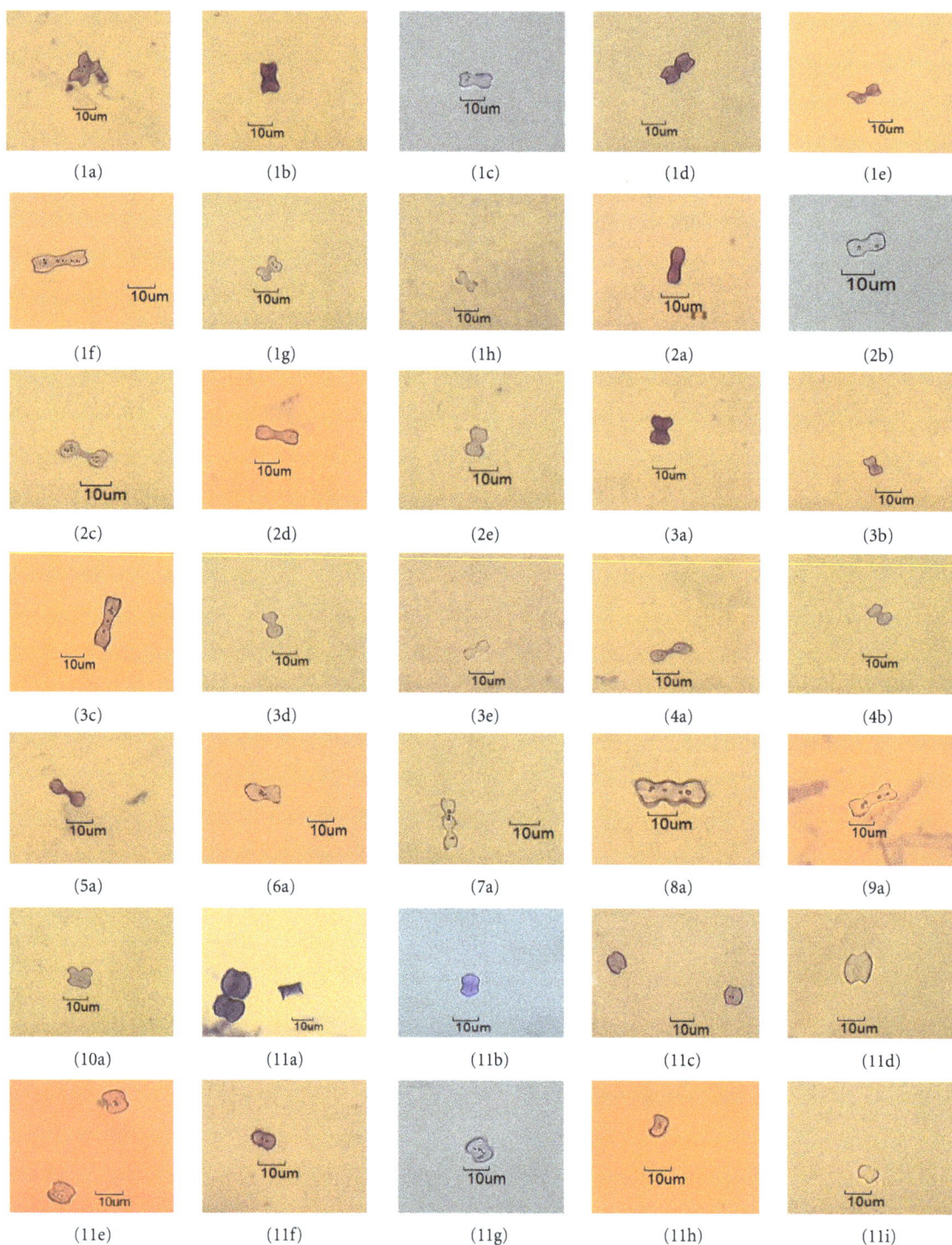

FIGURE 2: 1–6: Simple bilobates (1) Bilobate with concave margins. (1a) *Eleusine indica*, (1b) *Eragrostis tenella*, (1c) *E. viscosa*, (1d) *Leptochloa chinensis*, (1e) *Perotis indica*, (1f) *Sporobolus diander* (1g) *S. piliferus* (1h) *S. tenuissimus*; (2) bilobate with flattened margins. (2a) *Eragrostis tenella*, (2b) *E. viscosa*, (2c) *Perotis indica*, (2d) *Sporobolus diander*, (e) *S. piliferus*; (3) bilobate with flattened/concave margins. (3a) *Eleusine indica*, (3b) *Perotis indica*, (3c) *Sporobolus diander*, (3d) *S. piliferus*, (3e) *S. tenuissimus*; (4) Bilobate with flattened/concex margins. (4a) *Perotis indica*, (4b) *Sporobolus piliferus*; (5) (5a) Bilobate with convex margins of *Perotis indica*; (6) (6a) Bilobate with convex/concave margins of *Sporobolus diander*; (7) (7a) complex bilobate with concave margins of *S. piliferus*; (8) (8a) Complex bilobate with flattened margins of *Eleusine indica*; (9) (9a) nodular bilobate with concave margins of *Sporobolus diander*; (10) (10a) quadralobate phytoliths of *S. piliferus*; (11) saddles. (11a) *Chloris barbata*, (11b) *Cynadon arcuatus*, (11c) *C. dactylon* (11d) *Dactyloctenium aegyptium* (11e) *Eleusine corocana* (11f) *E. indica*, (11g) *Eragrostis viscosa*, (11h) *S. diander*, (11i) *S. tenuissimus*.

(1) Associated Epidermal Polygonal Cells. Polygonal shaped cells of the epidermal cells with smooth or sinuate margins. These were observed only in *Eragrostis nutans* with a frequency below 1%.

(2) Associated Long Cells. These are associated long cells of the epidermis either with stomata or prickle hairs. These are very common and are found in almost all the species. Their frequency in *Eragrostis unioloides* is greater than 60%.

(3) Associated Papilla. These are protrusions from the epidermal cells. Sometimes 2 or more papillae are seen associated together. These were observed only in *Chloris barbata* and *Cynadon dactylon* with a frequency less than 2%.

(4) Associated Short cells. These are long cells associated with one or more short cells. Frequency of this type was found to be always less than 3%. However they were observed in six species.

(5) Bulliform Elements. These are silicified bulliform cells associated with each other. These were observed only in two species, that is, *Dactyloctenium aegyptium* and *Sporobolus diander*, with a frequency below 4%.

(6) Stomatal Complex. These are silicified guard cells joined together. This is a common type and were observed in nine species studied. The highest frequency of about 40% was observed in *Eragrostis viscosa*.

(7) Subepidermal Elements. These are silicified mesophyll or other cells present beneath the epidermis. These were observed in five species but their frequency was less than 1%.

3.2. Isolated Forms. Phytoliths remain isolated without association with other cells. These include the following.

(8) Simple Bilobate. These are phytoliths with two equal lobes connected by a shank. Their margins can be concave, convex or flattened or the two lobes of the same phytolith can have different margins. At least one type of simple bilobate phytoliths is represented in all the species except *Chloris barbata* and *Cynadon arcuatus* where they were found to be totally absent.

(9) Complex Bilobate. These are phytoliths with 3 equal lobes. Their margins may be concave or flattened. This is a rare type in the subfamily and their frequency falls below 2%. They were observed only in *Leptochloa chinensis* and *Perotis indica*.

(10) Nodular Bilobate. These are bilobate phytoliths with a projection or a small lobe between the two major lobes. Margins can be concave or flattened. Convex margins are found to be totally absent in the subfamily.

(11) Quadralobate. They are characterised by almost similar length and width dimensions and in having four lobes. They were observed only in four species with a frequency of less than 2%.

(12) Normal Saddle. Saddle shaped phytoliths are arranged vertically on the epidermis. They are present in almost all the members of Chloridoideae except one, that is, *Leptochloa chinensis*.

(13) Elongate. These are silicified long cells of the epidermis with echinate or sinuate walls. They are represented in all the species except two. However their frequency does not exceed beyond 10%. The highest frequency of 8.8% was observed in *Cynadon dactylon*.

(14) Fan. These are silicified bulliform cells of the epidermis. They were found in *Eragrostis viscosa*, *Sporobolus diander* and *S. piliferus* with a frequency less than 1%.

(15) Horned Towers. Cylindrical short cells with two projections on one side. This type was observed only in *Sporobolus piliferus* with a frequency of 0.7%.

(16) Macro Hairs. These are elongated hairs present on the epidermis. They were observed in eight species studied. Their frequency in most cases falls below 5%. However in *Sporobolus diander* it is 20%.

(17) Microhairs. These are formed as a result of silicification of slightly curved two-celled blunt hairs of the epidermis. This type is rare and was observed in *Eragrostis nutans* and *E. viscosa*. Their frequency is less than 1%.

(18) Prickle Hairs. These are formed as a result of silicification of sharply pointed and slightly curved structures on the epidermis. This is a common type found in all the species except *Eleusine indica*. Highest frequency of about 15% was observed in *Eleusine corocana* and *Eragrostis nutans*.

(19) Papilla. Protrusions from the epidermis seen isolated. This is very rare in the subfamily and was observed only in *Cynadon arcuatus* with a frequency of 2.4%.

All the dimensions of phytoliths were studied. For taking measurements surface view of phytoliths were considered. Measurements of phytoliths of each species are shown in Table 3. and frequency assemblages in Table 4. Photographs of different types of phytoliths are given in Figures 1 and 2.

Key for the identification of species based on the frequency, size, and shape of phytoliths.

4. Key to the Species

(1) Saddles less than 1%, simple bilobates dominant with a frequency of 90–96%...2.

(1) Saddles greater than 1%, simple bilobates usually not dominant, if so, frequency below 80%...3.

(2) Simple bilobates with concave margins dominant with frequency greater than 70%...*Leptochloa chinensis*.

(2) Simple bilobates with flattened margins dominant with frequency less than 50%...*Perotis indica*.

(3) Saddles the most dominant type, simple bilobate phytoliths rare, if present only less than 1%...4.

(3) Saddles usually not the most dominant type, if so simple bilobates 5% or greater...8.

(4) Average length of saddles 11-12 µm...5.

(4) Average length of saddles 7–9 µm...6.

(5) Hook shaped prickles present, average width of saddles about 10.6 µm...*Dactyloctenium aegyptium*.

(5) Hook shaped prickles absent, straight sharply pointed structures seen, average width of saddles about 12.9 µm...*Chloris barbata*.

(6) Saddles constitute about greater than 75% of the total phytoliths...7.

(6) Saddles constitute less than 60% of the total phytoliths...*Eleusine corocana*.

(7) Sharply pointed structures rare, less than 2% saddles about 95%...*Cynadon arcuatus*.

(7) Sharply pointed structures present at about 5%, saddles about 76%...*Cynadon dactylon*.

(8) ALC and SC together constitute more than 50% of the total phytoliths...9.

(8) ALC and SC usually very less and will not exceed 20%...12.

(9) Average length of SC about 26 µm...*Eragrostis viscosa*.

(9) Average length of SC less than 25 µm...10.

(10) Proportion of ALC and SC nearly 1 : 1...*Eragrostis nutans*.

(10) Proportion of ALC and SC is about 4 : 1...11.

(11) SC about 12.5 µm wide, outline more or less rounded...*Eragrostis tenella*.

(11) SC about 16 µm wide, outline somewhat V-shaped, not rounded...*Eragrostis unioloides*.

(12) Frequency of saddle shaped phytoliths greater than 60%...13.

(12) Frequency of saddle shaped phytoliths less than 50%...14.

(13) ALC and SC together constitute more than 15% of the total phytoliths...*Eleusine indica*.

(13) ALC and SC together constitute less than 5% of the total phytoliths...*Sporobolus tenuissimus*.

(14) Frequency of simple bilobate phytoliths greater than 65%, average width about 12.09 µm...*Sporobolus piliferus*.

(14) Frequency of simple bilobate phytoliths less than 15%, average width about 20.9 µm...*Sporobolus diander*.

5. Discussion

The present study of grasses from the Indian subcontinent underscores what previous investigations have shown already elsewhere—the Poaceae family produces a diverse array of phytolith shapes yet certain shapes of phytoliths can consistently appear at least at the subfamily level.

Saddle shaped phytoliths characterises the subfamily Chloridoideae. Of the 15 species studied saddle shaped phytoliths were found to be the dominant type in seven species. Saddles were not observed only in *Leptochloa chinensis*. Highest frequency of saddles was noticed in *Chloris barbata*, *Cynadon arcuatus*, *Cynadon dactylon*, and *Dactyloctenium aegyptium*. In *Leptochloa chinensis*, *Perotis indica*, and *Sporobolus piliferus* bilobate phytoliths were found to be the dominant type. Associated long cells and stomatal complex were found in almost all the species studied. However their frequency was found to supersede in species of *Eragrostis*. More species has to be studied before this character could be attributed as diagnostic to the genus.

Frequency of phytolith assemblages and measurements of phytoliths are found to be consistent within a species and have been useful for developing the key. Even though phytolith multiplicity and redundancy occur in grasses, frequency assemblages reveal that a particular morphotype dominate over the other in a given species. Hence the present study shows that characteristics of phytoliths can be used as a tool for the identification of the species in the vegetative stage.

A brief survey on 15 species of Chloridoideae grasses has proved the usefulness of phytoliths in the identification of grasses. The present paper represents only a preliminary study towards developing an identification key for all South Indian grasses based on the foliar phytolith characteristics.

Acknowledgments

Thanks are due to Department of Science and Technology (DST) for providing the Women Scientist Scheme Project (Order no. SR/WOS-A/LS-238/2009 dated 06.09.2010). The authors also thank Dr. A. K. Pradeep, Herbarium Curator, CALI, for providing the specimens needed for the present study.

References

[1] L. H. P. Jones and K. A. Handreck, "Silica in soils, plants, and animals," *Advances in Agronomy*, vol. 19, pp. 107–149, 1967.

[2] E. Blackman, "Opaline silica in the range grasses of southern Alberta," *Canadian Journal of Botany*, vol. 49, pp. 769–781, 1971.

[3] I. Rovner, "Plant opal-phytolith analysis. Major advances in archaeobotanical research," in *Advances in Archaeological Method and Theory*, M. Schiffer, Ed., vol. 6, Academic Press, New York, NY, USA, 1983.

[4] J. Bartolome, S. E. Klukkert, and W. J. Barry, "Opal phytoliths as evidence for displacement of native Californian grassland," *Madrono*, vol. 33, pp. 217–222, 1986.

[5] P. C. Twiss, "Grass-opal phytoliths as climatic indicators of the Great Plains Pleistocene," in *Quarternary Environments of Kansas*, W. C. Johnson, Ed., pp. 179–188, Kansas Geological Survey, Lawrence, Kan, USA, 1987.

[6] D. Piperno, *Phytolith Analysis : An Archaeological and Geological Perspective*, Academic Press, New York, NY, USA, 1988.

[7] R. F. Fisher, C. N. Bourn, and W. F. Fisher, "Opal phytoliths as an indicator of the floristics of prehistoric grasslands," *Geoderma*, vol. 68, no. 4, pp. 243–255, 1995.

[8] A. J. Alexandre, D. Meunier, A. M. Lézine, A. Vincens, and D. Schwartz, "Phytoliths: indicators of grassland dynamics during the late Holocene in intertropical Africa," *Palaeogeography, Palaeoclimatology, Palaeoecology*, vol. 136, no. 1–4, pp. 213–229, 1997.

[9] J. A. Carter, "Late devonian, permian and triassic phytoliths from antarctica," *Micropaleontology*, vol. 45, no. 1, pp. 56–61, 1999.

[10] A. L. Carnelli, J. P. Theurillat, and M. Madella, "Phytolith types and type-frequencies in subalpine-alpine plant species of the European Alps," *Review of Palaeobotany and Palynology*, vol. 129, no. 1-2, pp. 39–65, 2004.

[11] L. Gallego and R. A. Distel, "Phytolith assemblages in grasses native to central Argentina," *Annals of Botany*, vol. 94, no. 6, pp. 865–874, 2004.

[12] D. Barboni and L. Bremond, "Phytoliths of East African grasses: an assessment of their environmental and taxonomic significance based on floristic data," *Review of Palaeobotany and Palynology*, vol. 158, no. 1-2, pp. 29–41, 2009.

[13] C. R. Metcalfe, *Anatomy of the Monocotyledons. 1. Gramineae*, Claredon Press, Oxford, UK, 1960.

[14] A. L. Ollendorf, S. C. Mulholland, and G. J. Rapp, "Phytolith analysis as a means of plant identification: arundo donax and Phragmites communis," *Annals of Botany*, vol. 61, no. 2, pp. 209–214, 1988.

[15] T. B. Ball, J. D. Brotherson, and J. S. Gardner, "A typological and morphometric study of variation in phytoliths from Triticum monococcum," *Canadian Journal of Botany*, vol. 71, pp. 1182–1192, 1993.

[16] S. S. Whang, K. Kim, and W. M. Hess, "Variation of silica bodies in leaf epidermal long cells within and among seventeen species of Oryza (Poaceae)," *American Journal of Botany*, vol. 85, no. 4, pp. 461–466, 1998.

[17] S. Krishnan, N. P. Samson, P. Ravichandran, D. Narasimhan, and P. Dayanandan, "Phytoliths of Indian grasses and their potential use in identification," *Botanical Journal of the Linnean Society*, vol. 132, no. 3, pp. 241–252, 2000.

[18] P. B. Kaufman, P. Dayanandan. Y. Takeoka, W. C. Bigelow, J. D. Jones, and R. K. Iller, "Silica in shoots of higher plants," in *Silicon and Silicious Structure in Biological System*, T. L. Simpson and B. E. Voliani, Eds., pp. 409–449, Springer, New York, NY, USA, 1981.

[19] L. H. P. Jones and K. A. Handreck, "Studies of silica in the oat plant—III. Uptake of silica from soils by the plant," *Plant and Soil*, vol. 23, no. 1, pp. 79–96, 1965.

[20] E. Blackman, "Observation on the development of the silica cells of the leaf sheath of wheat," *Canadian Journal of Botany*, vol. 47, pp. 827–838, 1969.

[21] J. A. Raven, "The transport and function of silica in plants," *Biological Reviews of the Cambridge Philosophical Society*, vol. 58, pp. 179–207, 1983.

[22] D. R. Piperno, "A comparison and differentiation of phytoliths from maize and wild grases. Use of morphological criteria," *American Antiquity*, vol. 49, pp. 361–383, 1984.

[23] D. M. Pearsall, K. Chandler-Ezell, and A. Chandler-Ezell, "Maize can still be identified using phytoliths: response to Rovner," *Journal of Archaeological Science*, vol. 31, no. 8, pp. 1029–1038, 2004.

[24] Y. Zheng, A. Matsui, and H. Fujiwara, "Phytoliths of rice detected in the neolithic sites in the valley of the Taihu Lake in China," *Environmental Archaeology*, vol. 8, no. 2, pp. 177–183, 2003.

[25] W. D. Clayton and S. A. Renvoize, *Genera Graminum, Grasses of the World*, Her Majesty's Stationery Office, London, UK, 1986.

[26] M. Madella, A. Alexandre, and T. Ball, "International code for phytolith nomenclature 1.0," *Phytolitharien*, vol. 15, pp. 7–16, 2005.

19

Photosynthesis and Nitrogen Metabolism of *Nepenthes alata* in Response to Inorganic NO_3^- and Organic Prey N in the Greenhouse

Jie He and Ameerah Zain

Natural Sciences and Science Education Academic Group, National Institute of Education, Nanyang Technological University, 1 Nanyang Walk, Singapore 637 616

Correspondence should be addressed to Jie He, jie.he@nie.edu.sg

Academic Editors: E. Collakova, M. Kwaaitaal, and D. Zhao

This study investigates the relative importance of leaf carnivory on *Nepenthes alata* by studying the effect of different nitrogen (N) sources on its photosynthesis and N metabolism in the greenhouse. Plants were given either inorganic NO_3^-, organic N derived from meal worms, *Tenebrio molitor*, or both NO_3^- and organic N for a period of four weeks. Leaf lamina (defined as leaves) had significant higher photosynthetic pigments and light saturation for photosynthesis compared to that of modified leaves (defined as pitchers). Maximal light saturated photosynthetic rates (P_{max}) were higher in leaves than in pitchers. Leaves also had a higher light utilization than that of pitchers. Both leaves and pitchers of plants that were supplied with both inorganic NO_3^- and organic prey N had a similar photosynthetic capacity and N metabolism compared to plants that were given only inorganic NO_3^-. However, adding organic prey N to the pitchers enhanced both photosynthetic capacity and N metabolism when plants were grown under NO_3^- deprivation condition. These findings suggest that organic prey N is essential for *N. alata* to achieve higher photosynthetic capacity and N metabolism only when plants are subjected to an environment where inorganic N is scarce.

1. Introduction

Carnivorous plants are restricted to environments with an abundant supply of water and light but are poor in nutrients [1]. Although plants are autotrophic with respect to reduced carbon, they must scavenge nitrogen (N) and other minerals from the environment, usually from the soil through uptake by their roots. On the other hand, plant carnivory is an alternate and efficient means to acquire nutrients in nutrient-poor habitats [2]. For instance, preys caught in the *Nepenthes* pitchers are digested in a pool of digestive enzymes in the pitcher where glands function to perceive chemical stimuli, secrete digestive enzymes, and absorb nutrients for plant growth and development [3].

Nepenthes are tropical pitcher plants, and there are approximately 90 species in the genus *Nepenthes*. Osunkoya et al. [4] suggested that most *Nepenthes* species are N-(but not P or K) limited, and thus have evolved the pitcher to assist in their uptake of N. The leaf morphology of the different

Nepenthes species is similar with a photosynthetic lamina and a tendril to which a pitcher is attached. *N. alata* is a pitcher plant that efficiently captures, retains, and digests predominantly insect prey in highly modified leaves, and pitchers [5]. The pitcher consists of the lid, peristome (upper rim of pitcher) which attracts prey, a waxy zone that is involved in trapping prey, and a digestive zone which digests prey [6]. Ellison and Gotelli [2] reported that both leaves and pitchers of the *Nepenthes* plants can photosynthesize. Clarke [7] found that *Nepenthes* fail to produce pitchers if the light or humidity is too low, or nutrient availability is too high.

In Singapore, *N. alata* is a popular ornamental plant that can be found in many home gardens and has commercial value. In nature, these large pitcher plants usually grow in soils consisting of low N, as the pitchers of these plants can obtain N from organic sources like insects or small animals. Usually, *N. alata* plantlets are obtained from tissue culture stock in the nurseries. Some growers of *Nepenthes* feed the pitchers of this plant with meal worms to provide the

Photosynthesis and Nitrogen Metabolism of Nepenthes alata in Response to Inorganic NO$_3^-$ and Organic Prey N in the Greenhouse

165

additional source of N. However, there are only a few studies that have examined directly the linkage between inorganic N uptake from the soil by the carnivorous plants and photosynthetic rate [5]. In addition, there is little information available on the overall prey and inorganic nutrient acquisition for the carnivorous plants such as *Nepenthes* [5].

This project focused on the relative importance of leaf carnivory and root nutrition mainly with inorganic NO$_3^-$ on photosynthesis and N metabolism of *N. alata* in the greenhouse. Using NO$_3^-$ and prey-derived organic N source, this project aimed to compare the photosynthetic characteristics and light utilization between leaf and pitcher and to study the effects of NO$_3^-$ and prey on the photosynthesis and N metabolism of leaf and pitcher. The parameters studied were photosynthetic O$_2$ evolution, chlorophyll (Chl) fluorescence, photosynthetic pigments, total reduced N content, and soluble protein content. Understanding the contributions of inorganic NO$_3^-$ and organic prey N to photosynthesis and N metabolism, horticulturalists can select the optimal fertilizer required for cultivation of *N. alata*.

2. Materials and Methods

2.1. Plant Material. *N. alata* plants with 7-8 leaves and 4-5 pitchers were obtained from a commercial nursery. They were transplanted to pots (15 cm diameter) containing sand and vermiculite (1 : 1), and each pot had only one plant. The pitchers were emptied and washed with distilled water. An amount of 10 mL of distilled water was added to each of the pitchers which were then plugged with glass wool to prevent colonization by common pitcher inhabitants and capture of prey. All plants were acclimatized for one month in the greenhouse under a maximal photosynthetic photon flux density (PPFD) of 600–700 μmol m^{-2} s^{-1}. The daily ambient temperature ranged from 24 to 33°C. All plants were watered daily with tap water and supplied with nutrient solution based on full-strength Netherlands Standard Composition every alternate day. This nutrient solution contains full NO$_3^-$.

2.2. Experimental Design for Different N Treatments. After the plants were acclimatized under the previously stated conditions for one month, all yellow leaves and dead pitchers were removed. Each plant had 6 fully expanded leaves with fully developed pitchers and two young leaves without pitchers. The pitchers were emptied and washed with distilled water. An amount of 10 mL of distilled water was again added to each of the pitchers which were then plugged with glass wool to prevent colonization by common pitcher inhabitants and capture of prey. In order to study the photosynthetic characteristics and N metabolism in response to organic prey and inorganic NO$_3^-$, for each plant, six alive meal worms (*Tenebrio molitor*) (6 × 0.4 g) were added to each of the three pitchers, while the other three pitchers of the same plant did not receive any prey. After adding the meal worms, plants were divided into two groups: one group was watered nutrient solution with full NO$_3^-$, while the other group was watered with nutrient solution without NO$_3^-$ every alternate day. Therefore, there are four treatments: (1) NO$_3^-$,

(2) NO$_3^-$ + prey, (3) prey, and (4) no NO$_3^-$, no prey. The durations of different treatments were two, three, and four weeks, respectively. Significant differences in responses to different treatments were observed after four weeks. Thus, only data obtained after four weeks were presented.

2.3. Measurement of Chl Fluorescence. Electron transport rate (ETR), photochemical quenching (qP), and nonphotochemical quenching (qN) of Chl fluorescence were determined from both leaves and pitchers using the Imaging PAM Chl Fluorometer (Waltz, Effeltrich, Germany) at 25°C under different PPFDs in the laboratory as described by He et al. [8].

2.4. Measurement of Photosynthetic O$_2$ Evolution. The photosynthetic O$_2$ evolution of leaf and pitcher were determined with a Hansatech leaf disc O$_2$ electrode (King's Lynn, Norfolk, UK). Each leaf and pitcher section was placed in saturating CO$_2$ conditions (1% CO$_2$ from 1 M carbonate/bicarbonate buffer, pH 9). Leaf or pitcher section was illuminated, starting from the lowest photosynthetic photon flux density, PPFD (34 μmol m^{-2} s^{-1}), to the highest (1000 μmol m^{-2} s^{-1}). The photosynthetic light response curve was obtained by plotting the O$_2$ evolution rates against respective light intensity. Maximal photosynthetic O$_2$ evolution rates (P_{max}) of both leaf and pitcher were measured after two weeks of treatments under a PPFD of 1000 μmol m^{-2} s^{-1} at 25°C.

2.5. Measurement of Photosynthetic Pigments. Fresh samples of leaf or pitcher of 0.05 g were weighed and cut into smaller pieces. Total Chl and carotenoid were extracted from these samples with dimethylformamide and quantified using a spectrophometer following the procedure of Wellburn [9] at wavelengths of 480, 647, and 664 nm.

2.6. Measurement of Total Reduced N Concentration (TRN). Dry samples of 0.05 g of leaf and pitcher were placed into a digestion tube with a Kjeldahl tablet and 5 mL of concentrated sulphuric acid according to Allen [10]. The mixture was then digested about 60 min until clear. After the digestion was completed, the mixture was allowed to cool for 30 min, and TRN concentration was determined by a Kjeltec 2030 analyser unit (Höganäs, Sweden).

2.7. Total Soluble Protein (TSP) Extraction and Determination. Samples of 1 g were rapidly frozen in liquid nitrogen after weighing and stored at −80°C until used. Each leaf and pitcher sample was ground to fine powder in liquid N with pestle and mortar. After which, 1 mL of 100 mM Bicine-KOH (pH 8.1), 20 mM MgCl$_2$, and 2% PVP buffer were added [11]. After centrifugation (100,000 g, 30 min at 4°C), 4 mL of acetone was then added to 1 mL of the supernatant collected and centrifuged further for 10 min at 4000 rpm. Total soluble protein was extracted using the method described by Lowry et al. [11].

2.8. Statistical Analysis. For Table 1 and Figures 1 and 2, a *t*-test was used to test for differences between leaves and

TABLE 1: Total Chl content, Chl a/b ratio, total carotenoid content, and Chl/carotenoid ratio of leaf and pitcher of *N. alata*. The means and standard errors of four readings are given for each pigment. Letters represent comparison between leaf and pitcher. Any two means having a common letter are not significantly different.

Photosynthetic pigments	Leaf	Pitcher
Total Chl content (μg/g FW)	1273.12 ± 46.80^a	266.39 ± 29.42^b
Chl a/b ratio	2.95 ± 0.04^a	2.01 ± 0.06^b
Total carotenoid (μg/g FW)	207.11 ± 8.28^a	44.80 ± 5.00^b
Chl/Carotenoid ratio	6.15 ± 0.11^a	5.95 ± 0.04^a

pitchers. For Figures 3 and 4, ANOVA was used to discriminate means across all four treatments, followed by using Tukey's multiple comparison test. The difference between treatment means was considered significant at $P < 0.05$. All statistical analyses were carried out using Minitab software (Minitab, Inc., release 15, 2007).

3. Results

3.1. Comparative Studies on Photosynthetic Characteristics between Leaves and Pitchers. To compare the photosynthetic characteristics between leaves and pitchers, all plants were grown under the conditions described in Section 2.1 for one month. The leaves had significant higher total Chl content, Chl a/b ratio, and total carotenoid content compared to that of the pitchers (Table 1, $P < 0.05$). However, there was no significant difference in the Chl/carotenoid ratio between leaves and the pitcher. Light saturation point of photosynthetic O_2 evolution for leaf was achieved at PPFDs of about 600–800 μmol m^{-2} s^{-1} (Figure 1). For the pitchers, however, light saturation point was much lower, at PPFDs of about 200–400 μmol m^{-2} s^{-1}. These results indicate that the leaves have higher photosynthetic capacities compared to those of pitchers. Light utilization of leaf and pitcher was determined by qP, qN, and ETR. The leaves had qP values of about 0.9 to 0.6 under PPFD of 15–200 μmol m^{-2} s^{-1}. A drastic decrease of qP in the leaf was observed at PPFD higher than 200 μmol m^{-2} s^{-1} with values reaching zero at 1585 μmol m^{-2} s^{-1}. The pitcher had lower qP values compared to the leaves of 0.88 to 0.42 at PPFD of 50–200 μmol m^{-2} s^{-1}. The qP of zero was observed at about PPFD of 900 μmol m^{-2} s^{-1}. The ETR values of the leaves increased sharply to a maximum of 60 μmol electrons m^{-2} s^{-1} at a PPFD of 500 μmol m^{-2} s^{-1} after which the ETR values decreased gradually (Figure 2(b)). Similarly, the ETR of the pitcher increased rapidly to a maximum of 22 μmol electrons m^{-2} s^{-1} at a PPFD of 200 μmol m^{-2} s^{-1} after which it decreased gradually. These results show that the pitchers have a lower light utilisation compared to that of the leaves. On the other hand, the leaves had a gradual increase in qN from PPFD of 25–400 μmol m^{-2} s^{-1} after which it plateaued to about 0.8. Similarly, the pitcher had a rapid increase in qN from PPFD of 25–400 μmol m^{-2} s^{-1} after which it plateaued to about 0.6 (Figure 2(a)). Compared to that of pitcher, the higher qN levels in leaves indicate that higher amount of light energy could be dissipated as heat.

3.2. Responses of Photosynthesis and N Metabolism of Leaves and Pitchers to Inorganic NO_3^- and Organic Prey. Four weeks after different N treatments, there were no significant

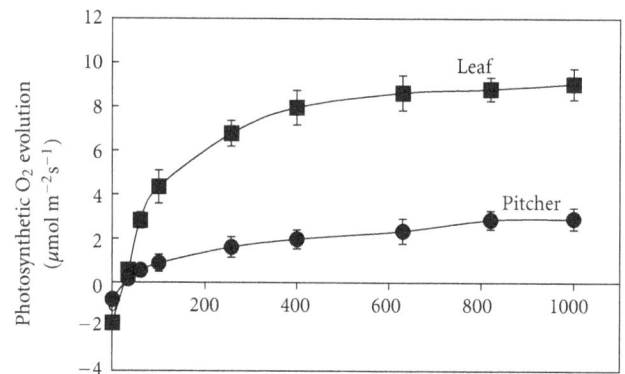

FIGURE 1: Photosynthetic light response curve of leaf and pitcher of *N. alata* under different PPFDs. Means of 4 measurements from 4 different leaves and 4 different pitchers. Vertical bars represent standard errors. When the standard error bars cannot be seen, they are smaller than the symbols.

differences in P_{max} and Chl content in the leaves of plants supplied with only NO_3^- to the roots and those supplied with both NO_3^- to the roots and prey added to the pitcher (Figures 3(a) and 3(c)). However, leaves that were supplied with prey to their pitchers only but without NO_3^- to the roots had significant lower P_{max} and Chl content ($P < 0.05$). Leaves that had neither NO_3^- supplied to the roots nor prey added to the pitchers exhibited the lowest P_{max} and Chl content (Figures 3(a) and 3(c), $P < 0.05$). After different N treatments, the changes of P_{max} in pitchers were very similar to those of leaves but the values of P_{max} were much lower compared to those of leaves (Figure 3(b)). However, lower total Chl content was only observed in pitchers of those plants without NO_3^- and prey (Figure 3(d), $P < 0.05$). Changes in TRN and TSP determined from the same leaves that were used to measure P_{max} and total Chl content are shown in Figure 4. The differences in TRN concentrations of both leaves and pitchers were very similar to those of P_{max} after different N treatments for 4 weeks (Figures 4(a) and 4(b)). It is also interesting to see that TSP concentrations of both leaves and pitchers showed the same patterns of Chl content after different N treatments for 4 weeks (Figures 4(c) and 4(d)).

4. Discussion

Carnivorous plants normally grow in moist, nutrient-poor soils. In order to adapt to an environment where critical

Photosynthesis and Nitrogen Metabolism of Nepenthes alata in Response to Inorganic NO₃⁻ and Organic Prey N in the Greenhouse

167

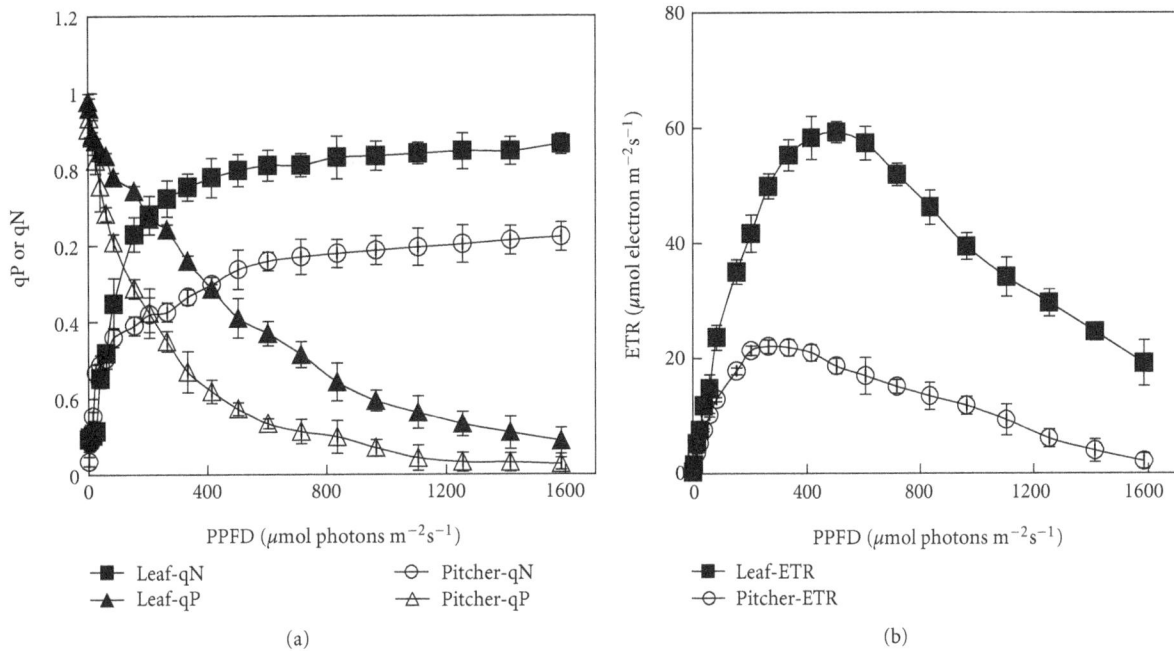

FIGURE 2: Changes in qP and qN (a) and ETR (b) of leaf and pitcher of *N. alata* under different PPFDs. Means of 4 measurements from 4 different leaves and 4 different pitchers. Vertical bars represent standard errors. When the standard error bars cannot be seen, they are smaller than the symbols.

nutrients are scarce and where light is not limiting, carnivorous plants have evolved modified leaves specialized for capturing animals and digesting the preys to acquire nutrients [1, 12–15]. In this study, using the carnivorous tropical pitcher plant, *N. alata*, it was demonstrated that the leaves have much higher P_{max} compared to that of the pitchers of the same plants (Figure 1). It was also reported by others that P_{max} of traps is usually lower than that of other noncarnivorous leaves of the same plants [1]. This could be due to the fact that leaves have higher levels of photosynthetic pigments compared to the pitcher (Table 1) and high efficiency of light energy utilisation and heat dissipation measured by qP, qN, and ETR (Figure 2). Pavlovič et al. [16] also reported that Chl content in two *Nepenthes* species, *N. alata* and *N. mirabilis*, was higher in the lamina (leaves) than in the pitcher. The red tint of *Nepenthes* pitchers suggests that they might not have much Chl, and this might lead to low P_{max}. Because P_{max} is positively correlated with N concentration and stomatal conductance, it is hypothesized that there is lower N concentration (Figure 4) and lower stomata density in pitchers than in the leaves. Most carnivorous plants exhibit very low rates of photosynthesis [5]. In the present study, although P_{max} was significantly higher in leaves than in pitchers, the value of P_{max} about $10 \, \mu mol \, m^{-2} \, s^{-1}$ was much lower compared to that of most C₃ plants. According to Ellison [5], photosynthetic rate of carnivorous plants was about 2 to 5 times lower than that of other noncarnivorous plants. Our finding of P_{max} of *N. alata* agrees with Ellison's report [5]. Low photosynthetic rate reflects the relatively low growth rate of *N. alata* plants (data not shown). Most carnivorous plants are at a competitive disadvantage in their

habitats due to a lower net photosynthetic rate when light is limiting, and the availability of soil nutrients is poor based on the cost-benefit model [1]. However, the relationship between photosynthetic performance of carnivorous plants and their carnivory is complex and ambiguous [3].

According to the ecological cost-benefit relationships, carnivory of carnivorous plants grown under their natural habitat of limited light and poor nutrients could lead to increasing photosynthetic rate if they were provided with a greater mineral nutrient availability [1]. *N. alata* is one of the most popular *Nepenthes* species in cultivation. According to our observation, *N. alata* plants that are cultivated in the local nursery under high light supplied with fertilizer grow well with numerous pitchers. These observations lead to the question of the relative importance of leaf carnivory and root nutrition mainly with inorganic NO₃⁻ on photosynthesis and N metabolism. In the present study, when full inorganic NO₃⁻ was supplied to the roots of *N. alata* plants, adding preys to the pitchers did not increase P_{max} and total Chl content of both leaves and pitcher (Figure 3). These results indicate that to improve plant growth, it seems sufficient to provide N in the form NO₃⁻ to the roots. This confirms the idea that carnivory is not indispensable for greenhouse growing carnivorous plants, but it is almost indispensable for carnivorous plants in natural habitats [17]. Prey captured in the pitcher could contribute 10–90% of the N budget of *Nepenthes* plants [16]. N in the end can be considered as limiting primary productivity [18]. For instance, when *N. alata* plants were grown under NO₃⁻ deprivation condition, adding preys to the pitchers increased P_{max} of both leaves and pitchers compared to those without preys. However, the

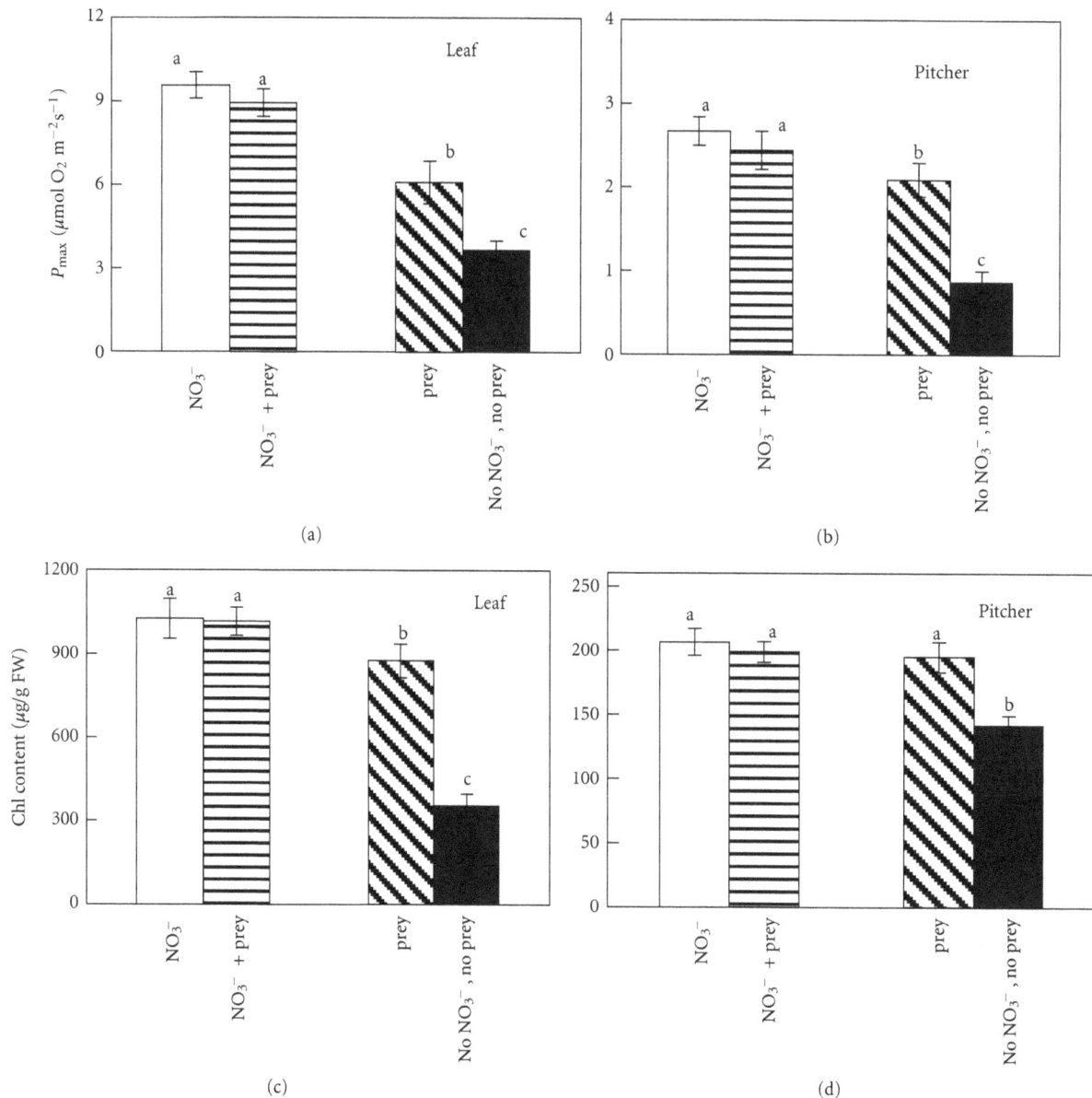

FIGURE 3: Changes in P_{max} (a), (b) and Chl content (c), (d) of leaves (a), (c) and pitchers (b), (d) of *N. alata* after 4 weeks of different N treatments. Means of four measurements were obtained from four different leaves from four different plants. Means with different letters above the columns are statistically different ($P < 0.05$) as determined by Tukey's multiple comparison test.

P_{max} values of plants supplied with only preys were significantly lower than those plants supplied with NO_3^- or both NO_3^- and preys (Figure 3). These findings suggest that in the event of low inorganic NO_3^- availability, the feeding of prey to the pitchers would bring about the same positive effect on growth (Figure 3). Chandler and Anderson [19] reported greater absolute growth in the presence of insects at low NO_3^- concentration in *Drosera whittakeri* and *Drosera binata* over a period of one growing season. In most *Sarracenia* species and in *Darlingtonia californica*, prey addition significantly increases P_{max} [1]. Ellison and Gotelli [20] concluded that there was an increase in photosynthetic rates followed by an addition of organic N in *Sarracenia*

purpurea. The result is also consistent with another study conducted by Wakefield et al. [21] on *Sarracenia purpurea*. However, benefits of carnivory including an increased rate of photosynthesis are often conflicting [20–22]. Obviously, the photosynthetic effect of prey addition is quite different in different carnivorous plant species [20–25]. It was reported that the total Chl content of the younger *Sarracenia* pitchers was positively and significantly correlated with the feeding level of prey. However, Chl content in *Sarracenia* was not correlated with either foliar N [23]. In the present study, even with additional prey N, leaves of *N. alata* plants had much higher P_{max} and total Chl content compared to their pitchers; feeding preys to the pitcher of plants that

Photosynthesis and Nitrogen Metabolism of Nepenthes alata in Response to Inorganic NO$_3^-$ and Organic Prey N in the Greenhouse

169

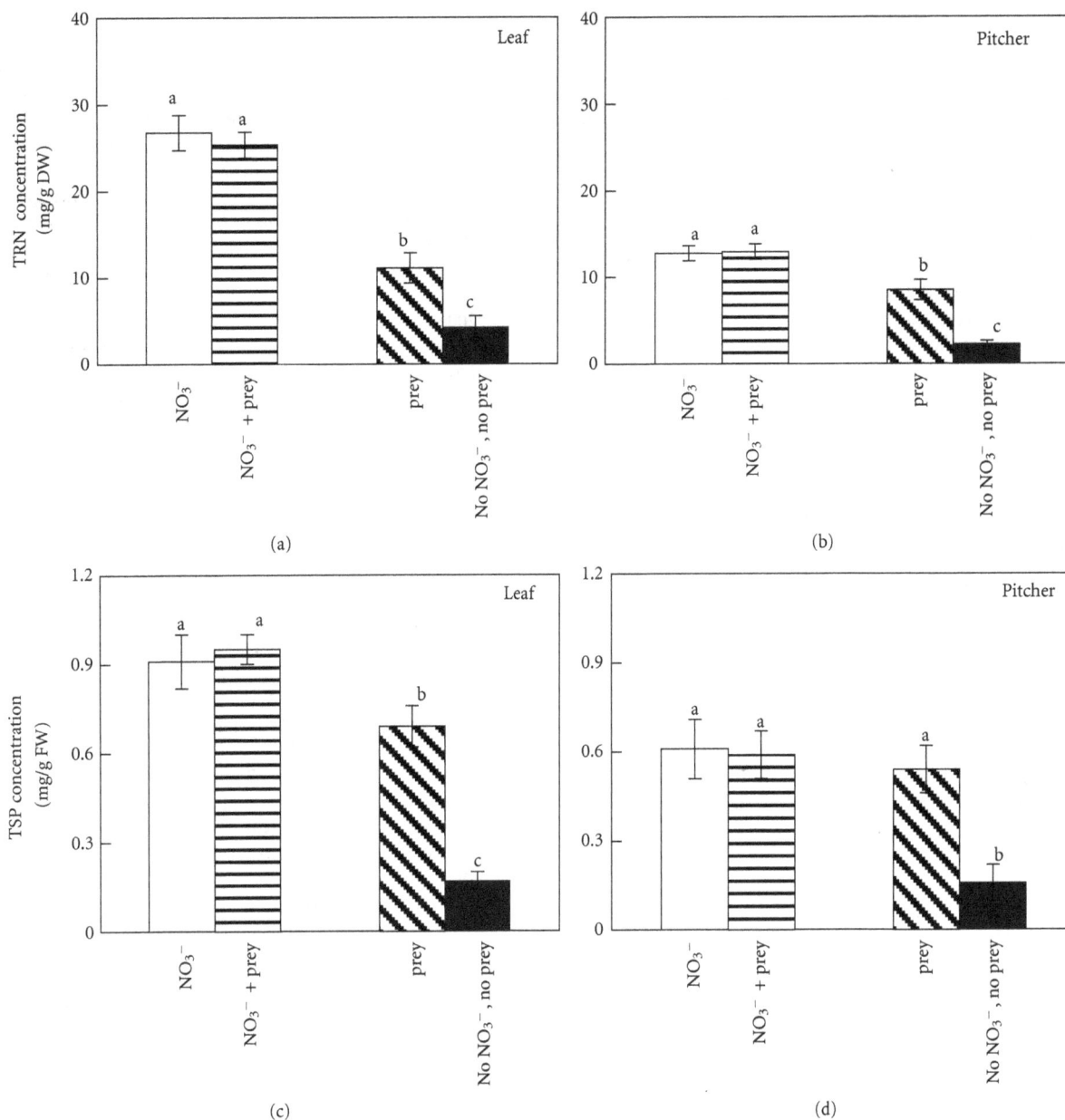

FIGURE 4: Changes in TRN (a), (b) and TSP (c), (d) concentration of leaves (a), (c) and pitchers (b), (d) of *N. alata* following 4 weeks of different N treatments. Means of 4 measurements were obtained from 4 different leaves from 4 different plants. Means with different letters above the columns are statistically different ($P < 0.05$) as determined by Tukey's multiple comparison test.

were supplied with full NO$_3^-$ did not enhance Chl further (Figures 3(c) and 3(d)). These results may suggest that the function of the pitchers is mainly for prey capture and not for photosynthesis, although they have photosynthetic pigments [2]. This could therefore explain why the leaves utilize light much more efficiently compared to the pitchers as demonstrated by the higher ETR, qP, and higher qN values of the leaves compared to that of the pitcher (Figure 2, Table 1).

It is well known that in noncarnivorous plants, with increasing inorganic NO$_3^-$ content, the photosynthetic capacity of leaves increases [26, 27]. However, there is very little information on the relationship between the amount

of organic N (prey), the rate of photosynthesis, and N metabolism in carnivorous plants. Typically, a reduction of TRN concentration in plants would lead to lower photosynthetic rates and vice versa. In this study, TRN was significantly higher in leaves than in pitchers (Figure 4). This could further explain why leaves had higher photosynthetic capacity and photosynthetic pigments compared to pitchers. In addition, differences in TRN concentrations of both leaves and pitchers among the different N-treated plants were very similar to those of P_{max}, indicating the close relationship between TRN and P_{max}. N availability obviously directly affects the amount of soluble proteins available in the plant, since N is a major element which makes up protein compounds [26, 27]. Plants require N (in the form of NO_3^-) to synthesize soluble proteins which in turn is required for the synthesis of RuBisCo that is the key enzyme in photosynthetic CO_2 fixation [27, 28]. This could explain why leaves, which had significantly higher TSP compared to the pitcher (Figure 4), had higher photosynthetic capacities (Figure 2). However, how much of the prey contributes to the N actually present in the foliage of *Nepenthes*? Studies like Ellison and Gotelli [20] suggest that in the field, the majority of foliar N is derived from substrate sources rather than carnivory. There are other conflicting findings like Moran et al. [18] who estimated that prey contributed from 54% to 68% of the total foliar N in *Nepenthes*. Schulze et al. [29] concluded that the total N concentration in leaves of *Dionaea muscipula* growing in areas where there was little insect availability was much lower than in plants which received a large amount of insect prey. When measuring NO_3^- reductase (NR) activity, it was shown that negative interactions can exist between organic and inorganic N sources. This was exhibited when insect feeding of greenhouse *Drosera binata* grown in complete solution led to 30–50% lower shoot NR activity compared to unfed plants [18]. Effects of organic prey on NR activity merit our future study.

5. Conclusion

The present study showed that the leaves of *N. alata* had higher photosynthetic capacity and light utilization compared to the pitchers. *N. alata* plants with matured pitchers supplied with both inorganic NO_3^- and organic prey N had a similar P_{max} compared to plants supplied with only inorganic NO_3^-, suggesting that carnivory is not indispensable for greenhouse growing carnivorous plants.

Abbreviations

Chl: Chlorophyll
ETR: Electron transport rate
N: Nitrogen
NO_3^-:   Nitrate
P_{max}:   Maximal light saturated photosynthetic rates
PPFD: Photosynthetic photon flux density
qN: Nonphotochemical quenching
qP: Photochemical quenching
TRN: Total reduced N
TSP: Total soluble protein.

Acknowledgment

This project was supported by teaching materials' vote of National Institute of Education, Nanyang Technological University, Singapore.

References

[1] T. J. Givnish, E. L. Burkhardt, R. E. Happel, and J. D. Weintraub, "Carnivory in the bromeliad *Brocchinia reducta*, with a cost/benefit model for the general restriction of carnivorous plants to sunny, moist, nutrient-poor habitats," *American Naturalist*, vol. 124, no. 4, pp. 479–497, 1984.

[2] A. M. Ellison and N. J. Gotelli, "Evolutionary ecology of carnivorous plants," *Trends in Ecology and Evolution*, vol. 16, no. 11, pp. 623–629, 2001.

[3] B. E. Juniper, R. J. Robins, and D. M. Joel, *The Carnivorous plants*, Academic Press, London, UK, 1989.

[4] O. O. Osunkoya, S. D. Daud, B. Di-Giusto, F. L. Wimmer, and T. M. Holige, "Construction costs and physico-chemical properties of the assimilatory organs of Nepenthes species in Northern Borneo," *Annals of Botany*, vol. 99, no. 5, pp. 895–906, 2007.

[5] A. M. Ellison, "Nutrient limitation and stoichiometry of carnivorous plants," *Plant Biology*, vol. 8, no. 6, pp. 740–747, 2006.

[6] E. Gorb, V. Kastner, A. Peressadko et al., "Structure and properties of the glandular surface in the digestive zone of the pitcher in the carnivorous plant *Nepenthes ventrata* and its role in insect trapping and retention," *Journal of Experimental Biology*, vol. 207, no. 17, pp. 2947–2963, 2004.

[7] C. M. Clarke, *Nepenthes of Borneo*, Natural History, Kota Kinabalu, Malaysia, 1997.

[8] J. He, B. H. G. Tan, and L. Qin, "Source-to-sink relationship between green leaves and green pseudobulbs of C3 orchid in regulation of photosynthesis," *Photosynthetica*, vol. 49, no. 2, pp. 209–218, 2011.

[9] A. R. Wellburn, "The spectral determination of chlorophylls a and b, as well as total carotenoids, using various solvents with spectrophotometers of different resolution," *Journal of Plant Physiology*, vol. 144, no. 3, pp. 307–313, 1994.

[10] S. E. Allen, "Analysis of vegetation and other organic materials," in *Chemical Analysis of Ecological Materials*, S. E. Allen, Ed., pp. 46–61, Blackwell, Oxford, UK, 1989.

[11] O. H. Lowry, N. J. Rosebrough, A. L. Farr, and R. J. Randall, "Protein measurement with the Folin phenol reagent," *The Journal of Biological Chemistry*, vol. 193, no. 1, pp. 265–275, 1951.

[12] V. A. Albert, S. E. Williams, and M. W. Chase, "Carnivorous plants: phylogeny and structural evolution," *Science*, vol. 257, no. 5076, pp. 1491–1495, 1992.

[13] R. J. Bayer, L. Hufford, and D. E. Soltis, "Phylogenetic relationships in Sarraceniaceae based on rbcL and ITS sequences," *Systematic Botany*, vol. 21, no. 2, pp. 121–134, 1996.

[14] K. M. Cameron, K. J. Wurdack, and R. W. Jobson, "Molecular evidence for the common origin of snap-traps among carnivorous plants," *American Journal of Botany*, vol. 89, no. 9, pp. 1503–1509, 2002.

[15] A. M. Ellison and N. J. Gotelli, "Energetics and the evolution of carnivorous plants—Darwin's "most wonderful plants in the world"," *Journal of Experimental Botany*, vol. 60, no. 1, pp. 19–42, 2009.

[16] A. Pavlovič, E. Masarovičová, and J. Hudák, "Carnivorous syndrome in Asian pitcher plants of the genus *Nepenthes*," *Annals of Botany*, vol. 100, no. 3, pp. 527–536, 2007.

[17] L. Adamec, "Photosynthetic characteristics of the aquatic carnivorous plant *Aldrovanda vesiculosa*," *Aquatic Botany*, vol. 59, no. 3-4, pp. 297–306, 1997.

[18] J. A. Moran, M. A. Merbach, N. J. Livingston, C. M. Clarke, and W. E. Booth, "Termite prey specialization in the pitcher plant *Nepenthes albomarginata*—evidence from stable isotope analysis," *Annals of Botany*, vol. 88, no. 2, pp. 307–311, 2001.

[19] G. E. Chandler and J. W. Anderson, "Studies on the nutrition and growth of *Drosera* species with reference to the carnivorous habit," *New Phytologist*, vol. 76, pp. 129–141, 1976.

[20] A. M. Ellison and N. J. Gotelli, "Nitrogen availability alters the expression of carnivory in the northern pitcher plant,

Photosynthesis and Nitrogen Metabolism of Nepenthes alata in Response to Inorganic NO₃⁻ and Organic
Prey N in the Greenhouse

171

Sarracenia purpurea," Proceedings of the National Academy of
Sciences of the United States of America, vol. 99, no. 7, pp. 4409–
4412, 2002.

[21] A. E. Wakefield, N. J. Gotelli, S. E. Wittman, and A. M. Ellison,
"Prey addition alters nutrient stoichiometry of the carnivorous
plant Sarracenia purpurea," Ecology, vol. 86, no. 7, pp. 1737–
1743, 2005.

[22] P. S. Karlsson, K. O. Nordell, B. Å. Carlsson, and B. M.
Svensson, "The effect of soil nutrient status on prey utilization
in four carnivorous plants," Oecologia, vol. 86, no. 1, pp. 1–7,
1991.

[23] E. J. Farnsworth and A. M. Ellison, "Prey availability directly
affects physiology, growth, nutrient allocation and scaling
relationships among leaf traits in 10 carnivorous plant
species," Journal of Ecology, vol. 96, no. 1, pp. 213–221, 2008.

[24] A. M. Ellison and E. J. Farnsworth, "The cost of carnivory
for Darlingtonia californica (Sarraceniaceae): evidence from
relationships among leaf traits," American Journal of Botany,
vol. 92, no. 7, pp. 1085–1093, 2005.

[25] M. Méndez and P. S. Karlsson, "Costs and benefits of carnivory
in plants: insights from the photosynthetic performance of
four carnivorous plants in a subarctic environment," Oikos,
vol. 86, no. 1, pp. 105–112, 1999.

[26] C. Field and H. A. Mooney, "The photosynthesis-nitrogen in
wild plants," in The Economy of Plant Form and Function, T.
J. Givinis, Ed., pp. 25–55, Cambridge University Press, Cambridge, UK, 1986.

[27] J. R. Evans, "Photosynthesis and nitrogen relationships in
leaves of C₃ plants," Oecologia, vol. 78, no. 1, pp. 9–19, 1989.

[28] D. Pankovic, M. Plesnicar, I. Arsenijevic-Maksimovic, N.
Petrovic, Z. Sakac, and R. Kastori, "Effects of nitrogen nutrition on photosynthesis in cd-treated sunflower plants," Annals
of Botany, vol. 86, pp. 841–847, 2000.

[29] W. Schulze, E. D. Schulze, I. Schulze, and R. Oren, "Quantification of insect nitrogen utilization by the venus fly trap
Dionaea muscipula catching prey with highly variable isotope
signatures," Journal of Experimental Botany, vol. 52, no. 358,
pp. 1041–1049, 2001.

20

Moss and Soil Substrates Interact with Moisture Level to Influence Germination by Three Wetland Tree Species

Alexander Staunch, Marie Redlecki, Jessica Wooten, Jonathan Sleeper, and Jonathan Titus

Department of Biology, Jewett Hall, SUNY-Fredonia, Fredonia, NY 14063, USA

Correspondence should be addressed to Jonathan Titus, titus@fredonia.edu

Academic Editors: M. Jullien and S. Satoh

To assess germination success in different microsites of a forested wetland environment, seeds of three common western New York wetland tree species, *Acer x freemanii, Fraxinus pennsylvanica*, and *Ulmus americana*, were sown into flats in the greenhouse with three substrates (mosses *Hypnum imponens* or *Thuidium delicatulum* or bare soil) and three hydrological conditions (wet, moist, or dry) in a factorial design. For the three species both treatment regimes and the interaction were highly significant, except for *Acer*, in which the substrate regime was not significant. *Fraxinus* germination had the highest tolerance for wet conditions and lowest for dry conditions followed by *Acer* and then *Ulmus*. Significant interactions showed that the effect of hydrological regime on germination is influenced by substrate type. Moss decreased germination under drier conditions and increased germination under wet conditions by lifting the seeds away from the soil and creating drier conditions than on bare soil. It is also possible that interspecific competition for moisture played a role in decreasing germination under dry conditions. By influencing the regeneration niche for three major tree species of swamps in the northeastern United States, the bryophyte layer plays an important role in determining community composition.

1. Introduction

Seed germination and seedling establishment are influenced by microsite environmental conditions. Favorable microsites or "safe sites" for seed germination and seedling establishment protect seeds and seedlings from extremes of temperature, humidity, moisture, and sunlight and affect nutrient availability, soil physical characteristics, seed predation, and a host of other factors. Thus, favorable microsites provide conditions beneficial for germination and establishment and their availability determines recruitment, which has a great impact on community composition, diversity, and succession [1–3]. An important aspect of a microsite, especially in wetlands, is microtopographic relief. In wetlands microsites with slight differences in elevation (centimeters) may differ in flooding duration, moisture retention, and availability, substrate composition and a host of other factors [4, 5]. Thus, minor differences in elevation in a swamp may have a powerful effect on germination and survival due to differences in inundation period and moisture availability.

Influencing differences in germination and survival elevation plays a role in tree seed germination and seedling survival and thereby influences species composition of the swamp forest [5]. That is, minor differences in elevation are different germination or regeneration niches [1, 6].

Bryophytes are a common substrate type on swamp microsites with strong effects on seed germination and seedling establishment. In general the presence of a bryophyte mat may positively [7–9] or negatively [10–16] affect seed germination and seedling survival. Seeds that get caught in a moss mat often remain too dry to germinate or are too far from the ground for the radicle to reach the soil surface and fail to establish [17–22]. Competition for water between the moss and the seedling is an important factor, especially in dry habitats [14, 23]. However, the presence of a bryophyte mat can lower temperatures and thereby reduce water stress on seeds and seedlings, and moss can reduce the rate of soil moisture loss, due to reduced evaporation and runoff positively influencing seedling survival especially in water-limited situations [10, 24, 25]. The position of a seed in the bryophyte mat and the size of the seed will interact to influence germination [3, 13, 19]. Species with larger seeds seem to be less influenced by the moss layer than small-seeded species [14].

Bryophyte effects on germination and establishment are species specific for the bryophyte and the vascular plant [4, 14, 20, 26]. For example, effects of bryophytes may be related to their turf structure with a thick moss cover affecting seed germination negatively and a thin moss cover promoting germination [4, 14, 27]. A moss mat may have a negative effect on germination and a positive effect on seedling survival [4, 15, 28, 29]. The effect of bryophytes on germination and seedling survival may differ between years depending on weather conditions [11].

A greenhouse study was conducted to investigate the effects of two moss species and bare soil under three hydrological regimes on the germination of seeds of the three dominant trees species, *Acer x freemanii*, *Fraxinus pennsylvanica*, and *Ulmus americana*, of western New York swamp forests. The three hydrological regimes represent elevation on microsites in a swamp and the substrates are the three substrates commonly found on microsites, that is, bare soil and the mosses *Hypnum imponens* and *Thuidium delicatulum*. These substrates may mediate the effect of inundation and moisture availability, which are controlled by elevation, on germination. The hypothesis is that substrate type and hydrological regime will affect the germination of the three tree species differently. This experiment would thereby clarify if minor elevational differences may be different regeneration niches for these three species.

2. Methods

The seeds of *Acer x freemanii* E. Murray (Freeman's maple), *Fraxinus pennsylvanica* Marsh. (green ash), and *Ulmus americana* L. (American elm) were used in this study. These species were used because they dominate the woody seedling, sapling, and canopy layers of Bonita Swamp and in many other swamps throughout western New York [5]. The seeds of these three species have epigeal germination. Most of the seeds of *Acer* and *Ulmus* germinate immediately after dispersal [30–32]. *Fraxinus* seed is reported to have dormancy mechanisms (e.g., [33, 34]) most likely because it is autumn-dispersed. Dirr [32] reported that *Fraxinus* seed requires 60 days at 20°C followed by 120 days at 0°C–5°C to break dormancy. However, we have found >50% germination with seed that was stored for 3 months under cool dry conditions and then planted in moist soil in the greenhouse (Titus unpubl. data). A cold stratification or other stratification treatments may increase germination beyond 50% [34]. In any case, the seeds of these species do not require freezing or scarification and are not major contributors to seed banks [35–38]. *Acer x freemanii* is a hybrid between *Acer rubrum* and *Acer saccharinum* and is identified by leaf and samara characteristics that are intermediate between the parent species [30, 39, 40]. These are large fruited species. *Acer* seeds with their samaras measured 4.3–5.8 cm long by 1.3–1.9 cm wide. *Ulmus* seeds,

in which the samara completely encircles the seed in an oval shape, were 1.0–1.3 by 1.2–1.5 cm for the two diameters. *Fraxinus* seeds have a narrower samara than *Acer* and were 3.8–4.4 cm long by 0.7–0.9 cm wide.

These tree species were selected because of their prevalence in Bonita Swamp, which is located along the Chadakoin River near Jamestown, Chautauqua County, NY (N 420730, W 0791709, 393 m asl). The 20.64 ha Bonita Swamp was purchased in 2004 by Chautauqua Watershed Conservancy and is part of the Chautauqua Lake Outlet Wetland Preserve (Chautauqua Watershed Conservancy 2009). Bonita Swamp is a palustrine, forested, deciduous, seasonally flooded (USUI, USGS 1996) class I wetland (NYSDEC 2007). The vegetation and hydrology of Bonita swamp are outlined in Blood and Titus [5]. Bonita Swamp is microtopographically complex with small differences in elevation dictating the time an area is inundated, which has a strong influence on species regeneration. Two moss species, *Hypnum imponens* Hedw. and *Thuidium delicatulum* (Hedw.) Schimp. occur on raised microtopographic features and cover 30% of the swamp floor [5]. Bare soil, wood, and litter also occupy raised microsites. These mosses form dense mats about 2 cm thick in Bonita Swamp. Blood and Titus [5] showed that raised moss microsites in Bonita Swamp contain more seedlings than bare soil or woody substrates presumably because the seeds are lifted above the saturated or inundated substrate.

Seeds and moss were collected on the ground in Bonita Swamp from eight locations separated by more than 10 m from one another. The mosses (*Hypnum* and *Thuidium*) were collected on 31 March 2007 and cultivated on potting soil in the greenhouse. *Acer* seeds were collected 11 June 2009 and planted 3 July 2009. *Fraxinus* seeds were collected 22 September 2007 and planted 15 February 2008. *Ulmus* seeds were collected 6 June 2008 and planted 18 September 2008. Seeds were stored under cool dry conditions.

Identical greenhouse experiments were carried out in the SUNY Fredonia campus greenhouse in Fredonia, NY, on three separate occasions using the seeds of the three species described above. Germination experiments were conducted consecutively for each species and not simultaneously due to greenhouse space limitations.

For each species a total of 72 26 cm × 26 cm flats were prepared with 100 seeds planted in each flat for *Fraxinus* and *Ulmus* and 75 seeds for *Acer*. Two treatment regimes, hydrologic regime (wet, moist, dry) and substrate (*Hypnum*, *Thuidium*, bare soil) were applied in a factorial manner across the flats for a sample size of eight for each treatment combination. Hydrologic treatments were wet, moist, and dry conditions. Wet conditions were established by placing the flats into larger flats without drainage holes allowing the soil to be continuously saturated, and if moss was present, the moss was wet. Moist flats were watered every three or four days and the soil was never allowed to dry out; however, if moss was present, the surface of the moss would become dry. Dry flats were watered weekly and dried out between watering and the surface of the moss would be quite dry. Two weeks before seeds were planted, moss was removed from the flats in which it was being cultivated and placed into treatment flats. A flat that received moss was covered

100% with a dense mat ~2 cm thick for each moss species. The seeds were then sprinkled onto the flats from above in order to simulate aerial dispersal in the swamp. The large size of the seeds meant that for these three species the seeds rested on the surface of the moss and did not penetrate into the mat. This is similar to the situation with these seeds in Bonita Swamp (unpubl. data).

Seedlings were counted weekly for a minimum of three months and the maximum number of seedlings obtained was used in the analyses. For *Acer* 17 days after initial seeding was used as the germinated seedling count for analysis, that is, after 17 days the number of seedlings did not increase. For *Fraxinus* 49 days after seeding was used, and for *Ulmus* 49 days after seeding was used. *Acer* seeds germinated much more rapidly than did seeds of the other species. Seedling survival was >95% for all three species and therefore was not assessed in this study. The greenhouse utilized ambient light which was occasionally supplemented with artificial lighting by fluorescent tubes at about 500 lux during the fall and winter. Thus, *Acer* received 14-15 hours of light per day where as *Fraxinus* and *Ulmus* received 11–13 hours. The greenhouse is heated in the winter and fan cooled in the summer. Greenhouse temperatures ranged from approximately 21°C–29°C for *Acer* in July and 18°C–24°C for *Fraxinus* and *Ulmus* in the fall and spring.

Germination data for each species were square root transformed to improve normality and homoscedasticity. Treatments were compared by two-way ANOVA and the post-hoc Tukey HSD test using SPSS [41]. Recall that for two-way ANOVA with a significant interaction post-hoc tests examine each treatment combination.

3. Results

The two-way ANOVA compares the two treatment regimes; however, the graphs show each treatment combination separately for greater visibility of the results (Figure 1). For the three species both treatment regimes and the interaction were highly significant, except for *Acer*, in which the substrate regime was not significant (Table 1). Overall the significant interactions showed that the effect of a hydrological regime on germination is influenced by substrate type. That is, for the three species under wetter conditions germination is higher on moss than on soil, but under drier conditions germination is higher on the soil.

Tukey's post-hoc results show that germination varied between the three species depending on both hydrology and substrate (Figure 1). *Acer* germination was the highest on wet moss and moist soil substrates and lowest on dry moss. *Fraxinus* germination was higher under wet conditions and lowest under dry conditions. *Ulmus* seeds had higher germination on moist soil substrate and lowest on dry moss and wet soil. Overall *Fraxinus* germination is most tolerant of wet conditions, followed by *Acer* and then by *Ulmus*.

A germinated seed with a radicle that was unable to penetrate the moss mat to reach the soil was not observed for any of these three relatively large seeded species.

TABLE 1: Two-way ANOVA and Tukey HSD results for greenhouse germination of three common wetland tree species under three moisture regimes (dry, moist, wet) on three substrates (*Hypnum*, *Thuidium*, bare soil).

Species	Treatment	F	Sig.
Acer x freemanii	Hydrology	26.808	0.000
	Substrate	1.473	0.237
	Interaction	11.879	0.000
Fraxinus pennsylvanica	Hydrology	721.365	0.000
	Substrate	16.124	0.000
	Interaction	22.581	0.000
Ulmus americana	Hydrology	33.651	0.000
	Substrate	20.030	0.000
	Interaction	23.761	0.000

4. Discussion

Based on the germination response to hydrology, the three species appear to have different regeneration niches for germination; that is, *Fraxinus* prefers the highest moisture levels for germination, *Acer* less moisture and *Ulmus* less. At each moisture level this effect is mediated by whether the substrate is moss or bare soil as is seen by the significant interactions and the post-hoc results (Table 1, Figure 1).

Fraxinus, under moist conditions, has higher germination on soil than on mosses. The mosses serve to lift the seed away from the soil making the germination site drier and the mosses may also compete for water with the germinating seed. Under wet conditions the saturated soil surface may reduce germination and so germination is slightly higher on the mosses which lift the seeds above the soil surface. *Acer* shows a similar pattern as *Fraxinus* but shifted slightly towards a drier moisture regime. Under wet conditions the moss lifts the seed away from the saturated soil and increases germination, whereas under moist and dry conditions the moss substrate is too dry and germination is higher on soil. For *Ulmus*, under moist conditions the soil substrate is strongly preferred for germination probably for the same reasons as above that the moss creates a drier substrate. The lack of difficulty for the radicles of these species to reach the soil supports the idea that the differences in germination are due to the moisture level at the point where the seed is located either on the surface of the soil or on the surface of the moss. No consistent differences were observed between the two moss species on germination.

Hörnberg et al. [4] observed "smothering" of germinating seeds by bryophytes. This was not observed in our study in which the seeds were perched on the surface of the moss and the radical appeared to have easily penetrated the moss mat. The species used on our study have larger seeds than the seeds of *Abies* in Hörnberg et al. [4].

Most *Acer* and *Ulmus* seeds germinate soon after dispersal and few enter dormancy [30–32]; however, the *Fraxinus* seeds used in this study may have been dormant at the time of planting [32–34]. Seed dormancy could potentially affect seed germination response to the substrate and moisture conditions in this study, particularly the wettest conditions.

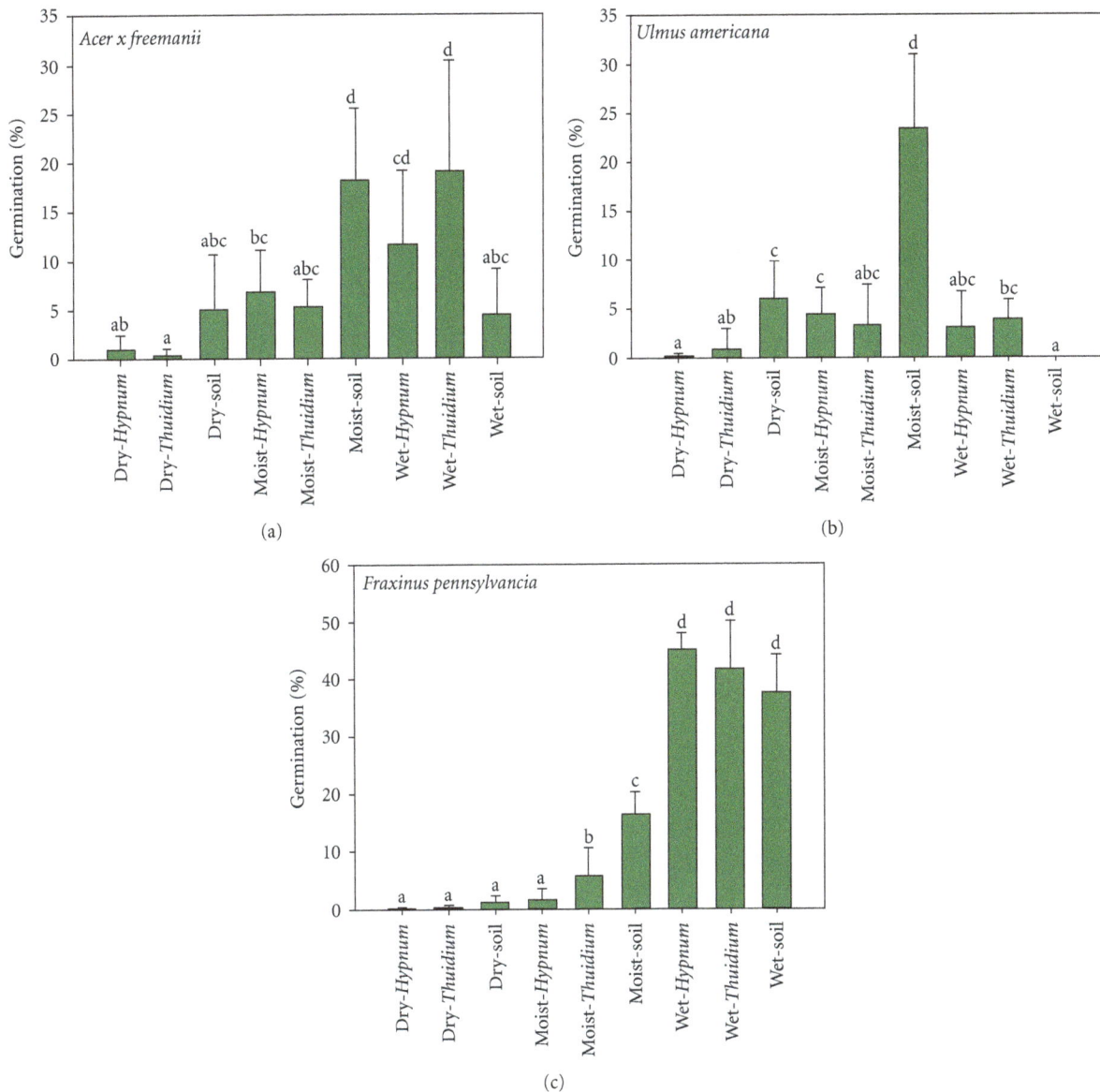

FIGURE 1: Mean + s.d. % germination for each of three wetland tree species (*Acer x freemanii, Fraxinus pennsylvanica, Ulmus americana*) under three hydrological regimes (wet, moist, dry) and three substrates (*Hypnum imponens, Thuidium delicatulum,* bare soil) in the greenhouse ($n = 8$ for each treatment combination). Seedlings were counted 17 days after seeding for *Acer*, and 49 days after seeding for *Fraxinus* and *Ulmus*. See Table 1 for statistical results. Bars with the same letter are not significantly different by Tukey's post-hoc test at $P \le 0.5$. Note that the vertical axis scale changes on each graph.

Fraxinus is a very tolerant species and has been observed both germinating in and inhabiting environments from moist to frequently flooded [5]; however, it is possible that dormancy affected the results obtained in this study.

It is possible that the greenhouse conditions influenced germination levels in these species (e.g., [16]). It was cooler (except on sunny days) and days were shorter for *Fraxinus* and *Ulmus* than for *Acer*. It is possible that given longer daylight hours and warmer conditions, for example, the seeds of *Fraxinus* or *Ulmus* could be more or less tolerant of wetter conditions. All of the greenhouse temperatures in this study are well within the normal field temperatures at the

time of germination for these species, thus, dramatic effects on germination dynamics are unlikely.

Moss cover can have chemical, physical, and mechanical effects on seeds. In this study the major effect appears to be physical with the presence of the moss lifting the germinating seed away from the soil surface decreasing the effective moisture level. These results are similar to those found in other studies where mosses affect moisture levels and thereby influence germination (e.g., [14, 17–19, 22]). Competition for water may also have decreased available water levels. In a swamp, where areas may be inundated for long periods of time, a small change in elevation, such as that provided by a moss mat, can have a major effect on moisture levels and therefore germination.

In Bonita Swamp the seeds of these three species are observed in large numbers perched on top of moss mats. By influencing the regeneration niche for the three major tree species of Bonita Swamp and many other swamps in the northeastern United States the bryophyte layer may play an important role in the forest dynamics of these systems.

Acknowledgments

A Holmberg Research Fellowship to A. Staunch funded this effort. The authors are indebted to E. Fuchs for identifying the mosses. P. Titus assisted in all aspects of the study. This study would not have been possible without the help of L. Blood, L. Martin, C. Sun, J. Wooten, K. Ludwig, and S. Strakosh. E. McCarrick and D. Hunt assisted with logistics. The authors would like to thank J. Jablonski, B. Nystrom, and the Chautauqua Watershed Conservancy for allowing the use of the study site.

References

[1] P. J. Grubb, "The maintenance of species-richness in plant communities: the importance of the regeneration niche," *Biological Reviews*, vol. 52, pp. 107–145, 1977.

[2] M. Fenner and K. Thompson, *The Ecology of Seeds*, Cambridge University Press, Cambridge, UK, 2005.

[3] T. W. Donath and R. L. Eckstein, "Effects of bryophytes and grass litter on seedling emergence vary by vertical seed position and seed size," *Plant Ecology*, vol. 207, no. 2, pp. 257–268, 2010.

[4] G. Hörnberg, M. Ohlson, and O. Zackrisson, "Influence of bryophytes and microrelief conditions on *Picea abies* seed regeneration patterns in boreal old-growth swamp forests," *Canadian Journal of Forest Research*, vol. 27, no. 7, pp. 1015–1023, 1997.

[5] L. E. Blood and J. H. Titus, "Microsite effects on forest regeneration in a bottomland swamp in western New York," *Journal of the Torrey Botanical Society*, vol. 137, no. 1, pp. 88–102, 2010.

[6] J. H. Titus, "Microtopography and woody plant regeneration in a hardwood floodplain swamp in Florida," *Bulletin of the Torrey Botanical Club*, vol. 117, pp. 429–437, 1990.

[7] G. Rusch and J. M. Fernandez-Palacios, "The influence of spatial heterogeneity on regeneration by seed in a limestone grassland," *Journal of Vegetation Science*, vol. 6, no. 3, pp. 417–426, 1995.

[8] L. S. Santiago, "Use of coarse woody debris by the plant community of a Hawaiian montane cloud forest," *Biotropica*, vol. 32, no. 4, pp. 633–641, 2000.

[9] E. G. Sedia and J. G. Ehrenfeld, "Lichens and mosses promote alternate stable plant communities in the New Jersey Pinelands," *Oikos*, vol. 100, no. 3, pp. 447–458, 2003.

[10] P. J. Keizer, B. F. van Tooren, and H. J. During, "Effects of bryophytes on seedling emergence and establishment of short-lived forbs in chalk grassland," *Journal of Ecology*, vol. 73, no. 2, pp. 493–504, 1985.

[11] B. F. van Tooren, "Effects of a bryophyte layer on the emergence of seedlings of chalk grassland species," *Acta Oecologica*, vol. 11, no. 2, pp. 155–163, 1990.

[12] I. Špačková, I. Kotorová, and J. Lepš, "Sensitivity of seedling recruitment to moss, litter and dominant removal in an oligotrophic wet meadow," *Folia Geobotanica*, vol. 33, no. 1, pp. 17–30, 1998.

[13] I. Kotorová and J. Lepš, "Comparative ecology of seedling recruitment in an oligotrophic wet meadow," *Journal of Vegetation Science*, vol. 10, no. 2, pp. 175–186, 1999.

[14] M. Zamfir, "Effects of bryophytes and lichens on seedling emergence of alvar plants: evidence from greenhouse experiments," *Oikos*, vol. 88, no. 3, pp. 603–611, 2000.

[15] M. Jeschke and K. Kiehl, "Effects of a dense moss layer on germination and establishment of vascular plants in newly created calcareous grasslands," *Flora*, vol. 203, no. 7, pp. 557–566, 2008.

[16] N. A. Soudzilovskaia, B. J. Graae, J. C. Douma et al., "How do bryophytes govern generative recruitment of vascular plants?" *New Phytologist*, vol. 190, no. 4, pp. 1019–1031, 2011.

[17] B. F. van Tooren, "The fate of seeds after dispersal in chalk grassland: the role of the bryophyte layer," *Oikos*, vol. 53, no. 1, pp. 41–48, 1988.

[18] H. J. During and B. F. van Tooren, "Bryophyte interactions with other plants," *Botanical Journal of the Linnean Society*, vol. 104, pp. 79–98, 1990.

[19] T. Nakamura, "Effect of bryophytes on survival of conifer seedlings in subalpine forests of central Japan," *Ecological Research*, vol. 7, no. 2, pp. 155–162, 1992.

[20] M. Ohlson and O. Zackrisson, "Tree establishment and microhabitat relationships in north Swedish peatlands," *Canadian Journal of Forest Research*, vol. 22, no. 12, pp. 1869–1877, 1992.

[21] I. Steijlen, M. C. Nilsson, and O. Zackrisson, "Seed regeneration of Scots pine in boreal forest stands dominated by lichen and feather moss," *Canadian Journal of Forest Research*, vol. 25, no. 5, pp. 713–723, 1995.

[22] J. W. Morgan, "Bryophyte mats inhibit germination of non-native species in burnt temperate native grassland remnants," *Biological Invasions*, vol. 8, no. 2, pp. 159–168, 2006.

[23] J. Czarnecka, "Seed longevity and recruitment of seedlings in xerothermic grassland," *Polish Journal of Ecology*, vol. 52, no. 4, pp. 505–521, 2004.

[24] C. D. Johnson and A. G. Thomas, "Recruitment and survival of seedlings of a perennial *Hieracium* species in a patchy environment," *Canadian Journal of Botany*, vol. 56, pp. 572–580, 1978.

[25] Y. Kameyama, N. Nakagoshi, and K. Nehira, "Safe site for seedlings of *Rhododendron metternichii* var. *hondoense*," *Plant Species Biology*, vol. 14, no. 3, pp. 237–242, 1999.

[26] H. E. Kirkpatrick, J. W. S. Barnes, and B. A. Ossowski, "Moss interference could explain the microdistributions of two species of monkey-flowers (*Mimulus*, Scrophulariaceae)," *Northwest Science*, vol. 80, no. 1, pp. 1–8, 2006.

[27] M. E. Harmon and J. F. Franklin, "Tree seedlings on logs in *Picea-Tsuga* forests of Oregon and Washington," *Ecology*, vol. 70, no. 1, pp. 48–59, 1989.

[28] G. Overbeck, K. Kiehl, and C. Abs, "Seedling recruitment of *Succisella inflexa* in fen meadows: Importance of seed and microsite availability," *Applied Vegetation Science*, vol. 6, no. 1, pp. 97–104, 2003.

[29] R. L. Eckstein, J. Danihelka, N. Hölzel, and A. Otte, "The effects of management and environmental variation on population stage structure in three river-corridor violets," *Acta Oecologica*, vol. 25, no. 1-2, pp. 83–91, 2004.

[30] O. M. Freeman, "A red maple, silver maple hybrid," *Journal of Heredity*, vol. 32, no. 1, pp. 11–14, 1941.

[31] M. Coladonato, *Ulmus americana*. In: Fire Effects Information System, U.S. Department of Agriculture, Forest Service, Rocky Mountain Research Station, Fire Sciences Laboratory,1992, http://www.fs.fed.us/database/feis/.

[32] M. A. Dirr, *Manual of Woody Landscape Plants: Their Identification, Ornamental Characteristics, Culture, Propagation, and Uses*, Stipes Publishing, Champaign, Ill, USA, 1998.

[33] G. P. Steinbauer, "Dormancy and germination of *Fraxinus* seeds," *Plant Physiology*, vol. 12, pp. 813–824, 1937.

[34] R. W. Tinus, "Effects of dewinging, soaking, stratification, and growth regulators on germination of green ash seed," *Canadian Journal of Forest Research*, vol. 12, pp. 931–935, 1982.

[35] L. A. Hyatt and B. B. Casper, "Seed bank formation during early secondary succession in a temperate deciduous forest," *Journal of Ecology*, vol. 88, no. 3, pp. 516–527, 2000.

[36] J. H. Lambers and J. S. Clark, "The benefits of seed banking for red maple (*Acer rubrum*): maximizing seedling recruitment," *Canadian Journal of Forest Research*, vol. 35, no. 4, pp. 806–813, 2005.

[37] J. H. R. Lambers, J. S. Clark, and M. Lavine, "Implications of seed banking for recruitment of southern Appalachian woody species," *Ecology*, vol. 86, no. 1, pp. 85–95, 2005.

[38] L. E. Blood, H. J. Pitoniak, and J. H. Titus, "Seed bank of a bottomland swamp in Western New York," *Castanea*, vol. 75, no. 1, pp. 19–38, 2010.

[39] A. E. Murray, "*Acer x freemanii*, E. Murray, *Hybrida Nova*," *Kalmia*, vol. 1, pp. 2–3, 1969.

[40] J. Schlegel, "A "new" tree in western New York: the Freeman maple," *Clintonia*, vol. 21, pp. 1–3, 2006.

[41] SPSS, *PASW for Windows Release 18.0.0*, SPSS, Chicago, Ill, USA, 2009.

21

Sequences Analysis of ITS Region and 18S rDNA of *Ulva*

Zijie Lin, Zhongheng Lin, Huihui Li, and Songdong Shen

College of Life Sciences, Soochow University, Suzhou 215123, China

Correspondence should be addressed to Songdong Shen, shensongdong@suda.edu.cn

Academic Editors: J. Elster and H. Sanderson

Ulva, as the main genera involved in green tides in the Yellow Sea, has attracted serious concern in China. Especially, *Ulva prolifera* is one of the causative species of the occurring. This paper focused on the complete sequences analyses of ITS, 18S, and the combined data to determine phylogenetic relationships among taxa currently attributed to *Ulva*, *Monostroma*, and some other green algal. The samples are all concluded in the area of Yellow Sea, China. The results showed the content of G+C in 18S was approximately concentrated upon 49% in average of 19 subjects while the ITS region content of base G and C is obviously higher than A and T. Comparing the ITS and 18S rDNA sequences obtained in this paper to other species retrieved from GenBank, the genetic distance and the ratio of sequence divergence reflect that *U. pertusa* and *U. prolifera* had closer genetic relationship with an 18S rDNA, which had genetic distance of 0.007 while ITS had further genetic distance. According to further comparison, *Ulva prolifera* has closest genetic distance with *Chloropelta caespitosa* (0.057) and *Ulva californica* (0.057), which is a reverification coincided *Chloropelta*, *Enteromorpha*, and *Ulva* are not distinct genera.

1. Introduction

Green tide is an ecological phenomenon that occurs globally in which fixed growth alga break away from the shallow beaches of calm bays, resulting in accumulation of extensive biomass of free-floating green alga [1, 2]. Eutrophication is the main reason for green tide [3, 4]. The main species involved in green tides are *Ulva* sp. and *Enteromorpha* sp. Large scale green tides have occurred continuously four times in the Yellow Sea, from 2007 to 2010 in China, [5, 6], which had serious influences on the local environment and the life of coastal residents. As well known, these green tides consist of *Enteromorpha* sp. [7, 8]; however, their sources are still subject to debate. It has been suggested that the southern coast of the Yellow Sea is the ultimate source; accordingly, this region has become the focus of many investigations into the subject [9]. *Ulva* sp. has been considered the main causative species of green tides, which have bloomed continuously in recent years in the central and southern Yellow Sea in China. *Enteromorpha* [10] belongs to *Ulvaceae*, *Ulvales*, *Chlorophyceae*, *Chlorophyta*. More than 100 species of this organism have been recorded worldwide; however, only about 30 species can be recognized based on their

morphology [11]. In addition, 23 species of *Enteromorpha* have been recorded in China [12–16]. Most of them are marine species, which are widespread along the coast of China.

Due to the divergence of *Enteromorpha* or *Ulva*, the *Ulva* sp. is named according to Tan et al. [20] and Hayden et al. [21] in this study. *Ulva* sp. has wide acclimatization and can grow well in a broad range of temperatures and salinities, but the morphological characteristics are changing easily in response to the environment [16]. Various morphological changes in the intraspecies and less differences that are difficult to identify, even among interspecies. Therefore, making an identification of *Ulva* sp. by using classical taxonomic methods is very difficult [22, 23]. Molecular biology and cross-hybridization methods [24, 25] have been introduced into this field to clarify the species. By this method, some reports have suggested that green tides are formed by an *Ulva linza-procera-prolifera* (LPP) complex instead of individual species [5, 26]. Analyses of nuclear ribosomal internal transcribed spacer DNA (ITS nrDNA; 29 ingroup taxa including the type species of *Ulva* and *Enteromorpha*), the chloroplast-encoded rbcL gene (for a subset of taxa), and a combined dataset were carried out. Combined with

TABLE 1: Primers used for PCR amplification and sequencing.

Gene	Region and direction	Primer name	Sequence (5'-3')
ITS region	5' end (F)	F	TCT TTG AAA CCG TAT CGT GA
	3' end (R)	R	GCT TAT TGA TAT GCT TAA GTT CAG CGG GT
18S rRNA	5' end (F)	NS1[1]	GTA GTC ATA TGC TTG TCT C
	~1150 (R)	NS4[1]	CTT CCG TCA ATT CCT TTA AG
	~1150 (F)	NS5[1]	AAC TTA AAG GAA TTG ACG AAA G
	3' end (R)	NS8[1]	TCC GCA GGT TCA CCT ACG GA
	5' end (F)	CRN5[2]	TGG TTG ATC CTG CCA GTA G
	~1137 (R)	1137[2]	GTG CCC TTC CGT CAA T
	~337 (F)	AB1[3]	GGAGGATTAGGGTCCGATTCC
	~1131 (R)	TW4[3]	CTTCCGTCAATTCCTTTAAG
	~1056 (F)	MonS	GCGGGTGTTTGTTTGA
	~1328 (R)	MonA	CTATTTAGCAGGCTGAGGT

[1] From White et al. [17], [2] from Booton et al. [18], [3] from Van Oppen [19].

earlier molecular and culture data, these data provide strong evidence that *Ulva*, *Enteromorpha*, and *Chloropelta* are not distinct evolutionary entities and should not be recognized as separate genera [21]. Partial sequences of the genes coding for rbcL and the 18S rRNA were used to determine the phylogenetic position of the order Prasiolales among other members of the Chlorophyta. Sequence divergence values within the Prasiolales for the rbcL gene (0–6.1%) and the 18S rRNA gene (0.4–3.8%) are both low compared to values among the other green algal sequences. Parsimony and distance analyses of the two subject genes sequences indicate that the Prasiolales is a well-delineated order of green algae containing both Prasiola and Rosenvingiella [27].

In this study, complete sequences of ITS and 18S are used to analyze the phylogenetic relationship among three common species (*Ulva prolifera*, *Ulva pertusa*, and *Monostroma grevillei*) in the Yellow Sea, China and some other green algal around the world. Especially, *Ulva prolifera* is the main causative species of green tides in China. Assessing the phylogenetic position of *Ulva* contributes to the investigation of the process of phylogenetic.

2. Material and Methods

2.1. Plant Material.
Ulva prolifera were collected from Lianyungang coast of Jiangsu Province. And *Monostroma grevillei* and *Ulva lactuca* were collected from Qingdao coast of Shandong Province. Thallus were cleaned up, dried, and stored at −20°C for further analysis. The samples were reanimated several hours to unfold entirely and then soaked in 0.7% KI solution for ten minutes, followed by scouring with ddH$_2$O.

2.2. DNA Extraction.
The cleaned-up samples were stored at −70°C overnight and then were triturated thoroughly with a chilled mortar and pestle. They were transferred into microcentrifuge tubes, which 2% sea snail enzyme with 2 M glucose was added into. The samples were digested for three

hours at 25°C in swing bed and then were filtered and collected. Total genomic DNA was extracted by CTAB method, which was modified according to the samples. In brief, each sample was suspended with DNA extraction solution (3% CTAB, 0.1 mol L^{-1} Tris-HCl, pH 8.0, 0.01 mol L^{-1} EDTA, 1.4 mol L^{-1} NaCl, 0.5% β-mercaptoethanol, 1% PVP) and digested with protease-K with a final concentration of 300 μg mL^{-1}. The mixture was incubated at 55°C for half an hour with shaking every ten minutes, and then it was cooled down to room temperature. One-third volume of 5 M KAc was mixed with the solution before extracting the protein with phenol : trichloromethane : isoamyl-alcohol (25 : 24 : 1 v/v/v). The extract was precipitated by isopropyl-alcohol at −20°C for two hours. The solution was centrifuged and then the sediment was collected and cleaned with 70% ethyl alcohol. Finally, the sediments were dissolved in TE buffer (10 mmol L^{-1} Tris-HCl, pH 8.0, 1 mmol L^{-1} EDTA) and could be used as template for the following PCR amplification.

2.3. PCR Amplification and Purification.
The primers F&R were designed according to ITS region sequence and MonS & MonA were designed according to 18S rDNA sequence retrieved from software Primer 5.0. And the primers were synthesized by Shanghai Sangon Biological Engineering Technology and Service Co., Ltd. (Table 1).

The reactions for PCR amplification were performed with a final volume of 50 μL, containing 27.5 μL ddH$_2$O, 10 μL 5×ExTaq Buffer (Mg^{2+}), 6 μL dNTPMix (2.5 mM), 5 μL of the DNA template, 2 μL of each PCR primer (50 pmol μL^{-1}), and 0.5 μL ExTaq DNA polymerase (5 U μL^{-1}).

The amplification of ITS region was performed with an initial denaturation at 94°C for 5 min, 35 cycles at 94°C for 1 min, 45°C for 2 min, and 65°C for 3 min. And the products of amplification were preserved in 4°C.

The amplification of 18S rDNA was performed with an initial denaturation at 95°C for 2 min, 35 cycles at 95°C

TABLE 2: List of species used in this study and GenBank accession numbers for ITS and 18S.

Taxon	ITS accession number	18S rRNA accession number
Ulva lactuca	AY422499	AF499666
Klebsormidium flaccidum	EU434019	M95613
Chlorella vulgaris	FM205855	X13688
Mantoniella squamata	FN562451	X73999
Ulva prolifera	**HQ902007**	**HQ850569**
Monostroma grevillei	**HQ902006**	**HQ850570**
Paulschulzia pseudovolvox	AF182428	U83120
Trebouxia asymmetrica	AF344177	Z21553
Kornmannia leptoderma	AF415168	AF499661
Monostroma nitidum	AF415170	AF499665
Blidingia minima	AJ000206	AF499659
Enteromorpha intestinalis	AJ000210	AF189077
Ulva fenestrata	AJ234316	AF499653
Chloropelta caespitosa	AY016309	AF499656
Ulvaria obscura var. blyttii	AY260571	AF499657
Ulothrix zonata	Z47999	AY278217
Ulva pertusa	**HQ902008**	**HQ850571**
Ulva californica	AY422518	AF499652
Percursaria percursa	AY016305	AF499658

for 1 min, 55°C for 1 min (while the AB1&TW4 was 50°C and the MonS&MonA is 44°C), 72°C for 4 min, and a final extension step at 72°C for 6 min. And the products of amplification are preservation in 4°C.

The PCR products were confirmed by electrophoresis in 1% agarose gel. The gels were stained with ethidium bromide and photographed by Bio-IMAGING System. The PCR products were purified by TaKaRa Agarose Gel DNA Purification Kit Ver.2.0.

2.4. Sequencing and Phylogenetic Tree Construction. The products were sequenced by Shanghai Sangon Biological Engineering Technology and Service Co., Ltd. The sequences were aligned using Clustal X, and further manually adjusted using BioEdit. The gained sequences should be ITS region and 18S rDNA by homology investigation with BLAST on the website of NCBI (http://www.ncbi.nih.gov/) by defining the boundary and length [28]. The sequences were aligned in order to observe the resemblance and to analyze the differences using the software DNAMAN with default parameters. With the multialignment analysis, we applied

the program MEGA 3.1 with Kimura's two-parameter model [29] to calculate the base composition, Kimura two-parameter distance and the ratio of sequence divergence. For phylogenetic analysis, we used MEGA3.1 (neighbor-joining method) to study the relationship of different species to construct the phylogenetic tree, and detect the degree of bootstrap confidence from 1000 replicates. For comparative analysis, sequences were taken from the GenBank, of which entrance numbers were shown in Table 4.

3. Results

3.1. Splicing of ITS Region and 18S rDNA Sequences Fragments. With the primers mentioned above, the genomic of the ITS region and 18S rDNA fragments were successfully amplified (Figure 1). We got the overall length of ITS region and 18S rDNA sequences about *U. prolifera*, *U. pertusa*, and *M. grevillei*.

3.2. ITS Region and 18S rDNA Sequences of the Three Species. The base size of *U. prolifera* was 536 bp in ITS and 1718 bp in 18S. And the base size of *M. grevillei* was 541 bp in ITS and 1755 bp in 18S. And the base size of *U. pertusa* was 567 bp in ITS and 1761 bp in 18S. The base sizes of the three species were almost the same in ITS region and 18S rDNA. After sequencing the ITS region and 18S rDNA, we got the accession numbers from GenBank. For comparative analysis, sequences of other sixteen species were also taken from the GenBank. All these data were shown in Table 2.

3.3. Base Composition of ITS Region and 18S rDNA Sequences. The base composition of ITS region and the content of G+C are shown in Table 3. And the base composition of 18S rDNA and the content of G+C are shown in Table 4. The ITS region content of G+C was about 58.45% in average in nineteen species. However, the content of G+C varied obviously from 51.65% to 65.78% in ITS region while 18S rDNA content of G+C was about 48.98% in average in nineteen species. Comparing with the content of G+C in ITS region, the content of G+C in 18S was approximately concentrated upon 49%. The ITS region content of base G and C in average in nineteen species is obviously higher than A and T.

3.4. Genetic Distance in Interspecies. Comparing the ITS region and 18S rDNA sequence obtained in this study and other species retrieved from GenBank, the results of the genetic distance and the ratio of sequence divergence are shown in Tables 5 and 6. *U. pertusa* and *U. prolifera* had closer genetic relationship with genetic distance of 0.007 in 18S rDNA. While, its had further genetic distance in ITS.

A comparison of the sequences of the ITS of the three strains evaluated in this study to those of other species of green algal retrieved from GenBank is shown in Table 5. Obviously, CC, UC, HT, UL, EI, SC, UF, UO, and PPE are subordinate to *Ulvaceae*. The region distance (ranged form 0.189 to 0.035) and the ratio of sequence divergence (ranged from 0.026 to 0.010) are lower than those comparisons within other 10 species from different families. Particularly prominent, CC, HT, and UC had the closest relationship

TABLE 3: Base composition of ITS region.

Taxon	A (%)	T (%)	G (%)	C (%)	G+C (%)
Ulva lactuca	21.37	15.88	28.24	34.51	62.75
Klebsormidium flaccidum	21.78	19.42	28.71	30.10	58.81
Chlorella vulgaris	20.00	23.45	25.32	31.22	56.55
Mantoniella squamata	22.76	21.27	28.17	27.80	55.97
Ulva prolifera	20.15	17.16	28.92	33.77	62.69
Monostroma grevillei	22.92	21.26	24.95	30.87	55.82
Paulschulzia pseudovolvox	24.26	24.09	24.75	26.90	51.65
Trebouxia asymmetrica	20.27	26.93	26.48	26.32	52.80
Kornmannia leptoderma	22.78	22.06	26.16	29.00	55.16
Monostroma nitidum	24.18	23.26	25.27	27.29	52.56
Blidingia minima	22.45	22.08	25.55	29.93	55.47
Enteromorpha intestinalis	18.85	18.85	28.84	33.46	62.29
Ulva fenestrata	20.04	17.12	28.99	38.85	62.84
Chloropelta caespitosa	20.51	19.15	29.06	31.28	60.34
Ulvaria obscura var. blyttii	20.42	18.13	28.44	33.02	61.45
Ulothrix zonata	23.98	24.18	24.56	27.27	51.84
Ulva pertusa	18.87	15.34	30.34	35.45	65.78
Ulva californica	20.41	19.11	28.39	32.10	60.48
Percursaria percursa	19.86	14.81	29.62	35.71	65.33
Average	21.36	20.19	27.41	31.31	58.45

TABLE 4: Base composition of 18S rDNA.

Taxon	A (%)	T (%)	G (%)	C (%)	G+C (%)
Ulva lactuca	26.37	24.50	28.78	20.35	49.13
Klebsormidium flaccidum	26.08	26.80	26.53	20.59	47.11
Chlorella vulgaris	24.97	25.42	27.64	21.97	49.61
Mantoniella squamata	26.01	26.80	26.74	20.45	47.19
Ulva prolifera	24.80	25.67	27.76	21.77	49.53
Monostroma grevillei	25.13	26.55	27.35	20.97	48.32
Paulschulzia pseudovolvox	24.91	25.49	28.02	21.58	49.60
Trebouxia asymmetrica	25.06	26.00	27.34	21.60	48.94
Kornmannia leptoderma	26.51	24.10	28.92	20.48	49.40
Monostroma nitidum	26.17	24.83	28.84	20.16	49.00
Blidingia minima	26.34	23.66	29.14	20.86	50.00
Enteromorpha intestinalis	24.87	25.74	27.78	21.61	49.39
Ulva fenestrata	26.30	24.43	28.84	20.43	49.27
Chloropelta caespitosa	26.44	24.43	28.84	20.29	49.13
Ulvaria obscura var. blyttii	26.44	24.30	28.97	20.29	49.27
Ulothrix zonata	25.09	27.01	27.07	20.83	47.90
Ulva pertusa	24.70	25.95	27.77	21.58	49.35
Ulva californica	26.30	24.57	28.84	20.29	49.13
Percursaria percursa	26.17	24.43	28.97	20.43	49.40
Average	25.72	25.30	28.11	20.87	48.98

among each other because of the lower region distance (HT/CC 0.057; HT/UC 0.057; CC/UC 0.035) and the ratio of sequence divergence (HT/CC 0.012; HT/UC 0.012; CC/UC 0.010). These data illustrate HT/UC/CC has extremely related species. According to other Ulva species, the sequence homology of these three species is closer which verified U. prolifera and C. caespitosa belonged to Ulva genera. On the other hand, JM, UZ, BM, MN, and KL had higher homology relationship within each other, the region distance ranged from 0.335 to 0.133, and the ratio of sequence divergence ranged from 0.033 to 0.021. The data are relatively stable, and in these five species, JM/UZ is the closest pair because of

TABLE 5: Kimura 2-parameter ITS region distance (below) and the ratio of sequence divergence (above).

Species	UL	JM	HT	SC	UZ	UO	CC	UF	EI	BM	MN	KL	TA	PP	MS	UC	PPE	CV	KF
UL		0.039	0.017	0.019	0.038	0.025	0.016	0.019	0.015	0.039	0.041	0.038	0.053	0.048	0.057	0.014	0.024	0.049	0.049
JM	0.377		0.037	0.040	0.021	0.037	0.037	0.038	0.037	0.025	0.032	0.033	0.051	0.048	0.048	0.039	0.035	0.045	0.044
HT	0.105	0.338		0.018	0.037	0.023	0.012	0.016	0.016	0.039	0.040	0.039	0.049	0.045	0.054	0.012	0.023	0.046	0.047
SC	0.118	0.369	0.102		0.041	0.026	0.019	0.015	0.018	0.042	0.044	0.042	0.053	0.049	0.059	0.017	0.023	0.046	0.051
UZ	0.364	0.133	0.355	0.398		0.036	0.037	0.037	0.039	0.022	0.028	0.033	0.046	0.049	0.052	0.038	0.037	0.043	0.045
UO	0.196	0.335	0.163	0.181	0.333		0.025	0.024	0.024	0.035	0.038	0.040	0.053	0.049	0.059	0.025	0.016	0.044	0.047
CC	0.085	0.341	0.057	0.116	0.344	0.189		0.018	0.016	0.039	0.039	0.038	0.054	0.048	0.057	0.010	0.024	0.048	0.048
UF	0.118	0.346	0.089	0.076	0.360	0.178	0.099		0.018	0.039	0.040	0.040	0.051	0.048	0.054	0.017	0.023	0.045	0.048
EI	0.070	0.348	0.085	0.112	0.360	0.185	0.079	0.109		0.039	0.040	0.038	0.050	0.045	0.056	0.015	0.022	0.048	0.048
BM	0.368	0.189	0.360	0.398	0.146	0.329	0.354	0.369	0.363		0.029	0.030	0.045	0.049	0.053	0.039	0.038	0.046	0.049
MN	0.399	0.257	0.370	0.416	0.219	0.349	0.378	0.374	0.374	0.243		0.036	0.048	0.046	0.053	0.040	0.039	0.039	0.046
KL	0.379	0.285	0.370	0.406	0.297	0.381	0.369	0.389	0.374	0.256	0.335		0.050	0.046	0.055	0.040	0.041	0.050	0.048
TA	0.510	0.486	0.464	0.508	0.438	0.476	0.510	0.485	0.474	0.442	0.449	0.503		0.036	0.053	0.051	0.051	0.048	0.049
PP	0.477	0.479	0.440	0.489	0.475	0.480	0.477	0.465	0.444	0.478	0.465	0.459	0.315		0.055	0.048	0.048	0.048	0.047
MS	0.549	0.474	0.527	0.571	0.520	0.558	0.551	0.537	0.548	0.534	0.529	0.565	0.519	0.554		0.055	0.056	0.059	0.042
UC	0.072	0.365	0.057	0.099	0.363	0.181	0.035	0.092	0.073	0.368	0.393	0.394	0.480	0.476	0.538		0.023	0.048	0.048
PPE	0.178	0.329	0.164	0.167	0.351	0.095	0.174	0.163	0.174	0.370	0.377	0.391	0.476	0.450	0.539	0.174		0.046	0.045
CV	0.476	0.423	0.441	0.448	0.397	0.434	0.460	0.443	0.456	0.429	0.383	0.483	0.480	0.458	0.611	0.470	0.435		0.049
KF	0.492	0.432	0.485	0.508	0.446	0.460	0.502	0.478	0.479	0.490	0.452	0.473	0.501	0.468	0.389	0.496	0.443	0.506	

Note: UL, *U. lactuca*; KF, *K. flaccidum*; CV, *C. vulgaris*; MS, *M. squamata*; HT, *U. prolifera*; JM, *M. grevillei*; PP, *P. pseudovolvox*; TA, *T. asymmetrica*; KL, *K. leptoderma*; MN, *M. nitidum*; BM, *B. minima*; EI, *E. intestinalis*; UF, *U. fenestrate*; CC, *C. caespitosa*; UO, *U. obscura var. blyttii*; UZ, *U. zonata*; SC, *U. pertusa*; UC, *U. californica*; PPE, *P. percursa*.

TABLE 6: Kimura 2-parameter 18S rDNA distance (below) and the ratio of sequence divergence (above).

Species	EI	CC	BM	UC	JM	SC	HT	CV	UL	PPE	UO	UF	TA	MS	PP	KF	UZ	MN	KL
EI		0.002	0.010	0.000	0.009	0.003	0.003	0.011	0.002	0.004	0.005	0.002	0.013	0.014	0.011	0.013	0.009	0.009	0.008
CC	0.003		0.010	0.002	0.010	0.002	0.002	0.011	0.002	0.004	0.005	0.002	0.013	0.014	0.011	0.014	0.009	0.009	0.008
BM	0.070	0.073		0.010	0.010	0.010	0.010	0.012	0.010	0.010	0.010	0.010	0.013	0.015	0..012	0.015	0.010	0.010	0.006
UC	0.000	0.003	0.070		0.009	0.003	0.003	0.011	0.002	0.004	0.005	0.002	0.013	0.014	0.011	0.013	0.009	0.009	0.008
JM	0.064	0.067	0.069	0.064		0.010	0.010	0.010	0.010	0.010	0.009	0.010	0.010	0.013	0.010	0.013	0.003	0.005	0.008
SC	0.007	0.004	0.078	0.007	0.072		0.003	0.011	0.003	0.005	0.005	0.003	0.013	0.015	0.011	0.014	0.009	0.009	0.009
HT	0.005	0.003	0.076	0.005	0.070	0.007		0.011	0.003	0.004	0.005	0.003	0.013	0.015	0.011	0.014	0.009	0.009	0.009
CV	0.086	0.089	0.108	0.086	0.067	0.093	0.092		0.011	0.011	0.011	0.011	0.009	0.011	0.009	0.010	0.009	0.009	0.011
UL	0.004	0.003	0.075	0.004	0.069	0.007	0.005	0.090		0.004	0.004	0.000	0.013	0.015	0.011	0.014	0.009	0.009	0.008
PPE	0.011	0.014	0.072	0.011	0.066	0.018	0.017	0.083	0.015		0.004	0.004	0.013	0.014	0.011	0.013	0.009	0.009	0.008
UO	0.018	0.017	0.073	0.018	0.064	0.021	0.019	0.091	0.017	0.012		0.004	0.012	0.014	0.012	0.014	0.009	0.009	0.008
UF	0.004	0.003	0.075	0.004	0.069	0.007	0.005	0.090	0.000	0.015	0.017		0.013	0.015	0.011	0.014	0.009	0.009	0.008
TA	0.113	0.116	0.115	0.113	0.075	0.121	0.119	0.054	0.114	0.108	0.105	0.114		0.011	0.011	0.011	0.010	0.011	0.012
MS	0.141	0.145	0.150	0.141	0.114	0.150	0.148	0.079	0.146	0.137	0.141	0.146	0.087		0.012	0.011	0.013	0.013	0.014
PP	0.086	0.089	0.104	0.086	0.078	0.093	0.092	0.064	0.090	0.086	0.094	0.090	0.081	0.102		0.012	0.010	0.010	0.011
KF	0.124	0.127	0.139	0.124	0.111	0.132	0.130	0.078	0.129	0.122	0.129	0.129	0.086	0.084	0.098		0.012	0.012	0.014
UZ	0.058	0.061	0.067	0.058	0.008	0.066	0.064	0.063	0.063	0.060	0.061	0.063	0.074	0.109	0.072	0.104		0.004	0.008
MN	0.060	0.063	0.069	0.060	0.018	0.067	0.066	0.063	0.064	0.061	0.066	0.064	0.078	0.114	0.073	0.104	0.011		0.008
KL	0.054	0.057	0.034	0.054	0.046	0.061	0.060	0.084	0.058	0.054	0.054	0.058	0.092	0.130	0.085	0.123	0.046	0.049	

Note: UL, *U. lactuca*; KF, *K. flaccidum*; CV, *C. vulgaris*; MS, *M. squamata*; HT, *U. prolifera*; JM, *M. grevillei*; PP, *P. pseudovolvox*; TA, *T. asymmetrica*; KL, *K. leptoderma*; MN, *M. nitidum*; BM, *B. minima*; EI, *E. intestinalis*; UF, *U. fenestrate*; CC, *C. caespitosa*; UO, *U. obscura var. blyttii*; UZ, *U. zonata*; SC, *U. pertusa*; UC, *U. californica*; PPE, *P. percursa*.

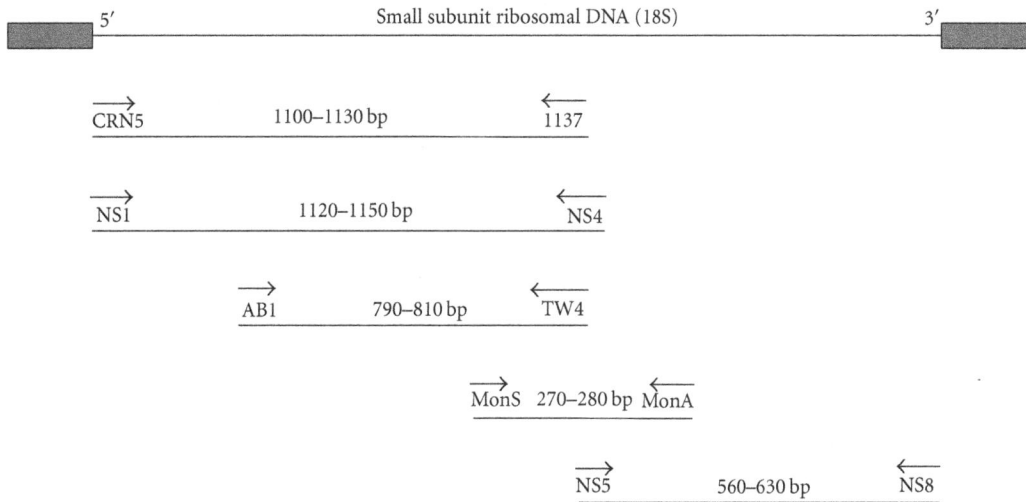

FIGURE 1: Splicing of 18S rDNA sequences fragments with the primers mentioned in Table 1.

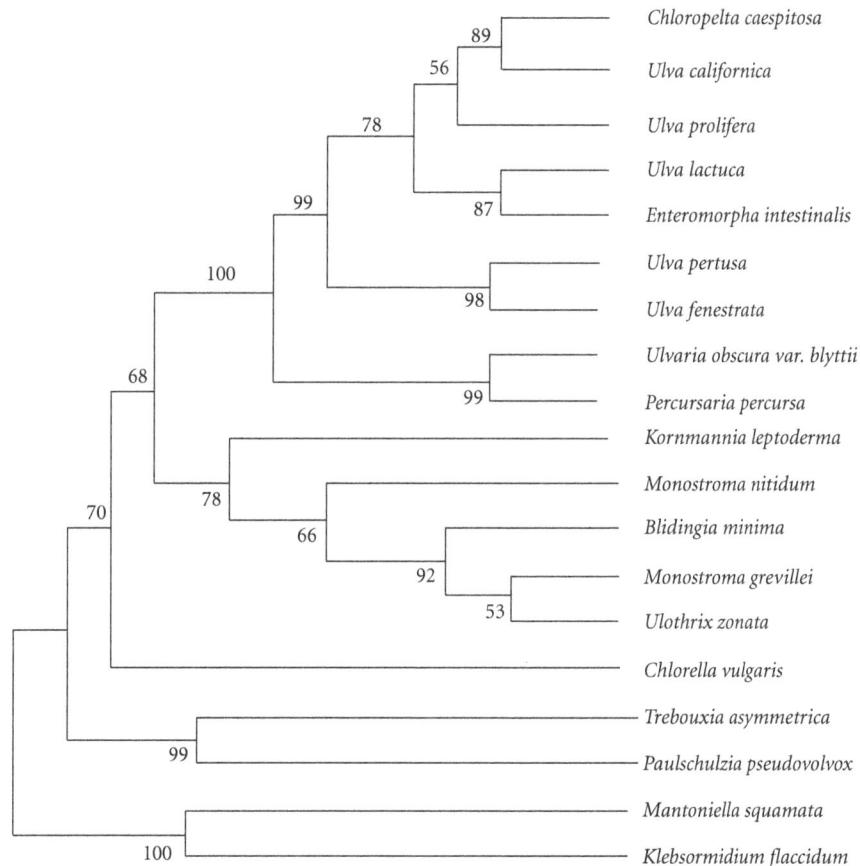

FIGURE 2: NJ phylogenetic tree constructed from ITS sequences.

the lowest region distance (0.133) and the ratio of sequence divergence (0.021).

A comparison of the sequences of the ITS of the three strains evaluated in this study to those of other species of green algal retrieved from GenBank is shown in Table 6. The same result with ITS, IE, CC, UC, SC, HT, UL, PPE, UO, and UF are subordinate to *Ulvaceae*. The region distance (ranged form 0.021 to 0.000) and the ratio of sequence divergence (ranged from 0.005 to 0.000) are lower than those comparisons within other 10 species from different families. Meanwhile, JM, UZ, BM, MN, and KL had higher homology relationship within each other, the region distance ranged

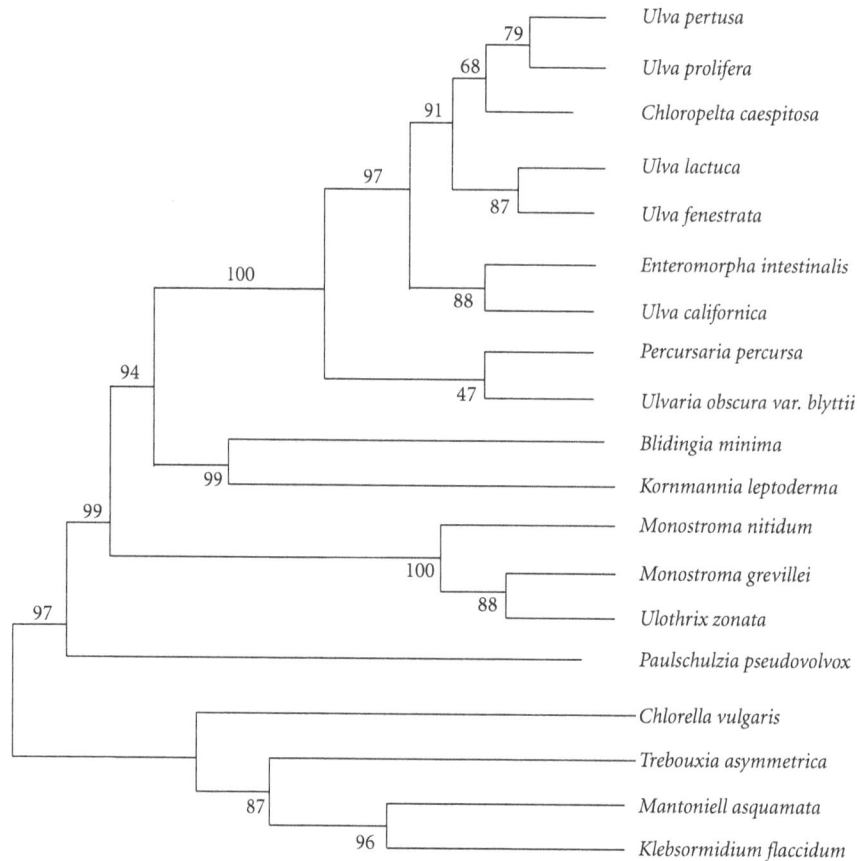

FIGURE 3: NJ phylogenetic tree constructed from 18S sequences.

from 0.069 to 0.008, and the ratio of sequence divergence ranged from 0.010 to 0.004.

3.5. Phylogenetic Tree Analysis. From the results of NJ phylogenetic tree constructed from ITS and 18S sequences (Figures 2 and 3), *U. pertusa* and *U. prolifera* belong to one branch. The homology of *U. pertusa*, and *U. prolifera* was obviously higher than *M. grevillei*.

The tree resulting from NJ analysis of the combined data of ITS is shown in Figure 2. As in previous analyses, the *Ulvaceae* and JM/UZ/BM/MN/KL were well supported. *C. caespitosa* and *U. californica* formed a tiny branch and *U. prolifera* joined them. *U. lactuca* and *E. intestinalis* composed another branch as *U. pertusa* and *U. fenestrata* did. There is strong support for the grouping of *C. caespitosa* with *U. californica* and *U. prolifera* in this phylogenetic tree, which coincided with the result from Hayden and Waaland [30]. *M. grevillei* and *U. zonata* formed one branch and then joined with *B. minima*, *M. nitidum*, and *K. leptoderma* forming one lager clade. The NJ tree and distance data both supported that *M. grevillei* has closer phylogenetic relation with *U. zonata* than *M. nitidum*.

The tree resulting from NJ analysis of 18S sequence is shown in Figure 3. In this tree, there is still strong support for the grouping of *C. caespitosa* with *U. californica* and *U. prolifera* while *U. pertusa* and *U. prolifera* formed a branch

and *C. caespitosa* joined with them, which is different from the result of ITS in Figure 2. Additionally, *E. intestinalis* came to the same branch with *U. californica* which confirmed well with the data of region distance and the ratio of sequence divergence in Table 6. *B. minima* and *K. leptoderma* became one clade. *M. grevillei* and *U. zonata* formed one branch again and *M. nitidum* joined them firstly. In the NJ tree of 18S, the smaller branch of *B. minima* and *K. leptoderma* is further away from the bigger one of *M. nitidum*, *M. grevillei*, and *U. zonata*, which is a more significant diversity compared with the NJ tree of ITS.

4. Discussions

During the past three decades, green tide has been gaining in scale and frequency in both marine and estuary environment all over the world. And in recent years, green tide massively occurred in China Yellow Sea. The main species involved in green tides are *Ulva* sp. and *Enteromorpha* sp.

The traditional taxonomy uses the morphology characters as the criterion. The thallus that consists of two layer cells is the genus *Ulva* while the monolayer cell form hollow tubular is *Enteromorpha*. But the *Enteromorpha* internal tubular thallus contained villiform protuberances and meshwork structures which were composed of glycoprotein. When the number of cells in cross-section of tubular thallus reached

30–50 h, the villiform protuberances disappeared and the meshwork structures became tight, then the mural cells of tubular thallus adhered and the foliolose thallus formed [31]. In culture studies of European *Ulva* species, Gayral [32] observed the development of both tubular and blade thalli from single populations of zoospores and parthenogenetic gametes.

Ulva sp. has wide acclimatization and can grow well in a broad range of temperatures and salinities, but change morphological characteristics easily in response to the environment [16]. Various morphological changes in the intraspecies and less differences among interspecies make identification of *Ulva* sp. using classical taxonomic methods very difficult [22, 23].

Molecular biology methods such as chloroplast *rbc*L, nuclear ITS, and 18S rDNA sequence analysis [5, 26, 27], alone or in combination with morphological methods, have been applied to species identification since green tides have occurred in the Yellow Sea. Despite multidisciplinary study, classification of the main causative species of green tides is still difficult. Several researchers have suggested that green tides are formed by an *Ulva linza-procera-prolifera* (LPP) complex instead of individual species [5, 17].

Although the thallus of *U. pertusa* is thick with many holes, it is different from *U. prolifera* in morphology. From the phylogenetic tree constructed from 18S sequences, it can be found that *U. pertusa* and *U. prolifera* group in one branch while *U. lactuca*, *U. fenestrate*, and *U. californica* are in another branch. From the phylogenetic tree constructed from ITS sequences, *U. prolifera* and *U. pertusa* are not in one branch. But *U. pertusa* and *U. fenestrate* grouped in one branch. No matter if the phylogenetic tree is constructed from ITS sequences or 18S sequences, it is clear that the clade of *Ulvaceae* is comprised of *Chloropelta*, *Enteromorpha*, *Percursaria*, *Ulva* and *Ulvaria*. These results are consistent with that of Hayden and Waaland [30] which proved that in different areas it had the same conclusion.

Hayden et al. [21] preferred that *Enteromorpha* should be transferred to *Ulva* and *Chloropelta*, *caespitosa* should be named *Ulva tanneri* according to the molecular phyletic evolution research. From the phylogenetic tree constructed from ITS and 18S sequences in our study, it is obviously showed that *Enteromorpha* and *Ulva* have closer phyletic evolution relationship.

Trees inferred from phylogenetic analysis of ITS and 18S both showed that *Monostroma nitidum* and *Monostroma grevillei* were not in one branch while *Monostroma grevillei* and *Ulothrix zonata* are in one branch. *Monostroma grevillei* has a closer 18S genetic distance with *Ulothrix zonata* (0.008) than with *Monostroma nitidum* (0.018). The same result is in ITS genetic distance with *Ulothrix zonata* (0.133), other than that with *Monostroma nitidum* (0.257). These results are also consistent with that of Hayden and Waaland [30].

So, we can make use of sequences not only in 18S rDNA to analyse genera, but also in ITS region and other sequences, such as 28S rDNA and chloroplast *rbc*L, all these together are the better way to analyze the whole phyletic evolution relationship.

Acknowledgments

This paper was supported by the National Scientific Foundations of China (Grant no. 30570125) and the Key Construction Laboratory of Marine Biotechnology of Jiangsu Province (Grant no. 2010HS03). L. Zijie and L. Zhongheng contributed equally to this work.

References

[1] D. Schories and K. Reise, "Germination and anchorage of *Enteromorpha spp.* in sediments of the Wadden Sea," *Helgoland Marine Research*, vol. 47, no. 3, pp. 275–285, 1993.

[2] R. L. Fletcher, "The occurrence of 'green tides'," in *Marine Benthic Vegetation: Recent Changes and the Effects of Eutrophication*, W. Schramm and P. H. Nienhuis, Eds., Ecological Studies, vol. 123, pp. 7–43, Springer, Berlin, Germany, 1996.

[3] T. A. Nelson, K. Haberlin, A. V. Nelson et al., "Ecological and physiological controls of species composition in green macroalgal blooms," *Ecology*, vol. 89, no. 5, pp. 1287–1298, 2008.

[4] T. A. Nelson, J. Olson, L. Imhoff, and A. V. Nelson, "Aerial exposure and desiccation tolerances are correlated to species composition in 'green tides' of the Salish Sea (northeastern Pacific)," *Botanica Marina*, vol. 53, no. 2, pp. 103–111, 2010.

[5] P. Jiang, J. F. Wang, Y. L. Cui, Y. X. Li, H. Z. Lin, and S. Qin, "Molecular phylogenetic analysis of attached Ulvaceae species and free-floating *Enteromorpha* from Qingdao coasts in 2007," *Chinese Journal Oceanology and Limnology*, vol. 26, no. 3, pp. 276–279, 2008.

[6] S. Sun, J. F. Wang, C. L. Li et al., "Emerging challenges: massive green algae blooms in the Yellow Sea," *Nature Precedings*, 2008.

[7] L. P. Ding, X. G. Fei, Q. Q. Lu, Y. Y. Deng, and S. X. Lian, "The possibility analysis of habitats, origin and reappearance of bloom green alga (*Enteromorpha prolifera*) on inshore of western Yellow Sea," *Chinese Journal Oceanology and Limnology*, vol. 27, no. 3, pp. 421–424, 2009.

[8] N. H. Ye, Z. M. Zhang, X. S. Jin, and Q. Y. Wang, "China is on the track tackling *Enteromorpha spp.* forming green tide," *Nature Precedings*, 2008.

[9] D. Y. Liu, J. K. Keesing, Q. G. Xing, and P. Shi, "World's largest macroalgal bloom caused by expansion of seaweed aquaculture in China," *Marine Pollution Bulletin*, vol. 58, no. 6, pp. 888–895, 2009.

[10] H. F. Link, "Epistola ad virum celeberrimum Nees ab Esenbeck. de algis aquaticis, in genera *Disponendis*," in *Horae Physicae Berolinense*, C. G. D. Nees von Esenbeck, Ed., pp. 1–8, Bonnae, Bonn, Germany, 1820.

[11] E. Nic Dhonncha and M. D. Guiry, "Algaebase: documenting seaweed biodiversity in Ireland and the world," *Biology and Environment*, vol. 102, no. 3, pp. 185–188, 2002.

[12] C. K. Tseng and J. F. Zhang, "Economic seaweeds of northern China," *Journal of Shandong University (Natural Science)*, vol. 2, pp. 57–82, 1952 (Chinese).

[13] C. K. Tseng and J. F. Zhang, "The economic seaweeds flora of the Yellow Sea and East Sea," *Oceanologia et Limnologia Sinica*, vol. 2, pp. 43–52, 1959 (Chinese).

[14] C. K. Tseng, D. R. Zhang, and J. F. Zhang, *China Economic Seaweeds Records*, Science Press, Beijing, China, 1962.

[15] C. K. Tseng and J. F. Zhang, "A preliminary analytical study of the Chinese marine algal flora," *Oceanologia et Limnologia Sinica*, vol. 5, no. 3, pp. 245–253, 1963 (Chinese).

[16] M. L. Dong, "A preliminary phytogeographical studies on Chinese species of *Enteromorpha*," *Oceanologia et Limnologia Sinica*, vol. 5, no. 1, pp. 46–51, 1963 (Chinese).

[17] T. J. White, T. Bruns, S. Lee, and J. Taylor, "Amplification and direct sequencing of fungal ribosomal RNA genes for phylogenetics," in *PCR Protocols: A Guide to Methods and Applications*, M. A. Innes, D. H. Gefland, J. J. Sninsky, and T. J. White, Eds., pp. 315–322, Academic Press, New York, NY, USA, 1990.

[18] G. C. Booton, G. L. Floyd, and P. A. Fuerst, "Polyphyly of tetrasporalean green algae inferred from nuclear small-subunit ribosomal DNA," *Journal of Phycology*, vol. 34, no. 2, pp. 306–311, 1998.

[19] M. J. H. Van Oppen, *Tracking trails by cracking codes*, Ph.D. dissertation, University of Groningen, Groningen, The Netherlands, 1995.

[20] I. H. Tan, J. Blomster, G. Hansen et al., "Molecular phylogenetic evidence for a reversible morphogenetic switch controlling the gross morphology of two common genera of green seaweeds, *Ulva* and *Enteromorpha*," *Molecular Biology and Evolution*, vol. 16, no. 8, pp. 1011–1018, 1999.

[21] H. S. Hayden, J. Blomster, C. A. Maggs, P. C. Silva, M. J. Stanhope, and J. R. Waaland, "Linnaeus was right all along: *Ulva* and *Enteromorpha* are not distinct genera," *European Journal of Phycology*, vol. 38, no. 3, pp. 277–294, 2003.

[22] C. Bliding, "A critical study of European taxa in *Ulvales*. I. *Capsosiphon, Percursaria, Blidingia, Enteromorpha*," *Opera Botanica*, vol. 8, pp. 1–160, 1963.

[23] J. Blomster, C. A. Maggs, and M. J. Stanhope, "Molecular and morphological analysis of *Enteromorpha intestinalis* and *E. compressa* (Chlorophyta) in the British Isles," *Journal of Phycology*, vol. 34, no. 2, pp. 319–340, 1998.

[24] M. Hiraoka, M. Ohno, S. Kawaguchi, and G. Yoshida, "Crossing test among floating *Ulva* thalli forming 'green tide' in Japan," *Hydrobiologia*, vol. 512, pp. 239–245, 2004.

[25] K. Niwa, A. Kobiyama, and T. Sakamoto, "Interspecific hybridization in the haploid blade-forming marine crop *Porphyra* (Bangiales, Rhodophyta): occurrence of allodiploidy in surviving F1 gametophytic blades," *Journal of Phycology*, vol. 46, no. 4, pp. 693–702, 2010.

[26] S. Shimada, M. Hiraoka, S. Nabata, M. Lima, and M. Masuda, "Molecular phylogenetic analyses of the Japanese *Ulva* and *Enteromorpha* (Ulvales, Ulvophyceae), with special reference to the free-floating *Ulva*," *Phycological Research*, vol. 51, no. 2, pp. 99–108, 2003.

[27] A. R. Sherwood, D. J. Garbary, and R. G. Sheath, "Assessing the phylogenetic position of the Prasiolales (Chlorophyta) using rbcL and 18S rRNA gene sequence data," *Phycologia*, vol. 39, no. 2, pp. 139–146, 2000.

[28] S. D. Shen, Y. Y. Li, X. J. Wu, and L. P. Ding, "Sequences and phylogeny analysis of *rbcL* gene in marine Chlorophyta," *Oceanic and Coastal Sea Research*, vol. 9, pp. 145–150, 2010.

[29] S. Kumar, K. Tamura, and I. B. Jakobsen, "MEGA2: molecular evolutionary genetics analysis software," *Bioinformatics*, vol. 17, no. 12, pp. 1244–1245, 2001.

[30] H. S. Hayden and J. R. Waaland, "Phylogenetic systematics of the Ulvaceae (*Ulvales, Ulvophyceae*) using chloroplast and nuclear DNA sequences," *Journal of Phycology*, vol. 38, no. 6, pp. 1200–1212, 2002.

[31] J. W. Wang, A. P. Lin, S. D. Shen, and B. L. Yan, "Microscopic observation on the development of *Enteromorpha prolifera* (Chlorophyta)," *Ecologic Science*, vol. 25, no. 5, pp. 400–404, 2006.

[32] P. Gayral, "Sur le démembrement de l'actual genre *Monostroma* Thuret (Chlorophycées, Ulotrichales s.l.)," *Comptes Rendus de l' Academie des Sciences*, vol. 258, pp. 2149–2152, 1964.

Permissions

The contributors of this book come from diverse backgrounds, making this book a truly international effort. This book will bring forth new frontiers with its revolutionizing research information and detailed analysis of the nascent developments around the world.

We would like to thank all the contributing authors for lending their expertise to make the book truly unique. They have played a crucial role in the development of this book. Without their invaluable contributions this book wouldn't have been possible. They have made vital efforts to compile up to date information on the varied aspects of this subject to make this book a valuable addition to the collection of many professionals and students.

This book was conceptualized with the vision of imparting up-to-date information and advanced data in this field. To ensure the same, a matchless editorial board was set up. Every individual on the board went through rigorous rounds of assessment to prove their worth. After which they invested a large part of their time researching and compiling the most relevant data for our readers. Conferences and sessions were held from time to time between the editorial board and the contributing authors to present the data in the most comprehensible form. The editorial team has worked tirelessly to provide valuable and valid information to help people across the globe.

Every chapter published in this book has been scrutinized by our experts. Their significance has been extensively debated. The topics covered herein carry significant findings which will fuel the growth of the discipline. They may even be implemented as practical applications or may be referred to as a beginning point for another development. Chapters in this book were first published by Hindawi Publishing Corporation; hereby published with permission under the Creative Commons Attribution License or equivalent.

The editorial board has been involved in producing this book since its inception. They have spent rigorous hours researching and exploring the diverse topics which have resulted in the successful publishing of this book. They have passed on their knowledge of decades through this book. To expedite this challenging task, the publisher supported the team at every step. A small team of assistant editors was also appointed to further simplify the editing procedure and attain best results for the readers.

Our editorial team has been hand-picked from every corner of the world. Their multi-ethnicity adds dynamic inputs to the discussions which result in innovative outcomes. These outcomes are then further discussed with the researchers and contributors who give their valuable feedback and opinion regarding the same. The feedback is then collaborated with the researches and they are edited in a comprehensive manner to aid the understanding of the subject.

Apart from the editorial board, the designing team has also invested a significant amount of their time in understanding the subject and creating the most relevant covers. They scrutinized every image to scout for the most suitable representation of the subject and create an appropriate cover for the book.

The publishing team has been involved in this book since its early stages. They were actively engaged in every process, be it collecting the data, connecting with the contributors or procuring relevant information. The team has been an ardent support to the editorial, designing and production team. Their endless efforts to recruit the best for this project, has resulted in the accomplishment of this book. They are a veteran in the field of academics and their pool of knowledge is as vast as their experience in printing. Their expertise and guidance has proved useful at every step. Their uncompromising quality standards have made this book an exceptional effort. Their encouragement from time to time has been an inspiration for everyone.

The publisher and the editorial board hope that this book will prove to be a valuable piece of knowledge for researchers, students, practitioners and scholars across the globe.

List of Contributors

Eri Adams, Minami Matsui and Ryoung Shin
RIKEN Center for Sustainable Resource Science, 1-7-22 Suehirocho, Tsurumi-ku, Yokohama, Kanagawa 230-0045, Japan

Celine Diaz
RIKEN Center for Sustainable Resource Science, 1-7-22 Suehirocho, Tsurumi-ku, Yokohama, Kanagawa 230-0045, Japan
Center for Research in Agricultural Genomics (CRAG), Universitat Aut`onoma de Barcelona, Cerdanyola del Vall`es, 08193 Barcelona, Spain

Joëlle Gérard and Ludwig Triest
Plant Biology and Nature Management (APNA), Vrije Universiteit Brussel, Pleinlaan 2, 1050 Brussels, Belgium

Soledad Ramos
Departamento de Ingenier´ia del Medio Agron´omico y Forestal, Escuela de Ingenier´ias Agrarias, Universidad de Extremadura, Avendia Adolfo Su´arez s/n, 06007 Badajoz, Spain

Francisco M. Vázquez
Departamento de Producci´on Forestal, Centro de Investigaci´on La Orden-Valdesequera, Consejer´ia de Infraestructura y Desarrollo Tecnol´ogico, Junta de Extremadura, Apartado de Correos 22, 06080 Badajoz, Spain

Trinidad Ruiz
Departamento de Biolog´ia Vegetal, Ecolog´ia y Ciencias de la Tierra, ´ Area de Bot´anica, Facultad de Ciencias, Universidad de Extremadura, Avendia de Elvas s/n, 06006 Badajoz, Spain

A. B. Nwauzoma
Department of Applied & Environmental Biology, Rivers State University of Science & Technology, PMB 5080, Port Harcourt 500001, Nigeria
Embrapa Agroenergia, PqEB-Final W3 Norte, Asa Norte, 7077091 Brasilia, DF, Brazil

K. Moses
Department of Applied & Environmental Biology, Rivers State University of Science & Technology, PMB 5080, Port Harcourt 500001, Nigeria

Ram Asheshwar Mandal and Siddhibir Karmacharya
Tirchandra College, Kathmandu, Nepal

Ishwar Chandra Dutta
Tribhuvan University Commission, Kirtipur, Kathmandu, Nepal

Pramod Kumar Jha
Central Department of Botany, Tribhuvan University, Kirtipur, Nepal

A. B. Nwauzoma
Department of Applied & Environmental Biology, Rivers State University of Science & Technology, PMB 5080, Port Harcourt 500001, Nigeria
Embrapa Agroenergia-PQEB-Final W3 Norte, Asa Norte, 7077091 Brasilia, DF, Brazil

Magdalene S. Dappa
Department of Applied & Environmental Biology, Rivers State University of Science & Technology, PMB 5080, Port Harcourt 500001, Nigeria

Alla I. Yemets, Galina Ya. Bayer and Yaroslav B. Blume
Institute of Food Biotechnology and Genomics, National Academy of Sciences of Ukraine, Osipovskogo Street 2a, Kiev 04123, Ukraine

Dominique (Niki) Robertson
Department of Plant Biology, North Carolina State University, P.O. Box 7612, Raleigh, NC 27695, USA

Edward Missanjo, Chikumbutso Maya, Dackious Kapira, Hannah Banda and Gift Kamanga-Thole
Malawi College of Forestry andWildlife, Private Bag 6, Dedza, Malawi

Jutta Papenbrock
Institute of Botany, Leibniz University Hannover, Herrenh"auser Straße 2, 30419 Hannover, Germany

Yury Kamenir and Zvy Dubinsky
The Mina and Everard Goodman Faculty of Life Sciences, Bar-Ilan University, Ramat-Gan 52900, Israel

Paola Ernandes and Silvano Marchiori
Laboratory of Systematic Botany and Plant Ecology, Di.S.Te.B.A. University of the Salento, 73100 Lecce, Italy

Michael P. Fuller, Hail Z. Rihan and Mohammad Al-Issaw
School of Biomedical and Biological Sciences, Faculty of Science and Technology, Plymouth University, Plymouth PL4 8AA, UK

Jalal H. Hamza
School of Biomedical and Biological Sciences, Faculty of Science and Technology, Plymouth University, Plymouth PL4 8AA, UK
Department of Agronomy, College of Agriculture, University of Baghdad, Baghdad, Iraq

Lassina Traoré, Amadé Ouédraogo and Adjima Thiombiano
Laboratory of Plant Biology and Ecology, University of Ouagadougou, 03 BP 7021, Ouagadougou 03, Burkina Faso

Jason T. C. Tzen
Graduate Institute of Biotechnology, National Chung Hsing University, Taichung 402, Taiwan

Koh-Ichi Takakura
Division of Urban Environment, Osaka City Institute of Public Health and Environmental Sciences, 8-34 Tojo-cho, Tennoji, Osaka 543-0026, Japan

P. I. Jattisha and M. Sabu
Department of Botany, University of Calicut, Kerala 673 635, India

JieHe and Ameerah Zain
Natural Sciences and Science Education Academic Group, National Institute of Education, Nanyang Technological University, 1 Nanyang Walk, Singapore 637 616

Alexander Staunch, Marie Redlecki, Jessica Wooten, Jonathan Sleeper and Jonathan Titus
Department of Biology, Jewett Hall, SUNY-Fredonia, Fredonia, NY 14063, USA

Zijie Lin, Zhongheng Lin, Huihui Li, and Songdong Shen
College of Life Sciences, Soochow University, Suzhou 215123, China

9 781632 391056